AF796465
WITHDRAWN

CERAMIC APPLICATIONS IN TURBINE ENGINES

CERAMIC APPLICATIONS IN TURBINE ENGINES

by

Harold E. Helms, Peter W. Heitman
Leonard C. Lindgren, Samuel R. Thrasher
General Motors Corporation
Allison Gas Turbine Division
Indianapolis, Indiana

np
NOYES PUBLICATIONS
Park Ridge, New Jersey, U.S.A.

TJ
778
.C38
1986

Copyright © 1986 by Noyes Publications
Library of Congress Catalog Card Number 86-5250
ISBN: 0-8155-1080-2
Printed in the United States

Published in the United States of America by
Noyes Publications
Mill Road, Park Ridge, New Jersey 07656

10 9 8 7 6 5 4 3 2 1

Library of Congress Cataloging-in-Publication Data

Ceramic applications in turbine engines.

Bibliography: p.
Includes index.
1. Gas turbines--Ceramic materials. I. Helms, Harold E.
TJ778.C38 1986 621.43'3 86-5250
ISBN 0-8155-1080-2

Foreword

This book documents a study of the use of ceramic components in turbine engines, for the purpose of improving cycle efficiency and raising operating temperatures. The study covers design, materials development, fabrication of ceramic components, and testing—including rig, engine, and vehicle demonstration tests. This book should prove to be of value to both ceramic and mechanical engineers.

The use of ceramic components in gas turbine engines allows significantly higher engine operating temperatures than metal components allow. This increased temperature capability yields a more efficient engine that could reduce heavy-duty truck fuel use per mile by more than 20%. Furthermore, ceramic components may be less expensive than complex air-cooled metal components and may offer reduced weight, improved erosion resistance, and reduced dependence on critical imported metals used in superalloys. The reduction in cost may make the gas turbine engine feasible for mass production vehicular applications (trucks or passenger cars).

As a part of the study described, a technology base was established which can be applied to automotive and other gas turbine engines. In addition, an understanding of ceramic design methodology was accomplished, design data and ceramic material characterization were provided, nondestructive testing technology was developed, ceramic component fabrication techniques were advanced, and chemical and structural stability of ceramic components in a gas turbine environment were evaluated.

The information in the book is from *Ceramic Applications in Turbine Engines—Final Report,* prepared by Harold E. Helms, Peter W. Heitman, Leonard C. Lindgren, and Samuel R. Thrasher of General Motors Corporation, Allison Gas Turbine Division, for the National Aeronautics and Space Administration, for the U.S. Department of Energy, October 1984.

The table of contents is organized in such a way as to serve as a subject index and provides easy access to the information contained in the book.

Advanced composition and production methods developed by Noyes Publications are employed to bring this durably bound book to you in a minimum of time. Special techniques are used to close the gap between "manuscript" and "completed book." In order to keep the price of the book to a reasonable level, it has been partially reproduced by photo-offset directly from the original report and the cost saving passed on to the reader. Due to this method of publishing, certain portions of the book may be less legible than desired.

NOTICE

The materials in this book were prepared as accounts of work sponsored by the National Aeronautics and Space Administration and the U.S. Department of Energy. Publication does not signify that the contents necessarily reflect the views and policies of the contracting agencies or the publisher, nor does mention of trade names or commercial products constitute endorsement or recommendation for use.

Acknowledgments

The authors would like to thank the United States Department of Energy, Office of Vehicle and Engine Research and Development, along with NASA, Lewis Research Center, for support of the work and for providing Allison Gas Turbine Division of General Motors Corp. the opportunity to participate in the development of a new structural material, ceramics, for use in gas turbine engines.

Special thanks are also due the subcontractors, especially the Carborundum Company, Corning Glass Works, GTE Laboratories Inc., Pure Carbon Company, and Norton Company. Thanks are also due the many contributors at Allison for the performance of the program, with special thanks to the Ceramic Materials Development Group headed by Dr. Peter W. Heitman.

Contents and Subject Index

I. Introduction and Summary

Program Objectives

The use of ceramic components in gas turbine engines allows significantly higher engine operating temperatures than metal components allow. This increased temperature capability yields a more efficient engine that could reduce heavy-duty truck fuel use per mile by more than 20%. Furthermore, ceramic components may be less expensive than complex air-cooled metal components and may offer reduced weight, improved erosion resistance, and reduced dependence on critical imported metals used in superalloys. The reduction in cost may make the gas turbine engine feasible for mass production vehicular applications (trucks or passenger cars).

The Ceramic Applications in Turbine Engines (CATE) program was initiated by Allison Gas Turbine Division of General Motors Corporation on 3 January 1978. The CATE program was sponsored by the United States Department of Energy (DOE) and monitored by the National Aeronautics and Space Administration (NASA), Lewis Research Center, under NASA contract No. DEN 3-17.

The program objectives included the following:

- advance the state of the art of highway vehicle gas turbine engines through utilization of ceramic components
- develop ceramic components that would make the Allison GT 404 engine acceptable for commercial application with regard to durability, maintenance frequency, operating characteristics, and noise and emissions
- develop structural ceramics technology, as shown in Table I, which would apply to other engines, such as the AGT 100 automotive gas turbine

The engine selected for the program was the Allison GT 404 industrial gas turbine engine (see Figure 1). Modifications, made by a step-by-step addition of ceramic components, were incorporated to permit operation at selectively higher temperatures.

The ceramic components were added to increase the gasifier turbine rotor inlet temperature from the all-metal engine value of 1002°C (1835°F) to 1241°C (2265°F), with intermediate temperature levels of 1038°C (1900°F) and 1132°C (2070°F). Program redirection due to funding constraints eliminated the 1241°C (2265°F) phase. The 1038°C (1900°F) and 1132°C (2070°F) levels were developed and were successfully demonstrated in the engine.

Table I.
Structural ceramics technology.

Development and Verification of Design Techniques
- Metallic interfaces
- Failure theory
- Analysis methodologies
- Material data base requirements

Development of Ceramic Materials/Component Fabrication Techniques

Development and Assessment of Ceramic Suppliers Capabilities
- Nine suppliers
- Source development

Establishment of Procedures
- Fixturing and machining
- Handling
- Assembling and installing
- Instrumentation
- Packaging

Development of NDE Techniques and Requirements

Development of Component Testing Procedures

Supporting Technologies
- Abradability
- Diffusion barriers
- Thermal barriers
- Insulation materials
- Regenerators (cores, seals, leakage)
- Engine control

Engine Testing and Vehicle Demonstration

Program Background and Schedule

The CATE program is the end product of the Heavy Duty Gas Turbine Engine (HDGTE) program conducted by Allison under contract No. NAS 3-20064 from the National Aeronautics and Space Administration, Lewis Research Center.

The HDGTE program started 1 July 1976 with engine demonstrations of available ceramic regen-

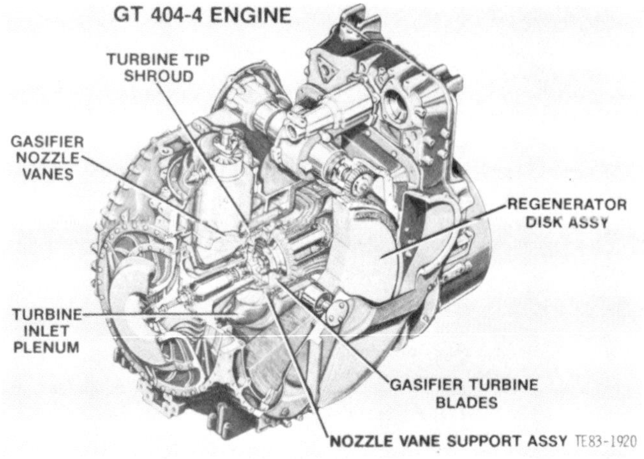

Figure 1. Ceramic components installed in an Allison GT 404 industrial gas turbine engine.

erator systems. It established a program plan that became the CATE program in January 1978. The HDGTE program objective was to study improvements to the Allison GT 404-4 engine that might result from incorporation of ceramic components. The benefits revealed by the study were impressive, so the CATE program approach was planned and included probabilistic design techniques, ceramic material behavior and characterization, development of nondestructive evaluation (NDE) techniques, component process development (manufacturing), assessment of ceramic suppliers' capabilities, component development testing, engine testing, and vehicle demonstration.

The original CATE program schedule is shown in Figure 2. The study phase shown represents the earlier HDGTE program, and the three temperature levels—1038°C (1900°F), 1132°C (2070°F), and 1241°C (2265°F)—represent the step-by-step introduction of additional ceramics, which reflects the CATE program approach to ceramic component development.

The program progressed through the 1038°C (1900°F) phase and into the 1132°C (2070°F) phase with successful engine and vehicular demonstrations. In 1981 and in 1982 changes in the Federal budget for energy programs forced redirection of DOE-funded work. The resulting CATE program

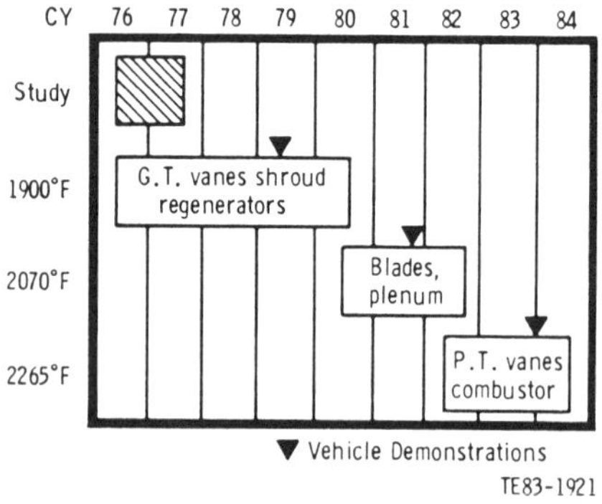

Figure 2. Original CATE program schedule.

schedule is shown in Figure 3. The plenum development work in the 1132°C (2070°F) phase was suspended, and the 1241°C (2265°F) phase of the program was deleted. The emphasis on vehicle demonstration was deleted and replaced by development of the 2070°F-configuration ceramic components, including an engine demonstration with all

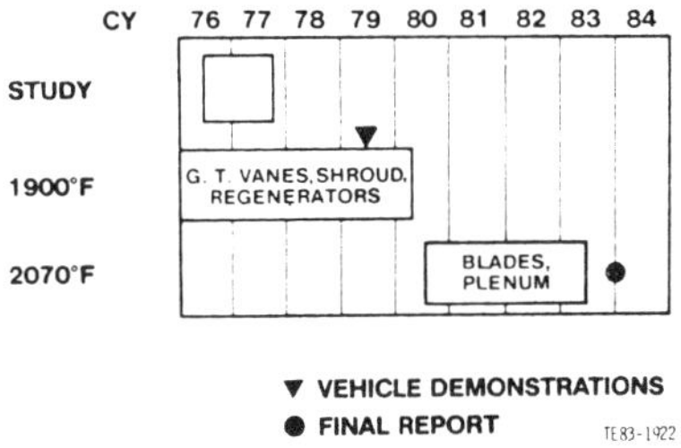

Figure 3. Revised CATE program schedule.

the 1132°C (2070°F) components except the ceramic plenum. During the summer and fall of 1982, the 2070°F-configuration ceramic components were successfully demonstrated at maximum engine power settings, 36,905 rpm gasifier speed and 1152°C (2106°F) burner outlet temperature, which is equivalent to 1132°C (2070°F) gasifier rotor inlet temperature.

This report will summarize the activities for each phase of the program, including the work accomplished on the 2265°F configuration prior to its deletion from the program. Detailed program progress was previously reported in semiannual progress reports (Ref 1-8); these will be referenced throughout this report to direct the reader to additional details.

Program Approach

The initial CATE program approach to developing ceramic components for the GT 404 industrial gas turbine engine was to introduce the components in a step-by-step manner, using the technology gained from each step to design and develop the following, more complex components (see Figure 3). This approach was rigorously followed and resulted in successful 1038°C (1900°F) and 1132°C (2070°F) ceramic component demonstrations.

The CATE program can be subdivided into four major categories: ceramic material behavior and characterization, ceramic component design, ceramic component fabrication, and ceramic component testing. A description of the approach used for each of these categories follows.

CERAMIC MATERIAL BEHAVIOR AND CHARACTERIZATION

The approach to ceramic material behavior studies was to evaluate the following properties:

- elastic properties
- thermal properties
- material strength
- fracture toughness
- abradable seal properties
- microstructure studies
- fracture surface topography

These properties were evaluated for the candidate materials proposed for use in fabricating components. These candidate materials included the following:

- SKT-B silicon carbide (reaction bonded) from Carborundum Company
- sintered alpha silicon carbide from Carborundum Company
- GTE 3502 sintered silicon nitride from GTE Laboratories
- Purebide Refel silicon carbide (reaction bonded) from Pure Carbon Company

This effort was intended to supply materials property information to support design activities and the characterization of ceramic components. Initially, test bar specimens were used to evaluate these properties; and as the program progressed, material cut from actual ceramic components was evaluated. This ensured that material properties that reflected actual component manufacturing processes were obtained.

All of the ceramic components received during the CATE program were characterized by use of NDE techniques. These NDE techniques included the following:

- visual examination at 30X magnification
- X-ray radiography
- fluorescent penetrant inspection (FPI) (Zyglo)
- ultrasonics
- dimensional evaluation

In addition, destructive evaluation was conducted on selected ceramic components and on test bars produced in conjunction with each ceramic component. The destructive evaluation included modulus of rupture (MOR) testing and microstructural evaluation of components (pore size and distribution, grain size, and contaminants).

Also included in the approach to ceramic material behavior and characterization was the development of new NDE techniques specifically designed for the unique ceramic material properties.

CERAMIC COMPONENT DESIGN

The approach to designing ceramic components acknowledged the brittle character and variability in strength of the ceramic materials. A unique design methodology was applied that was sensitive to known and newly discovered material characteristics, satisfying design and proof test reliability requirements, but was not unduly complex. Two specific activities were completed prior to component design work: (1) specific reliability objectives for each ceramic component were defined and (2) the Allison finite element models were coupled with appropriate volume and surface area probability integrals to introduce statistical techniques into design practices.

CERAMIC COMPONENT FABRICATION

The approach to ceramic component fabrication recognized that more than one candidate material was likely to be successful and that several ceramic suppliers currently are producing components made from similar ceramic materials. To address these considerations, both silicon-carbide (primary material for the CATE program) and silicon-nitride components were included for the gasifier turbine nozzle and blades. Each component was ordered from two or more suppliers, which might or might not have used the same manufacturing process (injection molding, isopressing, etc).

In all cases, a manufacturing process that offered cost-effective production potential was required for each ceramic component in the program. Each supplier was not required to use such a process, but at least one supplier producing each component was required to do so.

The following is a list of the CATE program ceramic component suppliers and the ceramic materials they used to produce components:

- Carborundum Company—reaction-bonded silicon carbide (SKT-B) and sintered alpha silicon carbide
- Corning Glass Company—alumina silicate and lithium alumina silicate
- GTE Laboratories—sintered silicon nitride
- Pure Carbon Company—Purebide Refel silicon carbide (reaction bonded)
- Norton Company—siliconized silicon carbide (NC430)
- NGK-Locke—sintered silicon nitride
- Kyocera International—sintered silicon carbide
- AiResearch Casting Company—sintered silicon nitride
- Annawerk—reaction-bonded silicon carbide
- Rosenthal—reaction-bonded silicon carbide and reaction-bonded silicon nitride

Another objective for procuring ceramic components from several suppliers was to assess supplier capabilities, both domestic and foreign, as well as to assess the different manufacturing processes used by the suppliers.

CERAMIC COMPONENT TESTING

To conduct a cost-effective and meaningful program, testing must be conducted on components that represent a production manufacturing process and a corresponding design that analysis predicts will survive. The manufacturing process for these components must be the highest quality possible, and NDE of components is needed to assess quality. However, flaws as small as 20 μm (0.0008 in.) in diameter are critical, and, for the CATE program, no existing NDE technique was available to detect these flaws. Therefore, the burden of finding flawed components fell on the test program. To assemble engines with more than 70 ceramic components and to conduct testing that could result in complete destruction of all the ceramic parts because of one flawed part is not cost effective. For that reason the concept of ceramic component proof testing was conceived and utilized extensively in the CATE program.

As ceramic components passed proof testing and subsequent inspection, they were considered qualified for engine testing. The proof testing was conducted in various rigs specifically designed for the individual ceramic components.

Engine testing is the only acceptable means to demonstrate ceramic component proof of concept. Once the ceramic components have been engine demonstrated, engine performance and endurance tests must be conducted to determine the acceptability of ceramics for the engine applications. The final demonstration of ceramic components is vehicle testing of the engine with ceramic components installed.

Program Summary

The development of the CATE ceramic components started in July 1976 as a study under the HDGTE program (Ref 9). The results of the study and the previously conducted ceramic regenerator development work led the way for the CATE program, which started in January 1978, under NASA contract No. DEN 3-17, for the U.S. Department of Energy, Conservation and Renewable Energy, Office of Vehicle and Engine Research and Development Programs. The CATE program continued through July 1983, when the demonstration of the 2070°F-configuration ceramic components was completed.

The schedule shown in Figure 3 identifies two ceramic component development efforts, the

1900°F and the 2070°F configurations. The 2070°F configuration is considerably more complex since it includes all of the 1900°F configuration plus additional components. A third, 2265°F configuration was originally planned but was deleted from the work scope due to redirection of DOE programs. Both the 1900°F configuration and the 2070°F configuration have been developed through all phases of activity: design, manufacturing, proof testing, and engine demonstration testing. The following subsections will summarize the activities for each configuration.

1900°F-CONFIGURATION SUMMARY

The 1900°F configuration consisted of three ceramic components and the necessary Allison GT 404-4 engine modifications to accept these ceramic components. The three ceramic components were regenerator disks, gasifier turbine vanes, and gasifier turbine tip shroud. A technical description and detailed discussion of the 1900°F-configuration design can be found in the HDGTE program progress report (Ref 9).

These three ceramic components successfully completed manufacturing process development at several suppliers, proof and development testing in rigs, engine demonstration and endurance testing, and, ultimately, vehicle demonstration in a heavy-duty over-the-road truck. Figure 4 summarizes the components, the ceramic material used to produce parts, the number of parts rig tested, and three columns of engine test data. The first column provides the total engine operating hours for each component; the second column lists the maximum engine test hours accumulated on each component. A check mark in the last column, labeled Truck Installed, means that the ceramic component has been vehicle demonstrated.

A total of 7475 engine test hours was accumulated, which represents more than 370,000 simulated heavy-duty truck miles. The vehicle demonstration with ceramic components logged 6656 miles of severe vibrational and thermal shock loading without incident. The truck demonstration was conducted over various city streets, railroad crossings, and local highways; then the vehicle was driven to the GM proving grounds in Detroit, Michigan, for a trip which included interstate highway operation. At the GM proving grounds, very severe vibrational and shock loading tests were conducted on the Belgian block and truck durability road courses. The following listing summarizes the vehicle demonstration ceramic hardware, conditions, and results:

Component	Material	Rig Tested (Parts)	Engine Tested Hours	High Time Part	Truck Installed
Vane 	Reaction Bonded Silicon Carbide	130	3138	1512	✓
	Reaction Bonded Silicon Nitride	67	81	81	
	Refel Silicon Carbide	4	435	435	✓
Turbine Tip Shroud 	Reaction Bonded Silicon Carbide	4	1546	985	
	Sintered Silicon Nitride	3	113	113	
	Lithium Alumina Silicate	3	11	11	
Regenerator 	Thick Wall 1000°C Alumina Silicate	5	2833	1808	✓
	Thin Wall 1000°C Alumina Silicate	16	3938	3050	✓
	Regenerator Inboard Seal	23	7475	2160	✓

Total Time: 7475 Hrs. Vehicle: 6656 Miles

TE83-1923

Figure 4. 1900°F-configuration ceramic component development summary.

- Hardware—alumina silicate regenerator disks and silicon carbide gasifier turbine vanes
- Conditions—city streets, railroad crossings, interstate highways, and GM proving grounds (Belgian block course and truck durability course)
- Results—ceramic parts withstood with no damage (1) severe shock and vibration and (2) severe thermal and mechanical operating environment; ceramic parts were capable of continued engine usage

The design, manufacturing, and test experience gained during the 1900°F-configuration development has attested to the structural and chemical stability of the ceramic material candidates in an engine environment. Furthermore, this development experience has served to support the promise of ceramics as a viable material for a gas turbine engine. In addition to encouraging a continued effort in structural ceramics, the development experience provided a strong background in ceramic material/component characterization, design methodology, manufacturing process development, and ceramic component proof and engine test techniques. This background was applied to the development of the 2070°F configuration discussed in the next subsection.

2070°F-CONFIGURATION SUMMARY

The 2070°F configuration features a totally ceramic gasifier turbine flow path and regenerator system and necessary modifications to the CATE GT

404 1900°F-configuration engine to accept these ceramic components. The turbine flow path is composed of 78 ceramic parts, 76 of which are in the gasifier turbine section, as shown in Figure 5. The remaining two ceramic components are the regenerator disks.

The preliminary 2070°F-configuration design was conducted as part of the HDGTE program (Ref 9) and finalized during the CATE program (Ref 1, 2, and 3).

Based on knowledge gained during the 1900°F-configuration development, considerable emphasis was placed on manufacturing process development throughout the 2070°F-configuration phase. Both ceramic gasifier nozzle and ceramic blade components were subjected to process development activities at Carborundum Company, the most extensive being the ceramic blade process development matrix experiment, which is discussed in Section IV of this report, in the gasifier turbine blade subsection. The other process development activities were ongoing throughout the program and are discussed in Ref 1 through 8.

All of the 2070°F-configuration ceramic components, with one exception, were successfully engine demonstrated at maximum speed (36,905 rpm) and rated turbine rotor inlet temperature (1132°C [2070°F]) conditions. The exception was the ceramic turbine inlet plenum, which was deleted from the program due to changes in work scope. Ceramic plenum components have been produced by three manufacturers: Carborundum Company, Norton Company, and NGK-Locke. Initial rig testing had started on the first Norton Company plenum when work-scope changes forced an end to plenum development.

The 2070°F-configuration ceramic component development is summarized in Figure 6. The information provided in Figure 6 includes the following data for each ceramic component:

- ceramic materials used to produce component
- number of each rig tested
- engine test hours for each component
- maximum engine test hours on any one part representing each component

The combination of 1900°F- and 2070°F-configuration ceramic component engine test experience is shown in Table II. The test experience is

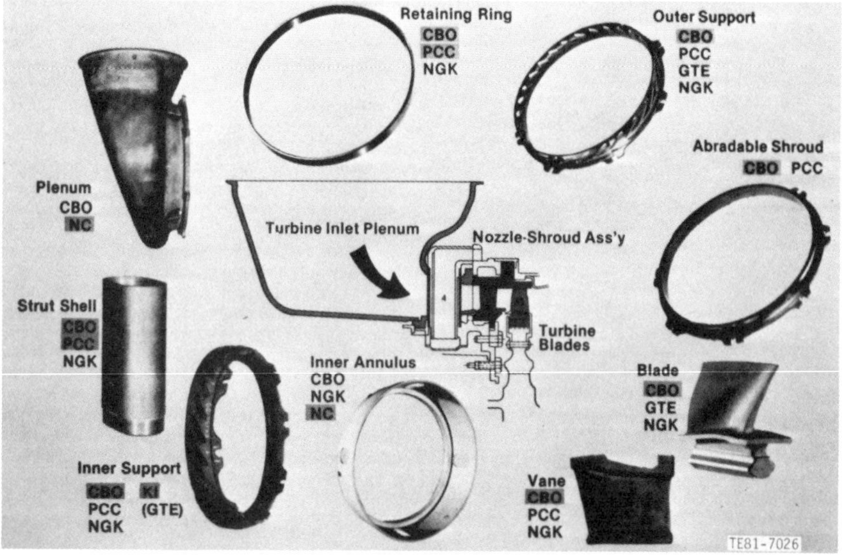

Figure 5. 2070°F-configuration gasifier turbine ceramic components.

Outer vane support

Material	Rig tested	Engine tested: Hours	Engine tested: High time part
Sintered alpha silicon carbide	1	4	4
Reaction bonded silicon carbide	9	896	890
Refel reaction bonded silicon carbide	1	—	—
Sintered silicon nitride	—	—	—

Vane

Material	Rig tested	Engine tested: Hours	Engine tested: High time part
Sintered alpha silicon carbide	24	894	890
Reaction bonded silicon carbide	164	900	890
Refel reaction bonded silicon carbide	7	—	—

TE82-6552(1)

Inner vane support

Material	Rig tested	Engine tested: Hours	Engine tested: High time part
Sintered alpha silicon carbide	1	—	—
Reaction bonded silicon carbide	9	900	896
Sintered silicon nitride	—	—	—

Vane retaining ring

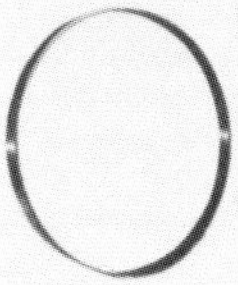

Material	Rig tested	Engine tested: Hours	Engine tested: High time part
Refel reaction bonded silicon carbide	5	900	890
Sintered silicon nitride	1	—	—
Reaction bonded silicon carbide	1	—	—

TE82-6552(2)

Regenerator disk

Material	Rig tested	Engine tested: Hours	Engine tested: High time part
1100°C rated alumina silicate thin wall matrix	6	1584	1267

TE82-6552(4)

Turbine tip shroud

Material	Rig tested	Engine tested: Hours	Engine tested: High time part
Reaction bonded silicon carbide	8	900	890
Sintered silicon nitride	—	—	—

Gasifier turbine blade

Material	Rig tested	Engine tested: Hours	Engine tested: High time part
Sintered alpha silicon carbide	327	20	16
Sintered silicon nitride	20	—	—

TE82-6552(3)

Strut shell

Material	Rig tested	Engine tested: Hours	Engine tested: High time part
Slip cast alpha silicon carbide	15	10	10
Refel reaction bonded silicon carbide	10	900	890
Sintered silicon nitride	1	—	—

Turbine inlet plenum

Material	Rig tested	Engine tested: Hours	Engine tested: High time part
Slip cast alpha silicon carbide	—	—	—
Siliconized silicon carbide	1	—	—
Slip cast sintered silicon nitride	—	—	—

Plenum inner annulus

Material	Rig tested	Engine tested: Hours	Engine tested: High time part
Slip cast alpha silicon carbide	—	—	—
Siliconized silicon carbide	1	—	—
Sintered silicon nitride	—	—	—

TE82-6552(5)

Figure 6. 2070°F-configuration gasifier turbine component.

Table II.
Combined 1900°F- and 2070°F-configuration ceramic component test experience.

Configuration	Hours 1900°F	2070°F	Total
Vanes (vane assy)	3406	900	4306
Tip shroud	1670	900	2570
Regenerator disk	7814	1584	9398
Blades	—	20	20
Engine	—	—	9398

summarized by component for each configuration and totaled for the engine CATE program. This total program test experience of 9398 hr on ceramic components represents a significant demonstration of ceramic materials as structural gas turbine engine components.

2265°F-CONFIGURATION SUMMARY

The final ceramic component configuration planned for the CATE program was the 2265°F configuration. Components for this configuration included all of those used in the 2070°F configuration plus a ceramic combustor and a two-stage power turbine with ceramic vanes and tip shroud. Initial design work was started on the combustor and the two-stage power turbine but was discontinued because of the program work-scope changes which eliminated the 2265°F configuration. The work completed on the 2265°F configuration is described in Sections IV and VI of this report. No hardware fabrication or development testing was conducted on the 2265°F configuration.

CERAMIC TECHNOLOGY DEVELOPMENT

The CATE program was the first of the major ceramic materials development programs sponsored by DOE and NASA. To accomplish the design, manufacturing, and testing previously described required a very significant effort in ceramic material technology development. The following is a summary of the ceramic material development activities that permitted the successful development of both 1900°F and 2070°F configurations:

- ceramic material behavior and characterization
- ceramic component characterization
- assessment of ceramic supplier capabilities
- evaluation of manufacturing processing and its influence on component strength and quality
- evaluation and development of abradable materials
- development and correlation (to test results) of NDE techniques
- failure analysis and design support
- support in planning and conducting the ceramic blade matrix experiment
- data analysis and verification of design methodology
- technical support for development of techniques used in the fixturing and machining of ceramic components
- development of supporting technologies such as diffusion and thermal barriers, insulating materials, and regenerator core and seal materials

Considerable development efforts in ceramic materials are still required, but the rate of progress and the successes enjoyed by the CATE program testify to the feasibility of commercial ceramic turbines.

Conclusions and Recommendations

The CATE program has enjoyed great success and represents a major step toward the goal of developing structural ceramic components for gas turbines and other types of engines. Several very significant conclusions have been made and have prompted the following recommendations to enhance the remaining DOE programs and other future ceramic development efforts:

1. Structural ceramic material development programs should concentrate on components for engines.
   - Transition from optimized test bar properties to equivalent properties in useful engine components is not yet possible.
   - Structural ceramic component design requires a well-defined engine environment that dictates the required ceramic material properties.
   - Ceramic components designed for some environment other than a "real life" engine do not help engine manufacturers apply ceramics to engines.
2. Basic ceramic materials research programs should be associated with an ongoing structural ceramic component development effort for an engine. This association will help ensure that the basic research work will have application to a useful product.
3. One of the objectives of ceramic material development programs should be to demonstrate ceramic components in an engine operating to its duty cycle.

- No laboratory scale test or rig test that simulates an engine can satisfactorily demonstrate ceramic component acceptability.
- Only engine testing to the most severe duty cycle expected for the engine application can demonstrate ceramic component acceptability.

4. A major part of any ceramic component development program should be a proof test for every ceramic part prior to its introduction to engine testing.
   - Engine testing of flawed ceramic components is extremely counterproductive because (1) engine failures are expensive and result in significant secondary damage and (2) engine failures of ceramic components leave little or no evidence to assist in understanding the cause or source of failure.
   - Proof testing of ceramic components is required to eliminate flawed parts.
   - Current NDE techniques are promising but require additional development to identify critical flaws in a cost-effective way.
5. NDE techniques have been developed and, in the laboratory, demonstrate the ability to find small flaws in ceramic components. However, these NDE techniques need further development to accomplish the following:
   - A correlation between the NDE indication of a flaw, the flaw size, and the location of the flaw within a component must be established and then related to stress distributions to predict if the flaw is critical to part survival.
   - The NDE techniques currently in the laboratory phase of development must be simplified to be used in a routine inspection environment.
6. Analysis techniques must be expanded to include time dependent failure modes. The probabilistic design techniques now developed are adequate to predict component reliability regarding the "fast fracture" failure mode. The following conclusions can be drawn relative to design methods:
   - Small static structural ceramic components such as vanes, shrouds, and rings can be designed, fabricated, and tested with reasonable confidence of success in a known environment.
   - Large static ceramic components such as plenums require more experience before a statement can be made regarding suitability as a gas turbine component. One conclusion, however, that can be drawn is that much more fabrication process development work is required by the ceramics suppliers if large duct-type structures are to be successful.
   - Rotating ceramic components, such as blades, can be successfully designed and their performance reliably predicted in a spin test rig environment.
   - Proof testing of ceramic components is required prior to engine testing. The proof test should duplicate or exceed the engine environment regarding mechanical and thermal loading.
7. The engine demonstrations conducted during the program illustrate the feasibility of using structural ceramic components in gas turbine engines.
8. The goal for all ceramic component testing should be to demonstrate ceramics in an engine operating to its duty cycle. To conduct cost-effective engine testing, ceramic components that have an acceptable probability of survival must be selected for use. Available inspection techniques are inadequate and advanced NDE techniques require additional development, so proof testing in rigs is required to eliminate the unacceptable components. Rig testing also provides a means to learn from failures. Because the evidence is not destroyed, it can be used to advance the understanding of brittle material failure mechanisms and to develop design methodology and nondestructive inspection techniques.

II. Engine Assembly and Test

Summary

Engine testing is the only acceptable means to demonstrate ceramic component proof of concept. The engine selected as the test bed to demonstrate the feasibility of structural ceramic components was the Allison 300-hp two-shaft GT 404 industrial gas turbine. The GT 404 was the baseline configuration engine which operated at 1002°C (1835°F) turbine inlet temperature. Figure 1 shows a GT 404 cutaway designating the ceramic components to be introduced.

This engine was initially modified from its all-metal configuration to accept individual inserted ceramic gasifier turbine vanes, a gasifier turbine tip shroud, and two ceramic regenerators. This configuration was referred to as the 1900°F-configuration engine, which signifies a gasifier turbine inlet temperature (TIT) of 1038°C (1900°F). The ceramic components are shown in Figure 7.

The 2070°F-configuration engine included a ceramic gasifier section consisting of 28 vanes, 3 flow-path rings, 4 flow-path strut shells, 40 rotor blades, a turbine tip shroud, and an improved ceramic regenerator system. Figure 8 is a cross section of the 2070°F-configuration gasifier turbine showing the ceramic components tested, except for the regenerator system.

The CATE program utilized three test bed engines for testing of ceramic components. These were designated as engines S/N C-1, C-2, and C-4. Engine S/N C-1 and C-2 were the 1900°F-configuration engines, and engine S/N C-4 was the 2070°F-configuration engine. In addition, engine S/N C-1 was installed and demonstrated in a 1979 GMC Astro 95 truck operating at 1038°C (1900°F) TIT. Figure 9 summarizes total hours accumulated on each engine configuration. Table II, shown in Section I, summarizes the combined 1900°F- and 2070°F-configuration engine test experience accumulated on structural ceramic components.

Engine S/N X-128 was the baseline configuration engine used under NASA contract No. NAS 3-20064. Results were reported in EDR 9346 (Ref 9). This EDR also included information on initial testing of engine S/N C-1.

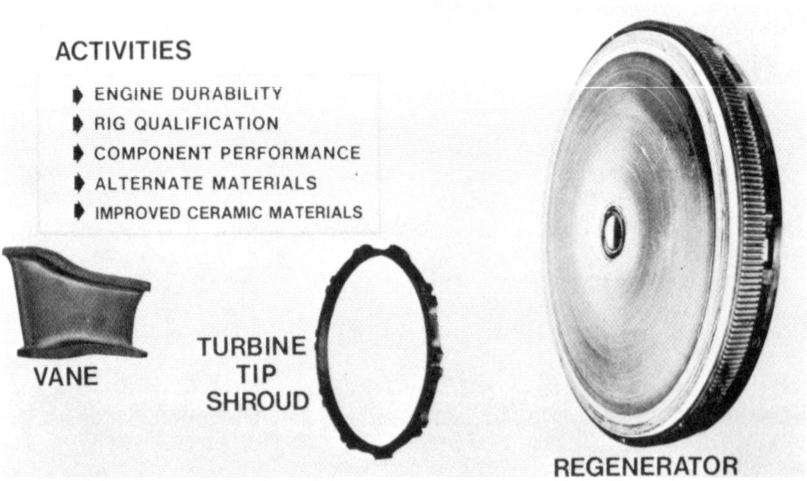

Figure 7. Initial ceramic component engine testing.

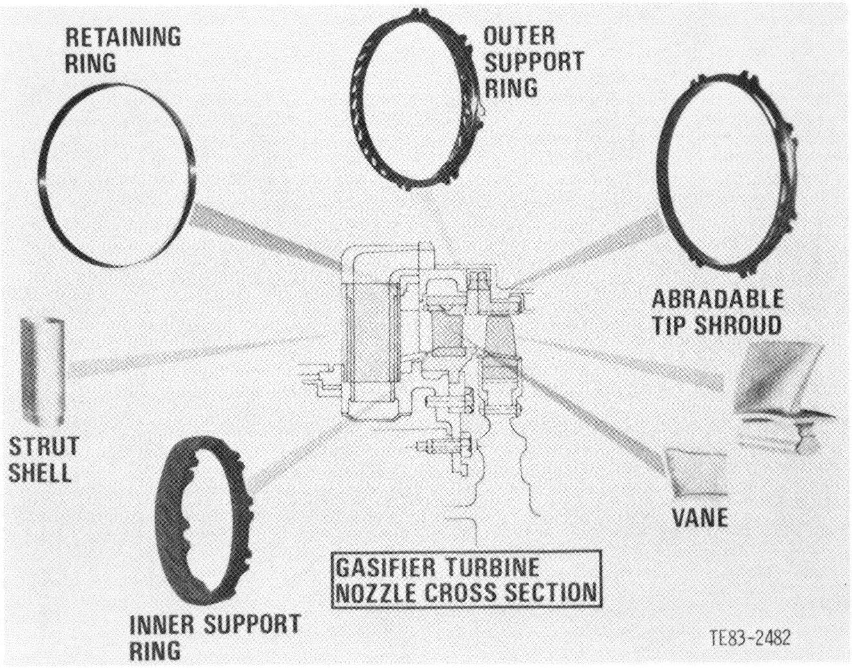

Figure 8. 2070°F-configuration gasifier turbine ceramic components tested.

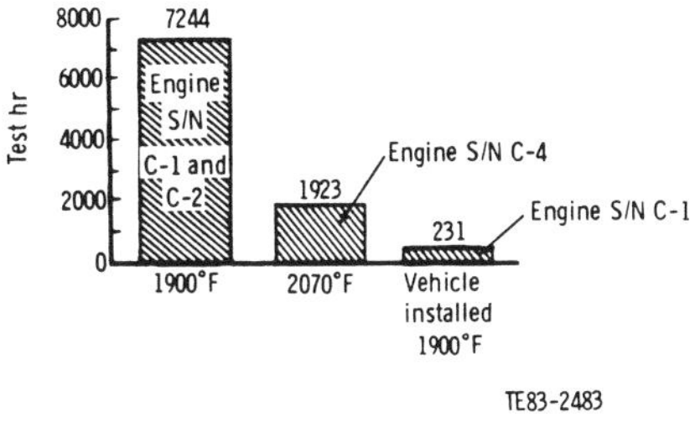

Figure 9. Test hours accumulated on each configuration.

Objective

The following are the objectives for the engine test portion of the CATE program:

1. evaluate ceramic components through exposure to actual engine environment and explore their durability capabilities
2. evaluate durability of ceramic materials and components operating with typical data cycles encountered in buses and trucks
3. demonstrate the feasibility of structural ceramic components in gas turbine engines

Approach

To successfully test ceramic components in an

engine environment, the concept of ceramic proof testing was used extensively before any ceramic component was installed in the engine. Proof testing consisted of a rig evaluation of ceramic components that induces a stress in the part equivalent to that expected in engine use. The critical stress developed in ceramic components is due to thermal gradients induced by the environment in the engine.

The concept of proof testing must be integrated with component characterization, research and development testing, engine testing, and design and manufacturing efforts. Figure 10 illustrates this relationship and identifies the proof test rigs used during CATE testing.

The following paragraphs briefly describe the rigs used to qualify ceramic components (proof testing) for engine test and the rigs used for research and development work.

Thermal Shock Rig—A thermal shock rig capable of accepting the gasifier nozzle assembly, plenum, and combustor was used to proof test ceramic nozzle and plenum components. The rig was capable of simulating the engine's most severe transient thermal environment to the extent that appropriate thermal stresses were induced in the parts.

Hot Regenerator Rig—A full-scale hot rig capable of evaluating the regenerator disk and associated seals at engine rated conditions was used to proof test ceramic disks and to measure seal leakage, disk pressure drop, and effectiveness. This rig is a practical means of evaluating regenerator system performance prior to engine test.

Spin Test Rig—A spin test facility was used to test metal rotors with either individual or up to 40 ceramic blades inserted into the rotor at speeds up to

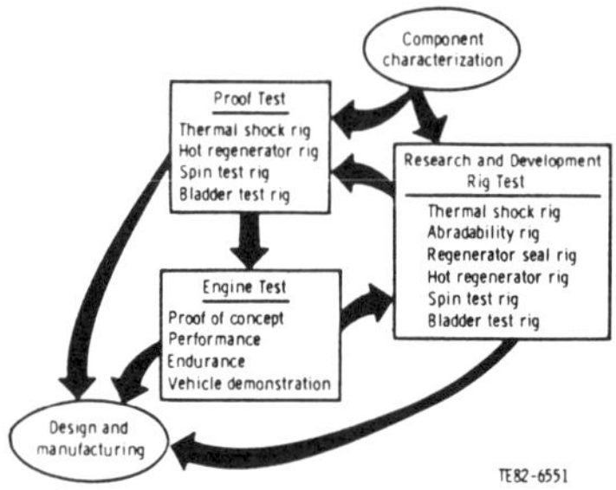

Figure 10. Ceramic component engine testing approach.

burst. For the CATE configuration, that is between 100% and 200% of rated speed (rated speed is 36,905 rpm).

Spin testing has been conducted at ambient temperatures and at temperatures up to 1250°F (677°C) in an evacuated (vacuum) environment. This test facility permits ceramic blade testing, which simulates the stresses due to centrifugal loads and at temperatures in the blade attachment expected during engine operation. All blades used for engine testing were first proof spin tested individually to 117% speed and then assembled into a rotor assembly (40 ceramic blades), which was proof tested to 108% speed.

Bladder Test Rig—The ceramic ring structures used in the nozzle assembly are fabricated to near net shape, which minimizes machining. However, some finish machining is required, which is a significant expense. To permit proof testing of ceramic rings before finish machining, a bladder rig was developed. This rig accepts the as-received ring and tests the ring by creating tensile stresses by pressurizing a bladder that fits into the inside diameter. Stresses similar to those created by thermal gradients during engine operation can be achieved. This test eliminates flawed components from the expensive machining and thermal shock rig testing operations.

Regenerator Seal Leakage Rig—The regenerator seal system used in the CATE program features rubbing seals between the disk and seal platform and thin foil leaf seals between the platform and the engine block structure. A special static seal leaf test rig has been developed to permit leaf leakage measurement while creating appropriate pressure differentials across the seal leaf structure. This rig operates at ambient temperature, with the rubbing side of the assembly sealed off to permit independent measurement of the leaf leakage.

Abradability Rig—For ceramic turbine assemblies to operate to their maximum efficiency potential, abradable turbine tip shrouds are required. This permits tolerances that are feasible for production ceramic components to be used with the knowledge that blade tip rub will occur without damage and the resulting turbine tip clearance will provide high turbine efficiencies. A laboratory scale rig, featuring a rotor driven metal wheel with provisions for attachment of ceramic blades, is used to evaluate candidate abradable materials. The abradable material can be attached to either metal and/or ceramic substrates, duplicating the desired engine configurations.

Proof testing was used to qualify ceramic components for the engine feasibility and durability evaluations. These engine tests are described in the following discussion for each configuration of ceramic components.

Discussion

1900°F CONFIGURATION

Engine test activities were completed for the 1900°F configuration during the first half of CY 1981. Figure 4 shows the ceramic components tested, the ceramic material evaluated, the number of parts rig tested, and a summary of engine experience. This figure also indicates if a ceramic component has been demonstrated in a vehicle by a check mark in the truck-installed column.

The vehicle demonstration was a significant test because, combined with the more than 7475 hr of dynamometer engine test experience, it demonstrated the feasibility of structural ceramic components in a vehicle engine operating under severe conditions. The vehicle test program, summarized in Section I, accumulated 231 hr of engine operation and covered 6656 miles over city streets, interstate highways, and the General Motors proving grounds.

The successful engine and vehicle demonstration of the 1900°F-configuration paved the way for the 2070°F-configuration development. Detailed discussions of the 1900°F-configuration test experience can be found in the following published semiannual reports covering a period from 16 January 1978 through 31 June 1981—EDRs 9519, 9722, 9951, 10156, 10383, and 10841 (Ref 1-5 and 7).

2070°F CONFIGURATION

Engine testing of 2070°F-configuration components was initiated with the buildup of engine C-4. The ceramic components for the 2070°F-configuration gasifier turbine section are shown in Figure 11. The only components not shown are the regenerator disks.

A total of 78 ceramic components make up the 2070°F configuration. The test experience of these components was shown earlier in Figure 6. All of the ceramic components in the 2070°F-configuration have been engine tested independently and in combination. That is, the gasifier nozzle assembly was tested with a metal gasifier turbine rotor and the ceramic gasifier turbine rotor was tested with a metal gasifier nozzle assembly. Discussion of this test phase (January 1980 through December 1981) can be found in EDR 10383, 10672, 10841, and 11006 (Ref 5-8).

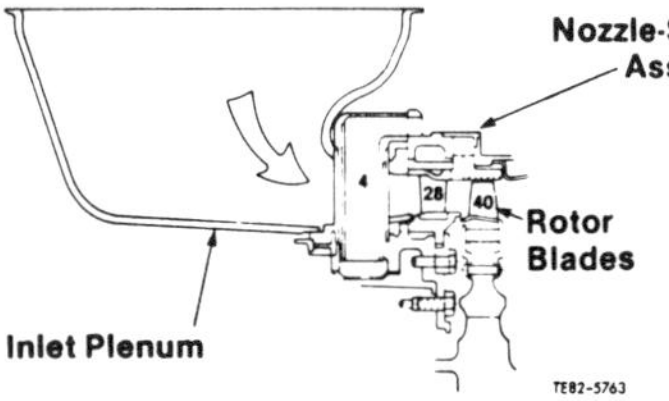

Figure 11. 2070°F-configuration gasifier turbine ceramic components.

A total of 890 hr of cyclic durability was accumulated on one ceramic gasifier nozzle assembly during this period.

During 1982, engine testing of the 2070°F configuration included testing of both the ceramic gasifier nozzle assembly and turbine rotor together. There were two engine builds of S/N C-4 (BU13 and BU14) for testing the ceramic gasifier nozzle and rotor assemblies. All of the ceramic components tested were rig qualified prior to engine build. The following paragraphs include a discussion of engine testing on the complete 2070°F configuration, including detailed discussions of BU13 and BU14. The tests were conducted in June and December 1982.

Engine C-4, BU13

Summary

During June 1982, the 2070°F-configuration ceramic gasifier section was tested in engine C-4. A summary of the ceramic components installed is provided in Table III.

The engine testing conducted can be described as two tests: a performance test over the entire speed range from idle to 100% and a durability test at the maximum rated speed and turbine inlet temperature. A total of 3 hr and 56 minutes was accumulated during both tests. The performance test included 1 hr and 12 minutes at 100% gasifier speed (36,905 rpm), which included 27 minutes at 1132°C (2070°F) average turbine inlet temperature (T_4). The remaining 45 minutes at 100% speed were at T_4 temperatures varying from 927°C (1700°F) to 1133°C (2070°F). The performance test was completed and a borescope inspection conducted with no evident problems. The second test consisted of a repeat schedule of the performance test up to the 100% speed and 1132°C (2070°F) T_4 condition. At that

Table III.
Ceramic components in engine C-4, BU13.

Component	Quantity	Material	Manufacturer
Gasifier turbine blades	40	Alpha SiC	Carborundum
Gasifier nozzle assembly vanes	7	Alpha SiC	Carborundum
	21	RBSiC	Carborundum
	28		
Outer vane support	1	Alpha SiC	Carborundum
Inner vane support	1	RBSiC	Carborundum
Turbine tip shroud	1	RBSiC	Carborundum
Vane retaining ring	1	Refel SiC	Pure Carbon Co.
Strut shells	1	Alpha SiC	Carborundum
	3	Refel SiC	Pure Carbon Co.
	4		
Regenerator disk	2	1100°C AS	Corning Glass

point, approximately 50 hr of testing was planned. During the acceleration from 95% to 100% speed, a failure occurred. The following conclusions were reached after the investigation:

- The primary cause of failure was not identifiable.
- All failure origins found on the parts appeared to be caused by impact damage.
- The possibility of engine structural distortion causing the failure has been eliminated.
- The most probable failure modes have been reduced to foreign object damage (FOD) induced blade failure caused by Fiberfrax insulation or nozzle outer vane support ring failure caused by vibration.

Objective

The test objective was to evaluate the 2070°F-configuration ceramic gasifier nozzle and rotor at rated speed and temperature, including engine performance and endurance testing. A secondary objective was continued evaluation of the block cooling design.

Approach

The configuration of the gasifier turbine assembly for BU13 is shown in Figure 12. All ceramic nozzle components were thermal shock rig tested (BR 95, BU11) prior to engine use. After 5 cycles simulating the most severe engine operating condition, the following parts were qualified for use in engine C-4, BU13:

- turbine tip shroud—reaction bonded silicon carbide (RBSiC), CBO with abradable layer (rig qualified to 1132°C [2070°F] twice, with one incident of rig overtemperature to 1571°C [2500°F])
- outer vane support—alpha SiC, CBO (rig qualified to 1132°C [2070°F])
- inner vane support—RBSiC, CBO (rig qualified to 1132°C [2070°F] with one incident of rig overtemperature to 1571°C [2500°F])
- vane retaining ring—Refel SiC, Pure Carbon (rig qualified to 1038°C [1900°F]; rig qualified to 1132°C [2070°F] with one incident of rig overtemperature to 1571°C [2500°F])
- strut shells—four parts (rig qualified to 1132°C [2070°F])
- vane—20 RBSiC and 8 alpha SiC, CBO (tested 1 to 7 times in rig; 1132°C [2070°F] qualified BU11)

Discussion: Engine Test

Testing of engine C-4 started with a performance test (Test 1) that was successfully conducted to 100% speed and 1148°C (2100°F) burner outlet temperature (BOT), which is equivalent to 1132°C (2070°F) rotor inlet temperature. The performance test required 2 hr and 38 minutes to complete and followed a carefully planned, step-by-step schedule, as shown in Figure 13. Each of the points shown on the curve in Figure 13 has a number in parentheses next to it; this is the time spent at that speed and temperature condition. The following is an outline of the performance test:

1. motor engine to 8000 rpm

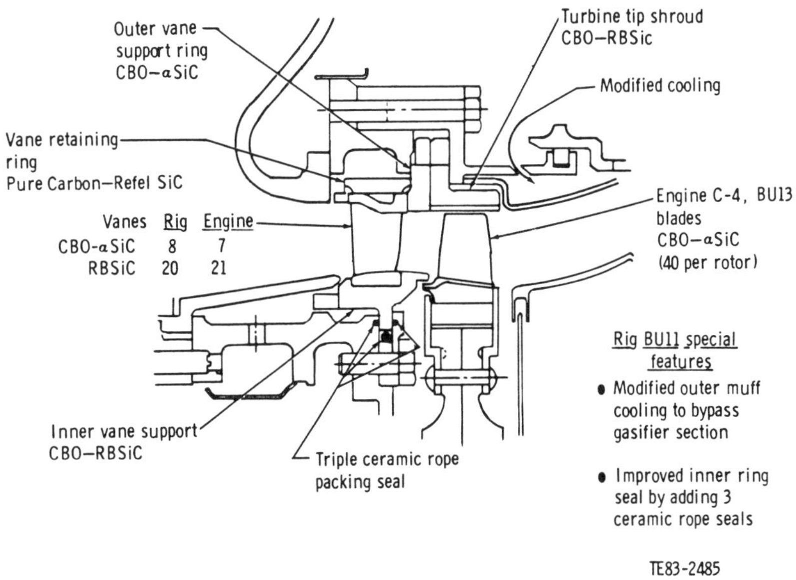

Figure 12. Engine C-4, BU13 gasifier turbine configuration.

2. borescope inspection—parts in excellent condition
3. start to idle, stabilize, record data, stop (running time: 30 minutes)

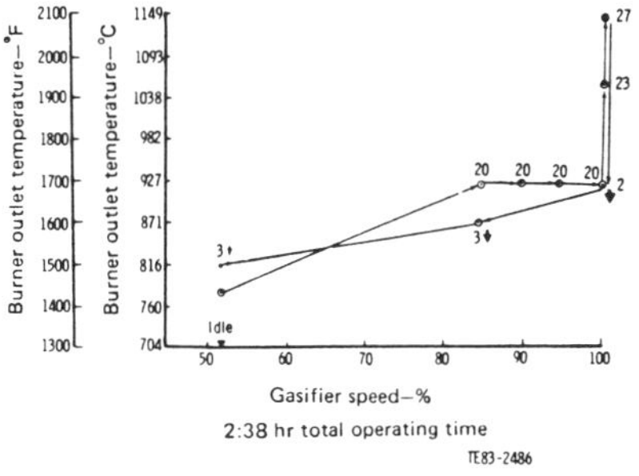

Figure 13. Engine C-4, BU13—Test 1 schedule.

4. borescope inspection—parts in excellent condition
5. start to idle, conduct performance test to 100% speed, 1144°C (2100°F) BOT conditions, stop (running time: 2 hr and 38 minutes)
6. borescope inspection—excellent (deposit noted on suction side of airfoils)

The ceramic materials, design, and project engineering personnel conducted the borescope inspections and concluded the engine was acceptable for continued testing. The next test planned was a durability test (Test 2) which was scheduled to accumulate approximately 50 hr at rated speed (36,905 rpm) and 1140°C (2100°F) BOT, which is equivalent to 1132°C (2070°F) gasifier rotor inlet temperature.

Test 2 included a start and acceleration to rated conditions similar to the schedule used in Test 1. Figure 14 shows the schedule and time at each speed condition up to 95% speed. During the acceleration from 95% to 100% speed, a failure occurred. The following is an overview of Test 2:

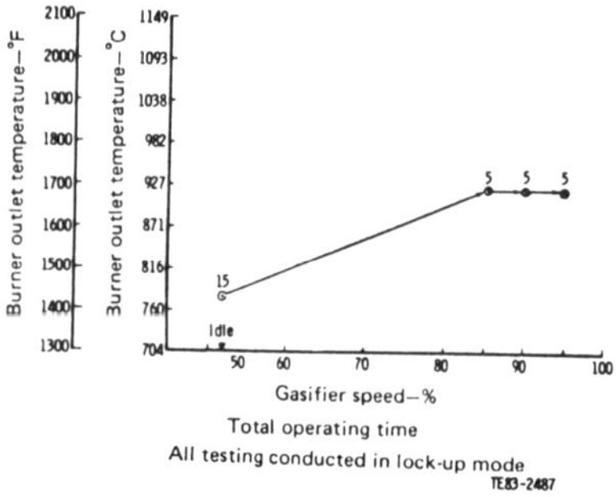

Figure 14. Engine C-4—Test 2 schedule.

- start to idle—discovered and repaired air leak restricting performance (running time: 12 minutes)
- start to idle—stabilized for 15 minutes, continued acceleration per schedule to 95% speed at 927°C (1700°F) BOT (running time: 30 minutes through to 95% speed)
- speed change initiated to accelerate to 100% speed, instantaneous vibration spike noted, loss of compressor discharge pressure (CDP) 61 kPa (18 in. Hg), negative torque (running time: 6 minutes to stop)
- Total time: 1 hr and 18 minutes

Data recorded during Test 2 are shown in Figure 15. Note the front lateral vibration spike and the low level of vibration that continued until shutdown. With all the other recorded parameters shown by the strip chart within limits, the loss in CDP and high negative torque was the only indication of a problem with the engine. Shutdown was initiated and the engine borescoped. Inspection revealed that a failure did occur in the gasifier turbine. The engine was left on the test stand while a failure investigation plan was formulated.

Prior to engine removal from the test stand, a failure investigation plan was formulated as follows:

1. Review all data before engine removal.
2. Identify potential failure modes.
3. Prepare engine disassembly checklists addressing each potential failure mode.
4. Direct engine disassembly in accordance with the checklists.

Engine disassembly was conducted based on the plan described. The data review resulted in the following conclusions:

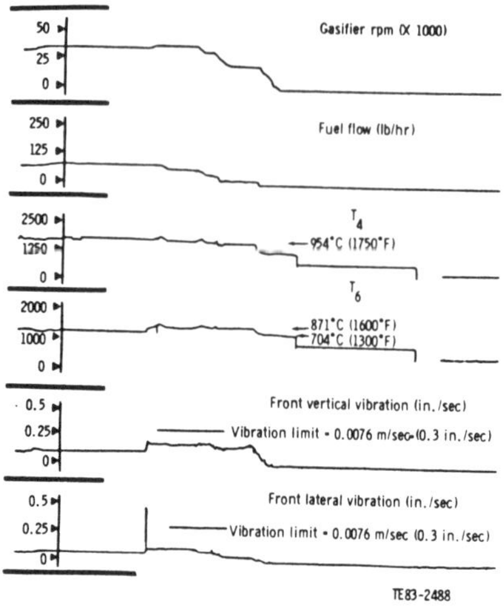

Figure 15. Strip chart data at time of failure for engine C-4, BU13.

- All four starts were similar and characterized by a slow temperature rise that did not exceed established limits.
- There was no abnormal engine vibration or noise other than at the moment of failure.
- All transients through the blade fundamental frequency response range were completed in less than 60 sec.
- All performance data were normal, and engine operation was considered acceptable up to the moment of failure.

The potential failure modes identified were the following:

- foreign object damage (nonceramic source)
- engine structural distortion causing mechanical loading
- ceramic nozzle failure
- ceramic rotor failure
- plenum/inner-annulus/combustor failure

The remaining part of the Engine Test Section concentrates on each of the potential failure modes. The mode of plenum/inner-annulus/combustor failures was eliminated because disassembly revealed those parts to be in good condition. Each potential mode of failure was analyzed individually, and each will be discussed in the following subsections.

Discussion: Foreign Object Damage

Any airborne particles entering the engine must pass through the regenerator disks before they reach the combustor and gasifier section. Since the regenerators act as very good filters for debris (particles larger than 0.508 mm [0.020 in.] in diameter cannot pass), investigation of FOD was restricted to an area downstream of the regenerator and upstream of the gasifier rotor assembly. This area includes the T_3 cavity, combustor, plenum, plenum inner annulus, and gasifier nozzle assembly. Items such as nuts, bolts, washers, locktabs, shim stock, etc, were checked to determine if any of these had dislodged and passed through the turbine. The combustor was intact as was the igniter and fuel nozzle. No distress was noted on the plenum or inner annulus. All shim stock on the instrumentation leads and instrumentation itself was intact. However, a piece of insulation was missing in T_3 cavity near the combustor. Figures 16 and 17 show the area where the insulation is missing. The size of this missing piece of insulation can fit through the vane throat and could impact the gasifier turbine rotor. The insulation is a Fiberfrax moldable material that is wet when applied and, after curing, becomes quite hard. There was also the possibility of a segment from the outer vane support ring failing due to vibration and impacting the gasifier turbine rotor blades.

To determine what this Fiberfrax insulation and a segment of the outer vane support ring could do to a ceramic blade on impact, a special test using high-speed photography was conducted in the blade spin pit. This special test is described in the following paragraphs.

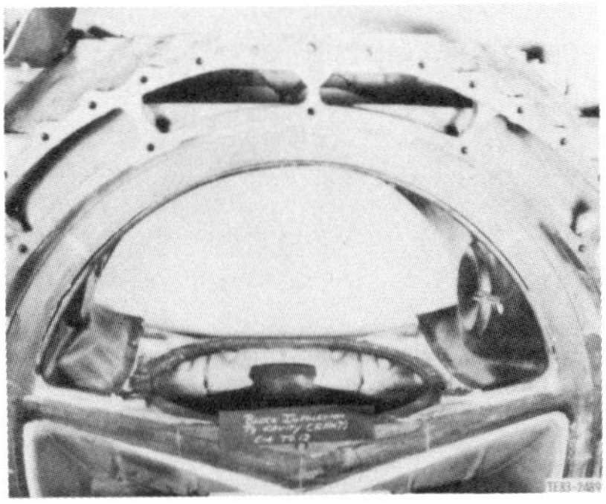

Figure 16. Fiberfrax insulation missing (T_3 cavity).

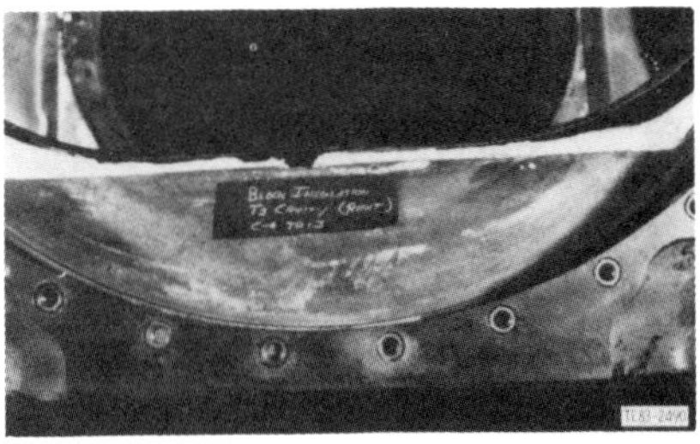

Figure 17. T_3 cavity area where Fiberfrax insulation could have entered.

Based on the information analyzed and the possibility of FOD, two probable causes of the ceramic gasifier turbine failure, engine C-4, test 13, were determined:

- loose Fiberfrax insulation (hard material) from the T_3 cavity
- a vibration failure of a segment (finger) from the outer vane support ring

One of these pieces probably passed through the gasifier nozzle/turbine components and caused the failure. Furthermore, the laboratory post-failure investigation of the turbine blade remnants revealed contact (impact) failure origins. Based on these observations, it was decided to conduct turbine blade FOD spin tests. The specific objectives of the spin tests were the following:

- with suspect particles, attempt to duplicate the engine blade failures
- attempt to demonstrate the successful impact of small particles

Three spin tests were conducted addressing these objectives utilizing the following:

- Fiberfrax piece of insulation from engine
- ceramic outer vane support ring finger
- one steel shot, 0.5 mm (0.020 in.) diameter

The test conditions were as follows:

- single blade previously proof tested at 117% N_G
- vacuum
- direction of rotation the same as the engine
- 97% N_G speed (35,800 rpm)—speed at time of engine failure

Release of the particles was accomplished by electric current burning a support wire, as illustrated in Figure 18. A high-speed (7500 frames per second) film record of the impact was taken with equipment shown in Figure 19.

Test 1. This test was conducted with a sugar-cube block of Fiberfrax (hard) insulation

Figure 18. FOD spin test configuration—0.5 mm (0.020 in.) steel shot.

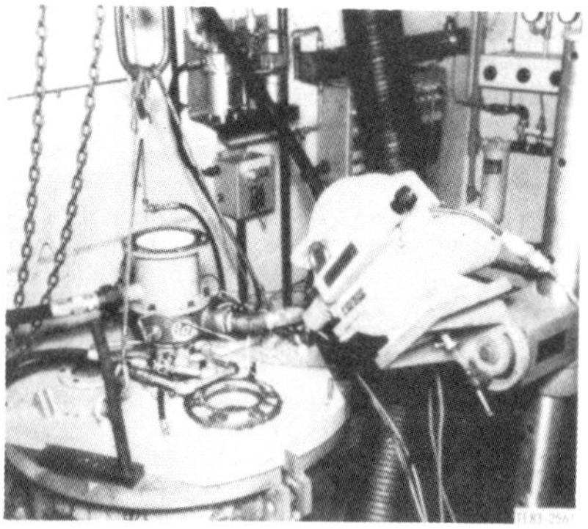

Figure 19. High-speed camera configuration, FOD impact tests.

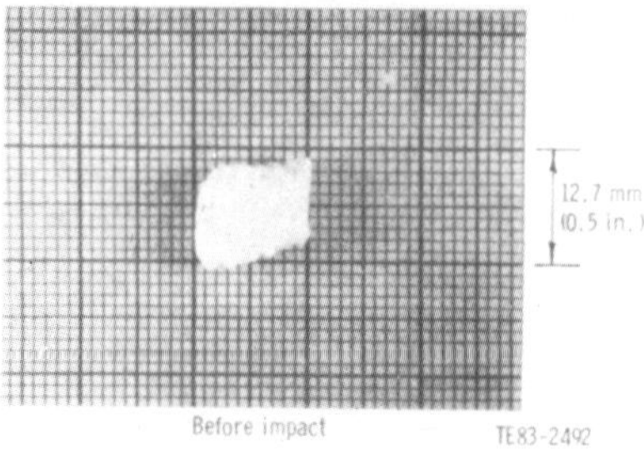

Figure 20. Fiberfrax FOD particle before impact, turbine blade test No. 1.

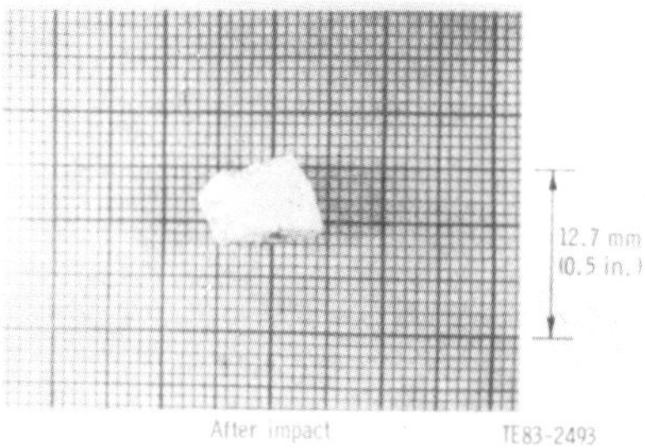

Figure 21. Fiberfrax FOD particle after impact, turbine blade test No. 1.

Figure 22. Failed spin test blade, Fiberfrax FOD test (top view).

weighing 0.715 g and illustrated in Figure 20. (This size will pass the combustor air dilution holes and the gasifier nozzle vane throat.) The bulk of the sample survived the impact test, as shown in Figure 21. The remnant weighed 0.57 g or 80% of the weight before impact. The insulation was positioned to strike the blade approximately 6 mm (0.24 in.) below the airfoil tip. This piece failed the airfoil near the root similar to the engine failure blades. The failed blade is shown in Figures 22 and 23. The laboratory identified the failure origin at the suction side crown, again similar to the engine.

Test 2. A segment from the ceramic outer vane support ring, weighing 2.415 g and illustrated in Figure 24, was utilized in the second FOD impact

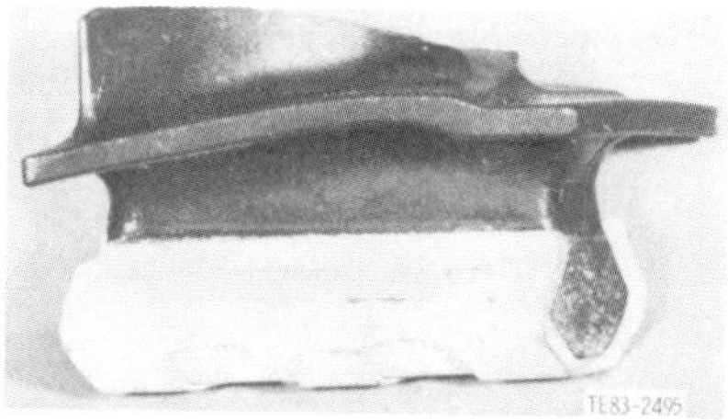

Figure 23. Failed spin test blade, Fiberfrax FOD test (side view).

test. Figure 24 shows the segment before impact and after impact, when the segment broke into multiple pieces. Again, this size particle will pass the nozzle vane throat. The segment was positioned to strike the blade approximately 6 mm (0.24 in.) below the airfoil tip. In the high-speed film, two

Figure 24. Outer vane support ring FOD segment before and after impact, turbine blade Test No. 2.

impacts of the outer vane support ring (OVSR) segment are evident. This piece failed the airfoil (completely obliterated) near the root similar to the engine failure blades. Furthermore, the forward acute corner of the platform was also failed similar to 12 engine blades. This is shown in Figures 25 and 26. Laboratory analysis identified the origin of the platform corner failure to be in the stalk, as was the case with the engine blades.

Test 3. The engine is configured with two fine-mesh ceramic regenerators that act as filters screening the compressor discharge air prior to entering the combustor. Trials demonstrated that 0.5 mm (0.020 in.) and smaller shot would pass through the regenerators. Thus FOD particles of this size could enter the engine and impact the turbine blades. The final impact test released one 0.5 mm

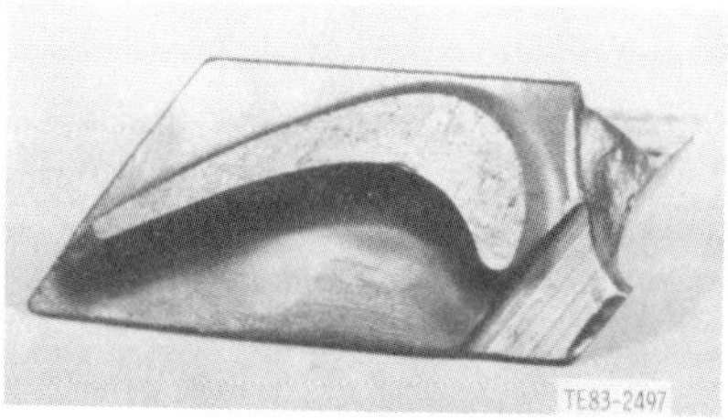

Figure 25. Failed spin test blade, OVSR segment FOD test (top view).

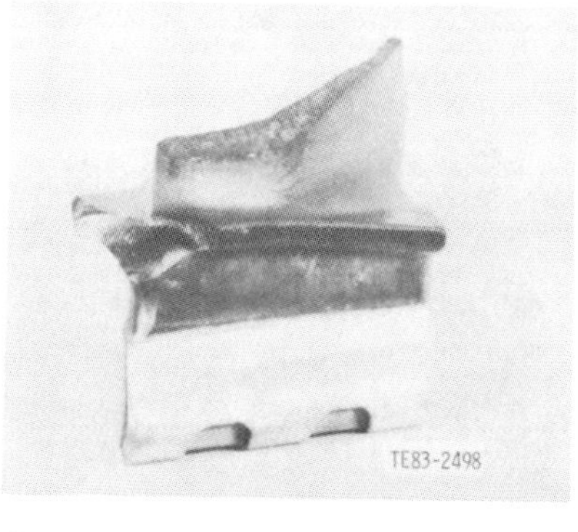

Figure 26. Failed spin test blade, OVSR segment FOD test (side view).

(0.020 in.) diameter steel shot (0.53 mg) into the path of a blade spinning at 97% N_G speed. The blade successfully withstood the impact.

The overall results of the blade impact tests are summarized as follows:

- FOD could occur downstream of the regenerator disks (Fiberfrax insulation and OVSR segment) that did fail the spin test blade in impact at 97% N_G speed.
- The blade successfully survived impact of a 0.5 mm (0.020 in.) diameter steel shot, which could pass the regenerator disks.

An overall summary of the turbine blade failure in engine C-4, BU13 is as follows:

1. Failure due to steady-state stress, vibratory response, rub, and block distortion was remote.
2. No blade that could be considered the primary engine failure was found.
3. Engine blade fracture surface appearance was duplicated on spin test blades that failed due to FOD.

Due to the number of impact type fractures, the number of failed pieces missing, and the special spin test results, the following conclusions were reached from engine C-4, BU13 failure analysis:

- The primary cause of failure cannot be identified.
- All failure origins found on the parts appear to be caused by impact damage.
- The possibility of engine structural distortion causing the failure has been eliminated.
- The most probable failure modes have been reduced to the following:
  - FOD induced blade failure caused by Fiberfrax insulation
  - nozzle outer vane support ring failure caused by vibration

The special spin tests conducted indicate that both the insulation and ceramic ring segment can cause ceramic blade airfoil failure when impact occurswith rotor speed at 97%. No failure occurred on impact with a 0.508 mm (0.020 in.) diameter steel shot.

Discussion: Gasifier Turbine Nozzle Failure

Upon removal of the gasifier nozzle assembly from the engine, some initial observations were made. (Figure 27 shows the gasifier section from the rear.) Approximately the top half of the nozzle section was missing. There are pieces of the rings and some vanes resting at the 180-deg position. Some vanes were cracked but still remained in place.

Figure 28 shows the nozzle after removal of the gasifier turbine and the shroud assembly. Some additional pieces have fallen out of place during handling, but the retaining ring was continuous through about a 260-deg arc. The portion of the outer vane support ring that remained in place was cracked in four places through a 180-deg arc. Twelve vanes remained firmly in place, three fewer than were seen in Figure 27. The inner vane support ring was in one piece with large chips removed from the trailing edge. As was mentioned earlier, very few pieces of the shroud remained in place.

Figure 27. Gasifier turbine assembly.

Figure 28. Gasifier nozzle after removal of rotor of engine C-4, TD14.

Figure 29. Power turbine nozzle assembly.

Figure 29 is a view into the inlet of the power turbine nozzle. This shows large pieces of the retaining ring and outer vane support ring, some pieces of the shroud, and a few pieces of vanes lodged against the power turbine nozzle. Some pieces are wedged between the vanes while most of the pieces are loose. The size of the pieces remaining indicated that the gasifier turbine blades were already

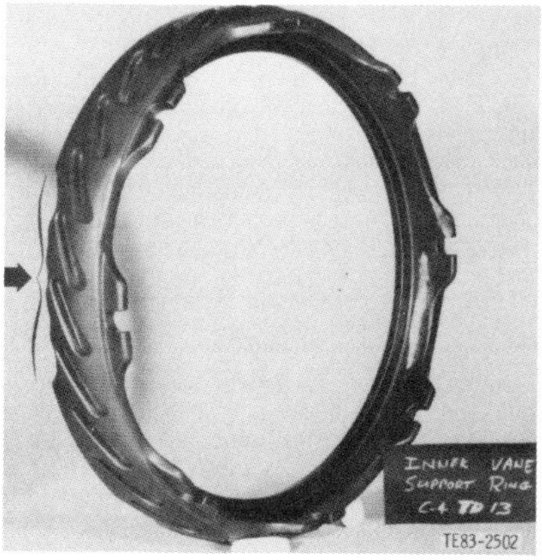

Figure 30. Inner vane support ring after disassembly of engine C-4 (aft view).

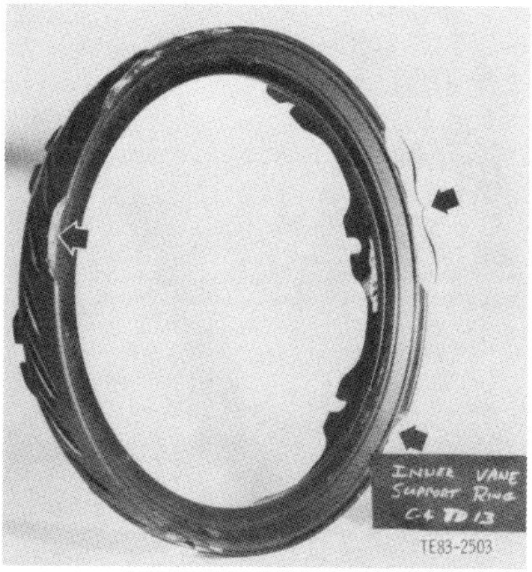

Figure 31. Inner vane support ring after disassembly of engine C-4 (forward view).

gone when these pieces passed through the rotor.

A nozzle failure could have been precipitated by a failure of any of the rings or strut shells. Each component is discussed in the following paragraphs.

Figures 30 and 31 show the inner vane support ring after removal from the assembly. The ring is not cracked through but has six areas of large chips at the trailing edge. Closer examination by the Materials Laboratory revealed that the inner vane support ring was primarily intact, with some minor damage on the lip at six locations. The area of the inner ring that was chipped was the back lip adjacent to vane pockets 3, 4, 7-10, 15, 20, and 26. All of these chips originated from the inside diameter of the ring near the vane pockets. A typical example of the damage is shown in Figure 32, pocket 20. Pocket 28 was also observed to be chipped at the forward section of the vane pocket. The damage to the inner ring is believed to have been caused by impact with platforms of the blades.

Figure 33 shows the remaining pieces of the outer vane support ring after disassembly. One vane was wedged in place with small particles of silicon carbide between the vane slot and the vane foot. All pieces were sent to the Materials Laboratory for closer evaluation.

The lower half of the outer vane support ring (pockets 9 to 25) was retained even though it was

Figure 32. Inner vane support ring fracture origin caused by impact.

broken in several places. In addition, pieces from pockets 26 to 2 and 3 to 6 were recovered. The partially reconstructed ring is shown in Figure 34. The

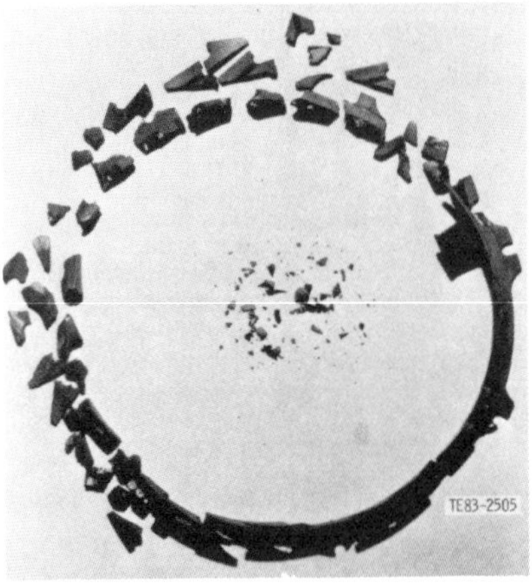

Figure 33. Remaining pieces of outer vane support ring after disassembly of engine C-4.

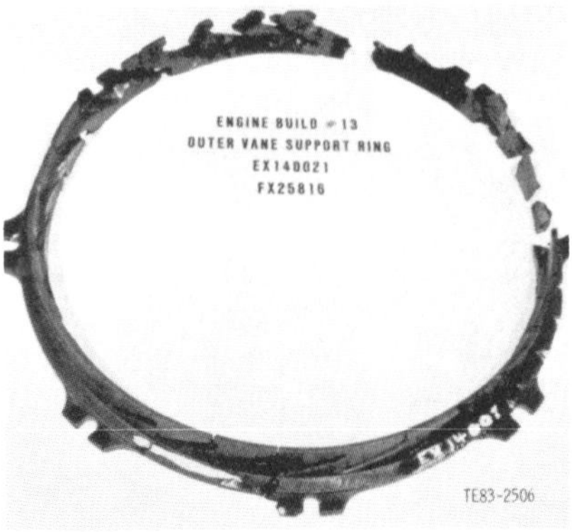

Figure 34. Outer vane support ring after reconstruction by Materials Laboratory.

majority of fractures in this component were contact or impact failures that initiated from the aft portion of the vane pocket surface. A typical example of this contact fracture is shown in Figure 35. Several pieces of the ring were fractured from the aft portion of the flange, inside diameter (see lug chart in Figure 36).

The failures at the first and fifth lugs were identified as contact fractures. These lugs failed from contact on both sides of the lug. The third lug also failed as a result of contact, but only on the top section. The bottom section of the seventh lug failed, but this was not a contact fracture. The top section of the lug was missing.

Since some pieces of the outer vane support ring were missing, it is still unknown whether or not this ring was the primary failure. One theory is that vibration caused failure of a cantilevered vane retention feature, which subsequently passed through the nozzle and failed the blades. This possibility is supported by the fact that the engine previously ran 890 hr with an unslotted outer vane support ring.

Prior to slotting a ring for engine testing, a vibration test was conducted on a slotted ring segment to determine its frequencies in the engine operating range. The segment was tested with various vane loads with and without damping. Figure 37 shows a frequency speed diagram indicating a number of modes do occur in the operating range. Response with very light damping was found to be minimal. The nozzle assembly features rope packing

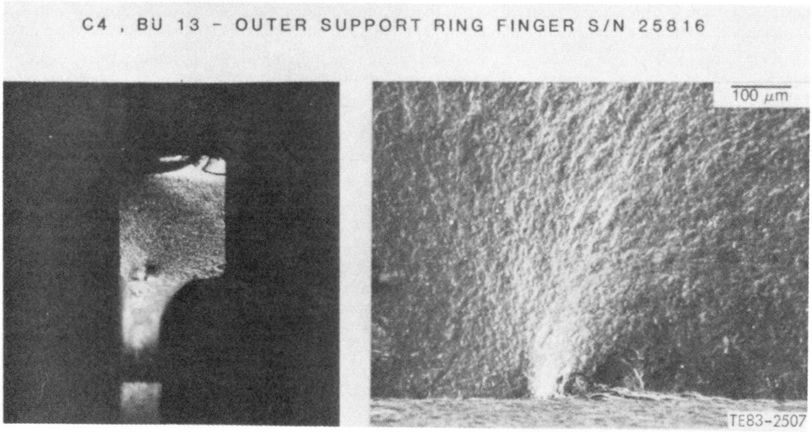

Figure 35. Outer vane support ring contact fracture.

between the outer vane support ring outside diameter and the vane retaining ring inside diameter. This rope packing provides vibration damping and acts as a seal.

The ring was slotted based on the original three-dimensional (3-D) stress analysis which indicated high stresses in the vane pockets. These results were subsequently determined to be erroneous. In EDR 11006 (Ref 8) modifications to the 3-D analysis and a two-dimensional (2-D) analysis that correlated with the 3-D results were shown. The 2-D analysis, along with revised temperature predictions from the instrumented nozzle testing, was used to determine the effect of slotting the leading edge of the ring. With slots at the leading edge, the maximum principal stress was 136.9 MPa (19.9 ksi) with a P_s of 0.99137 for 12 sec into deceleration.

Figure 38 shows that the principal stress in the

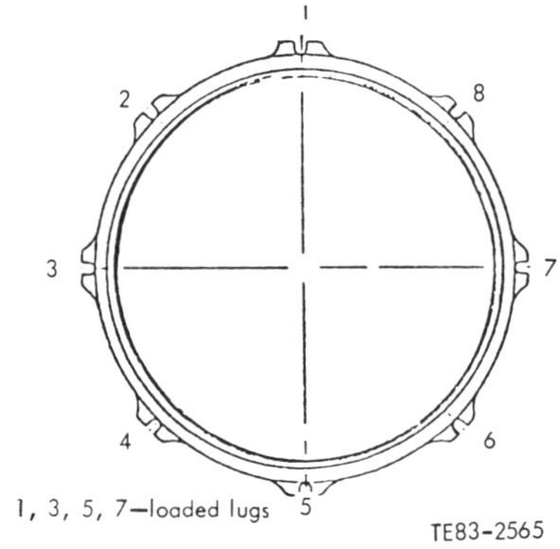

Figure 36. Lug chart, looking aft.

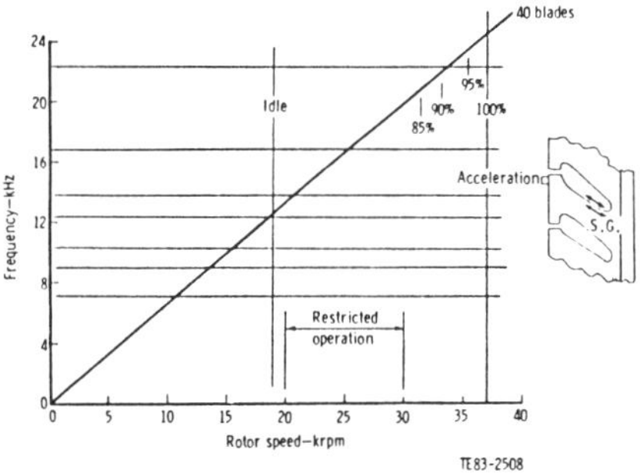

Figure 37. Ceramic outer vane support slotted segment frequencies.

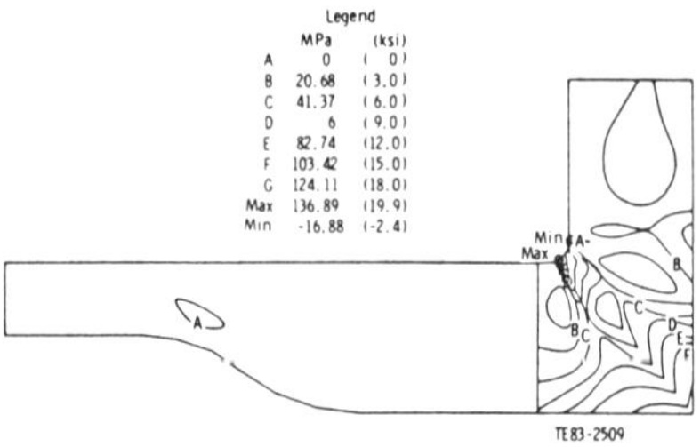

Figure 38. Plot of maximum principal stress at 12 sec into deceleration for support ring with slots at leading edge of each vane pocket.

pocket inside diameter is approximately 62.0 MPa (9.0 ksi). This 2-D analysis does not account for a stress concentration factor in that area. With no slots at the leading edge, the maximum principal stress was reduced to 125.7 MPa (18.2 ksi) with a P_s of 0.99535 for the same conditions. Figure 39 shows that the pocket stress increased to 96.5 MPa (14.0 ksi) even though the maximum principal stress decreased. The P_s stated for these cases assumes σ_{os} = 774.1 MPa (50.0 ksi), σ_{ov} = 626.3 MPa (27.0 ksi), and Weibull modulus = 8. Even with a stress concentration, the predicted pocket stresses would not cause failure in an unslotted ring. Therefore, to reduce the possibility of an outer vane support ring failure, the next ring to be engine tested was unslotted.

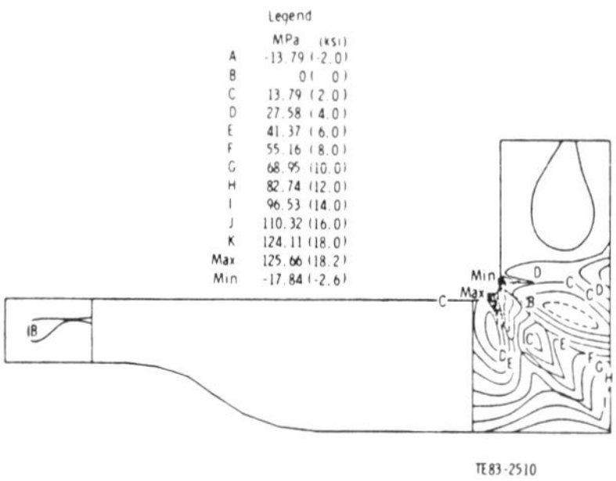

Figure 39. Plot of maximum principal stress at 12 sec into deceleration for outer vane support ring with no slots at leading edge of vane pockets.

As mentioned earlier, 13 of 28 vanes were missing. Twelve RBSiC vanes were intact and appear suitable for future testing. One other RBSiC vane had a chip out of the trailing edge. One vane had a foot broken off while another was split down the middle. Several pieces of other vanes were found lodged against the power turbine nozzle. Thirteen of the 28 vanes (vanes 9-11, 14-19, and 21-24) were found to be intact, although the surfaces on the trailing edge of the suction side, primarily near the outer vane support ring, were blasted by debris and exhibited superficial damage. Vane 12 was broken on the foot of the trailing edge by the outer support ring, probably a result of a surface crystal layer. Vane 13 was chipped on the trailing edge, presumably as a result of impact damage. Vane 20 was fractured axially into two halves. Traces of the abradable layer of the shroud were detected on the surface.

The vane retaining ring was approximately 75% intact. Figure 40 shows that most of the pieces of the ring were found and were able to be placed back in position. Materials Laboratory evaluation found the vane retaining ring fractured in 11 places located over about 30% of the circumference, which coincided with the area of maximum damage to the outer vane support ring. With the exception of one piece, which showed a typical tensile failure, all of the fractures in this ring were secondary fractures resulting from impact damage, primarily at the aft portion of the ring.

Figure 41 shows the few remaining pieces of the shroud ring. Most of the pieces are part of the flange that was trapped between the outer vane sup-

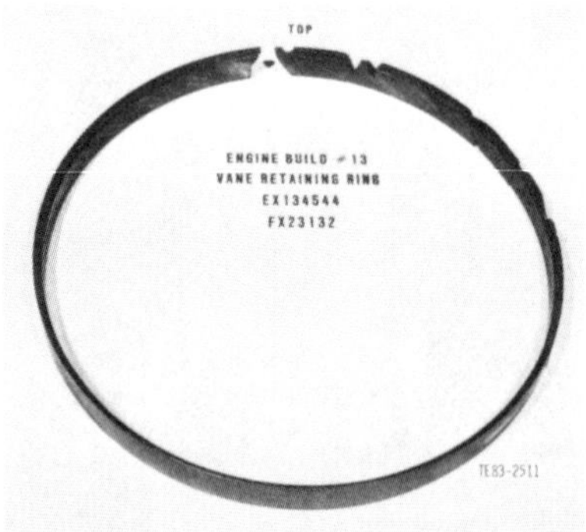

Figure 40. Vane retaining ring after reconstruction by Materials Laboratory.

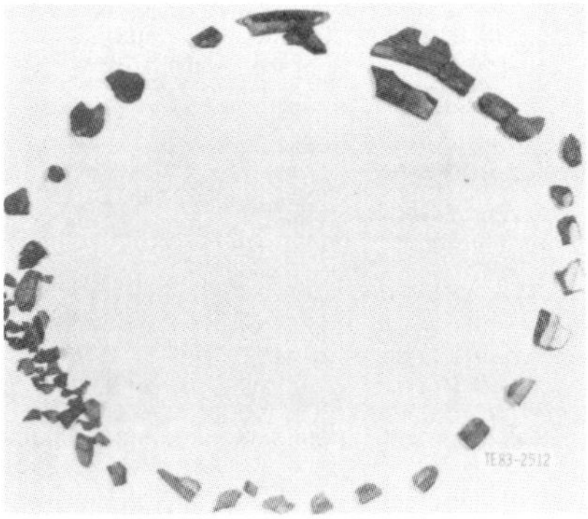

Figure 41. Remaining piece of shroud ring after disassembly of engine C-4, TD13.

port ring and the metal shroud assembly. The pieces were sent to the Materials Laboratory for closer evaluation.

Although insufficient pieces of the shroud remained to perform a meaningful evaluation, a few fragments of the abradable layer appeared to have been blasted by debris. No obvious cause of failure could be determined from the pieces remaining. The radial clearance between the shroud and the blade tip was an average of 1.12 mm (0.044 in.) cold and a calculated average of 0.889 mm (0.035 in.) hot. The calculated minimum hot radial clearance was 0.559 mm (0.022 in.). Previous test experience with metal blades and a ceramic shroud indicated that with an average hot radial clearance of 0.419 mm (0.0165 in.) the rotor would just rub the shroud. Therefore, with the ceramic blades growing less than the metal blades and with a larger clearance, it was thought that no rub was possible. This same magnitude of radial clearance was measured between the inner vane support ring lip and the blade platform, and no evidence of rub was visible.

Discussion: Gasifier Turbine Rotor

Engine C-4's gasifier rotor assembly was assembled with twice proof tested, alpha SiC gasifier turbine blades. The blades were individually proof spin tested at 117% (43,179 rpm) N_G speed, including a 5 minute dwell at maximum speed. The following discussion will address the postfailure investigation of the turbine section with the emphasis on the ceramic bladed gasifier rotor.

The following points are the results of the blade failure investigation:

- No blade was found that could be considered the primary engine failure.
- The common failure origin was the suction side, crown, near the root.
- The origins were contact (impact) in nature.
- Primary failure due to steady-state stress, vibratory response, rub, and block distortion is concluded to be remote.

The condition of the gasifier and power turbine rotor at engine teardown is illustrated in Figures 28, 29, 42, and 43. Figure 28 is a view of the outlet (exit) side of the gasifier turbine rotor with the rotor still installed in the mount pedestal. Considerable upstream ceramic component debris accumulated at the 6 o'clock position. No blade remnants could be identified. However, the most significant observation was the release of all airfoils at the root attachment just above the platform; there were no dovetail attachment or stalk region failures as typically observed in proof spin tests. This is not completely evident in Figure 29 although it was readily observed at teardown. Additional photos that illustrate this observation will be presented as part of the failure in-

Figure 42. Power turbine nozzle assembly, outlet side, engine C-4, TD13.

Figure 43. Power turbine, engine C-4, TD13.

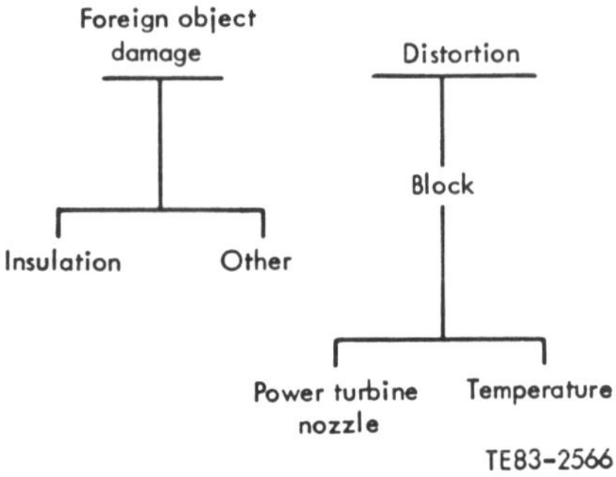

Figure 44. Potential failure modes.

vestigation. Note that no rub marks, rivet distress, or evidence of excessive temperatures are seen on the exit side of the turbine wheel.

Figure 29 is a view of the inlet side of the power turbine nozzle. Sizable pieces of upstream ceramic debris were identified as shroud and vane outer support ring pieces. These could not have passed through the turbine blades.

Figure 42 is a view of the outlet side of the power turbine nozzle. General abrasion by the SiC debris is evident, particularly in the blade tip track. In this photograph, some evidence of labyrinth seal rub is evident at the inside diameter. However, substantial local rub was observed at the 6 o'clock position.

Figure 43 is a view of the inlet side of the power turbine rotor. Extensive abrasion of the blades is evident, and one piece of ceramic debris is trapped between blades. Visual inspection of the rotor revealed bright (polished) rotoring labyrinth seals compatible with the rub evidence on the corresponding stationary seal member in the power turbine nozzle. No other distress was noted. This completed the teardown observations relative to the turbines.

To organize the failure investigation, potential failure modes were identified, as shown in Figure 44. This outline was then utilized to guide the logical investigation of each mode.

In support of the investigation, the failed blades were subjected to detailed laboratory analysis for definition of the failure nature and origin. Shown in Figures 45 through 48 are close-up views of the failed blades. The release of all airfoils at the root attachment (noted earlier) is evident. An interesting feature is the forward corner failure of the lip seal (platform) on multiple blades. No evidence of contact (rub) of the platform lip seal with the vane inner support ring is seen. The cover plates were removed and the blades withdrawn from the wheel. No bending was noted, and the contact (bedding) pattern on the compliant layer was uniform for every layer. This was consistent with no attachment failures.

Laboratory investigation noted that the airfoils of each of the gasifier blades were lost just above the platform. Analysis of the fracture surfaces indicated that all fractures originated as a result of impact damage at the leading edge of the airfoil and then propagated in several different modes. One failure mode involved the crack front shearing through the entire airfoil with no distortion, as shown in Figure 49. The other modes featured cracks propagating from the crown (near the leading edge) along the suction side of the airfoil, as shown in Figure 50, or along the pressure side, as shown in Figure 51. Several of the blades had double fracture origins, with the primary origin at the suction side of the airfoil near the leading edge (crown) and the secondary origin located on the suction side of the airfoil nearer the trailing edge, as shown in Figure 52. Several blades had more complicated multiple fractures, not

Figure 45. Gasifier turbine rotor C-4, TD13, blades No. 1 through 10 of 40 total.

Figure 46. Gasifier turbine rotor C-4, TD13, blades No. 10 through 20 of 40 total.

Figure 47. Gasifier turbine rotor C-4, TD13, blades No. 20 through 30 of 40 total.

Figure 48. Gasifier turbine rotor C-4, TD13, blades No. 30 through 40 of 40 total.

fitting the previous categories, in which the origin was either missing or badly damaged by debris. In all cases, the remaining airfoil stubs were blasted by debris with resultant chipping, especially at the fracture origin and adjacent areas.

Eleven blades had secondary fractures (not relevant to the airfoil fracture) at the upper leading edge of the stalk, on the curvature just below the platform. These were fractures also apparently impact related, i.e., not pure tensile fractures. Three of these blades had penny-shaped flaws at the fracture origin on the platform, as in Figure 49, similar in shape to flaws observed from shot blasting or as a result of hardness indentations.

Although no conclusive feature or explanation for the rotor blade failure has emerged, the impact failure is typically observed. This observation is a guide in the following review of potential modes per Figure 44. Table IV describes three possible modes of blade failure. Each table summarizes supporting and nonsupporting data for each mode.

Discussion: Engine Block Distortion

The concern during this portion of the investigation was to look for evidence of block distortion that could result in the power turbine nozzle imposing loads on the ceramic gasifier nozzle. The gasifier nozzle slips into the power turbine nozzle and seals a nominal radial gap of 0.686 mm (0.027 in.) with a piston ring. Measurements were taken at teardown that suggest block distortion was typical and did not result in line-to-line contact at the piston ring seal. The measurements were taken in two steps: (1) the diameter and runout of the gasifier and power turbine nozzle relative to the centerline of the gasifier rotor and (2) the diameter and runout of the

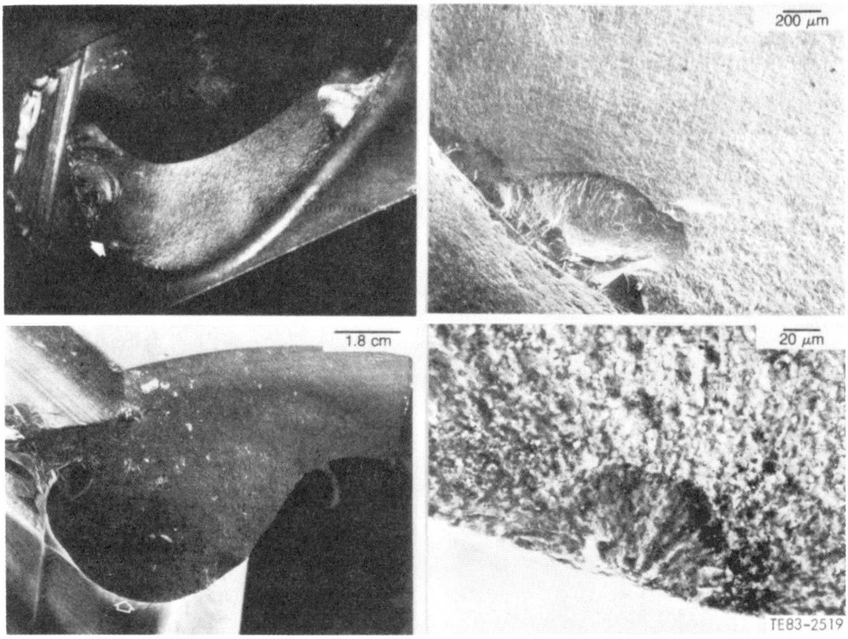

Figure 49. Example of blade primary fracture shearing through airfoil (note penny-shaped crack at secondary origin.)

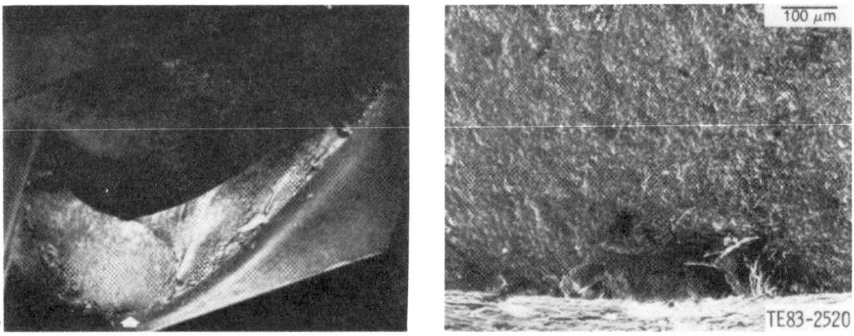

Figure 50. Blade fracture propagating along suction side of airfoil.

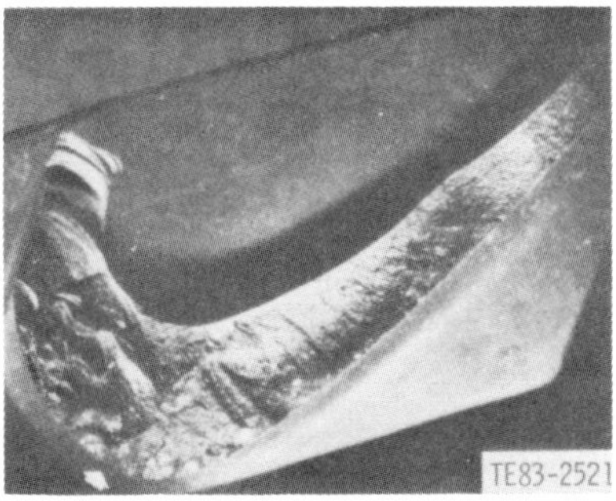

Figure 51. Blade fracture propagating along pressure side of airfoil.

power turbine nozzle relative to the centerline of the power turbine rotor. Each of the steps is subsequently described.

The diameter and runout relative to the centerline of the gasifier turbine at locations A, B, and C per Figure 53 were taken. These served to define the radial engagement of the gasifier nozzle into the power turbine nozzle. The results are shown in Figure 54. Note that the center of the power turbine nozzle is displaced 0.46 mm (0.018 in.) from the center of the gasifier nozzle. The piston ring seal accommodates this with no line-to-line contact. This offset also suggests seal 7 (see Figure 53) would interfere at assembly as shown in Figure 55 and prevent rotation of the power turbine. However, the power turbine spun freely at assembly, and this observation led to step 2 in the measurement investigation.

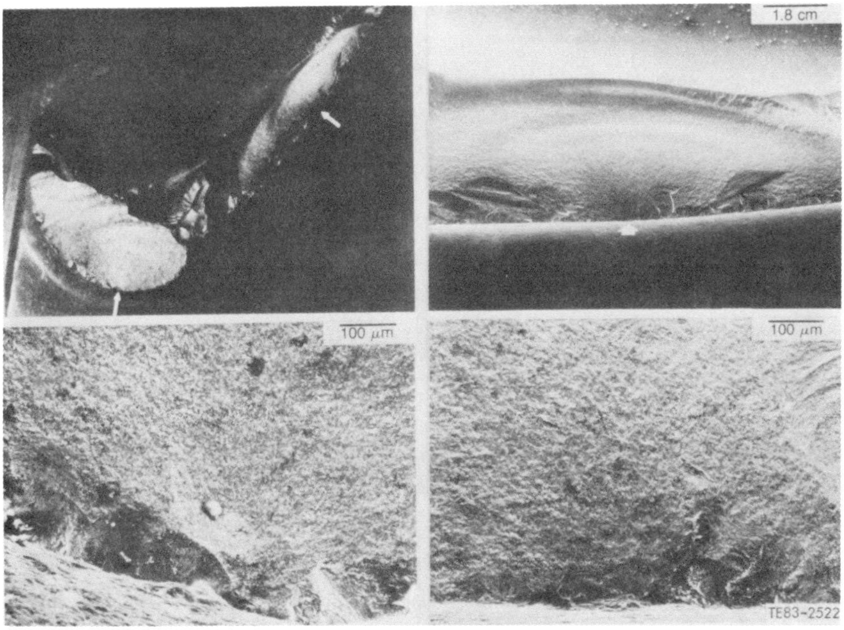

Figure 52. Blade fracture with double fracture origins.

Table IV.
Possible blade failure modes.

	Steady-state stress	Vibration	Rub
Supporting data	None	Analysis predicts blade first mode (bending) response to vane passage in the 60% to 85% N_G range Bench measured first mode frequency range substantiates calculations ~10,800-12,000 Hz	None (No. 8 labyrinth rub was typical of all prior experience)
Nonsupporting data	Blades individually spin tested to 117% of rated speed Rotor assembly proof tested to 108% of rated speed Engine C-4 tested to 100% speed Failure occurred at 97% N_G or 94% stress during second engine run Airfoil calculated maximum stress (see sketch) Calculated blade worst condition: 3 sec accel after braking, σ airfoil = 170.2 MPa (24.68 ksi) P_s airfoil = 0.9998 After 117% N_G proof test: P_s airfoil = 0.9998 (2 in 10,000) P_s dovetail = 0.99998 P_s blade = 0.9996 Conclude that primary airfoil failure remote	Failure occurred at 97% N_G 78.5% N_G (first mode coincident with vane passage) and with ± 10 ksi vibratory assumed and after 117% N_G proof test 1 blade P_s = 0.99966 Conclude that airfoil failure in vibration mode is remote	No rub witness marks at blade lip seal—airfoil tip unknown Airfoil tip build with 1.12 mm (0.044 in.) cold average clearance to shroud (no prior evidence of rub at this clearance) 0.89 mm (0.035 in.) average hot running 0.559 mm (0.022 in.) min hot running clearance

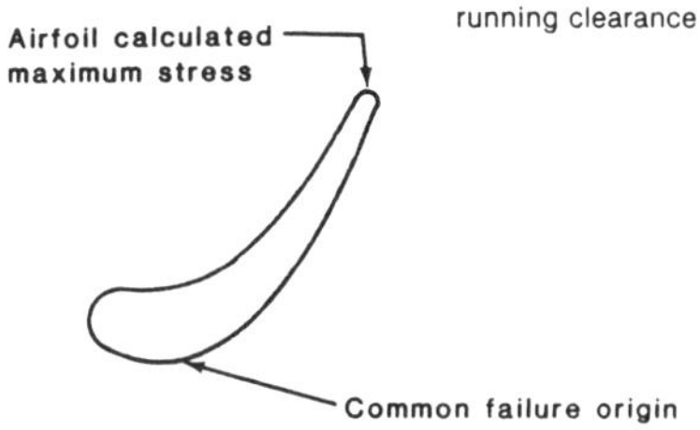

The diameter and runout relative to the centerline of the power turbine rotor at locations B and C, per Figure 55, were taken. Thus, the same surface was measured from two references (centerline of gasifier rotor and centerline of power turbine rotor). This identified an offset in the centerlines of 0.2 mm (0.008 in.). The position of both centers is identified in Figure 54. The offset in the centerlines relieves the apparent interference at labyrinth seal 7 (within measurement accuracy). The offset is typical of the GT engine. No witness marks were detected at teardown, so it was concluded that forcible distortion of the ceramic nozzle by the power turbine nozzle/block was unlikely.

Engine C-4, BU14

Summary

During November 1982, the second 2070°F-configuration ceramic gasifier section was tested in engine C-4. A summary of the ceramic components installed is provided in Table V.

The engine test conducted included three starts, a slow controlled acceleration to 100% engine speed at a constant temperature, and a temperature increase to 1132°C (2070°F) rotor inlet temperature (RIT) conditions at a constant 100% engine speed. A total of 6 hr and 11 minutes was accumulated. Scheduled borescope inspections

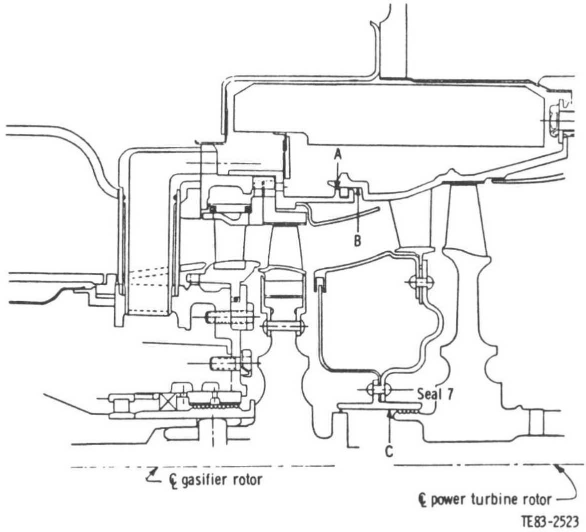

Figure 53. Measurement locations for checking engine block distortion.

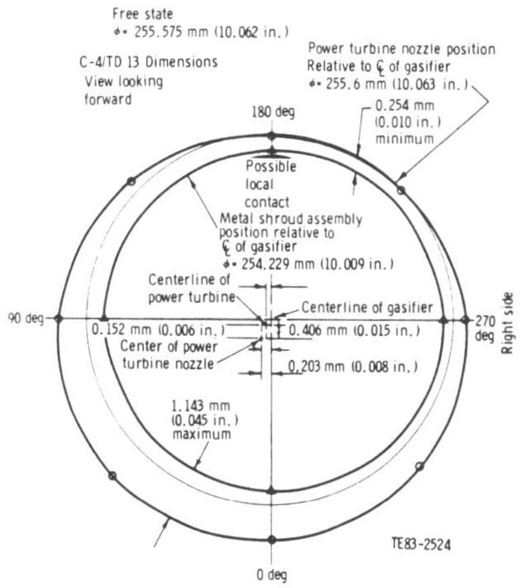

Figure 54. Engagement of the gasifier nozzle into the power turbine nozzle.

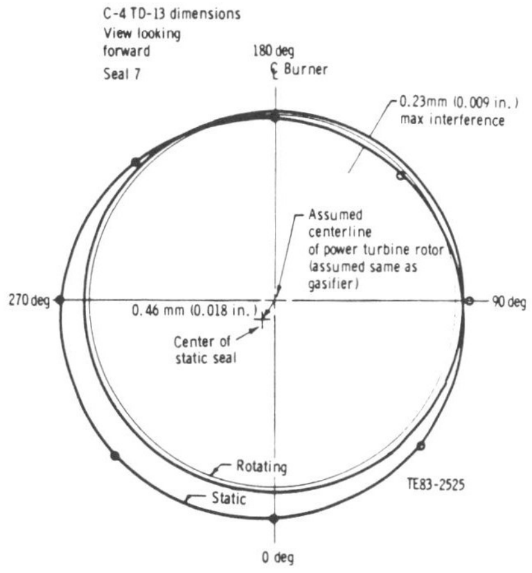

Figure 55. Apparent interference at labyrinth seal 7 assuming the centerline of the power turbine is coincident with the centerline of the gasifier turbine.

Table V.
Ceramic components in engine C-4, BU14.

Component	Quantity	Material	Manufacturer
Gasifier turbine blades	40	Alpha SiC	Carborundum
Gasifier nozzle assembly vanes	28	SiC	Carborundum
Outer vane support	1	Alpha SiC	Carborundum
Inner vane support	1	RBSiC	Carborundum
Turbine tip shroud	1	RBSiC	Carborundum
Vane retaining ring	1	Refel SiC	Pure Carbon Co.
Strut sheets	4	Alpha SiC	Carborundum
Regenerator disk	2	1100°C AS	Corning Glass

between starts showed no evident problems. At 2 hr and 39 minutes into the 10 hr endurance cycle at rated speed at 36,904 rpm and temperature at 1132°C (2070°F), the test was terminated by a failure of the gasifier turbine. A failure investigation was initiated. The conclusions reached after the investigation were the following:

- The primary cause of failure cannot be identified.
- The possibility of engine structural distortion causing the failure was eliminated.
- The most probable failure modes have been reduced to (1) ceramic blade failure, (2) foreign object damage, and (3) gasifier turbine nozzle assembly (metal) component failure.

The test objective was to evaluate the 2070°F-configuration ceramic gasifier nozzle and rotor at rated speed and temperature, including engine performance and 10 hr of endurance testing. A second objective was continued evaluation of the block cooling design.

The configuration of the gasifier turbine assembly is shown in Figure 56. After five cycles simulating the most severe engine operating condition (dynamic braking), the following parts were qualified for use in engine C-4, BU14:

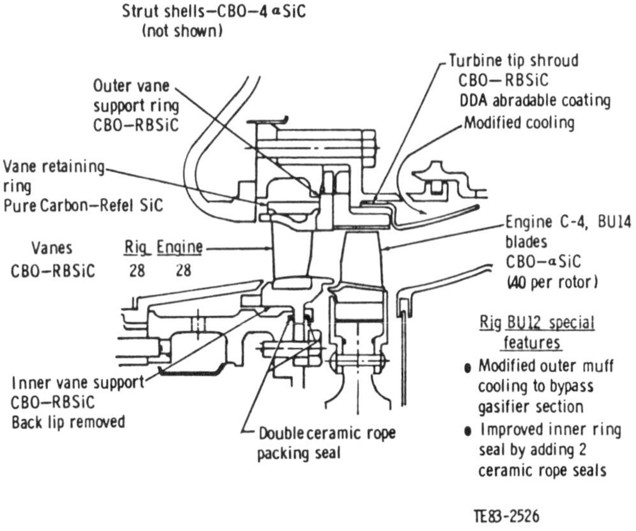

Figure 56. Engine C-4, BU14 gasifier turbine configuration.

- turbine tip shroud—RBSiC, CBO with abradable layer added by Allison (rig qualified to 1132°C [2070°F]; Allison abradable coating [80% yttria stabilized zirconia/14% eccospheres])
- outer vane support—alpha SiC, CBO (rig qualified to 1132°C [2070°F]; no leading edge slots)
- inner vane support—RBSiC, CBO engine qualified 890 hr; back lip removed)
- vane retaining ring—Refel SiC, Pure Carbon (rig qualified to 1132°C [2070°F])
- strut shells—4 alpha SiC, CBO (rig qualified to 1132°C [2070°F])
- vane—28 RBSiC, CBO (rig qualified to 1132°C [2070°F])

Discussion: Engine Test

Testing of engine C-4 was conducted in accordance with the planned test schedule, shown in Figure 57. Each of the points shown on the curve has a number in parentheses next to it. This number represents the time spent at that speed and temperature condition. Table VI provides an overview of the test.

Start No. 3 was scheduled to accumulate approximately 10 hr of durability testing at rated speed (36,905 rpm) and 1140°C (2100°F) BOT (which is equivalent to 1132°C [2070°F] gasifier rotor inlet temperature).

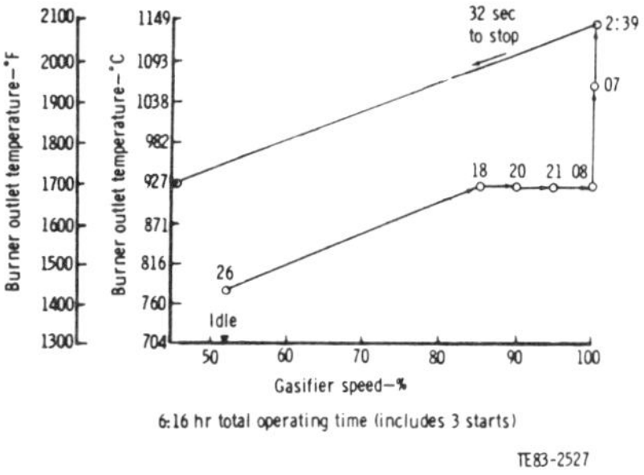

Figure 57. Engine C-4, BU14 Test 1 schedule.

Data recorded during the test are shown in Figures 58 and 59. The front lateral vibration spike at the time of failure indicates a time lapse of 17 sec

Table VI.
Overview of testing of engine C-4.

Engine test	Engine condition	Running time—minutes
Motored to 8000 rpm	Borescope inspection: parts in excellent condition	—
Start No. 1 (start-to-idle): engine stabilized and data recorded	Borescope inspection: parts in excellent condition	30
Start No. 2 (start-to-idle): performance testing to 95% gasifier turbine speed, 927°C (1700°F) BOT conditions, shutdown	Air leak: loss of gasket to the air transfer tube, cover gasket replaced Borescope inspection: deposit noted on suction side of blade airfoils (Spectrographic analysis was performed showing high content of calcium plus traces of other minerals. Probable sources are the test cell steam heaters.) Conclusion: acceptable for continued testing	82
Start No. 3 (start-to-idle): performance testing to 100% gasifier turbine speed, 1140°C (2100°F) BOT conditions, stabilized	At 159 minutes into the stabilization run, a loud pop was heard, followed by high vibration (30 sec delay), large loss of compressor discharge pressure (CDP), and torque; engine stop time was 32 sec	259

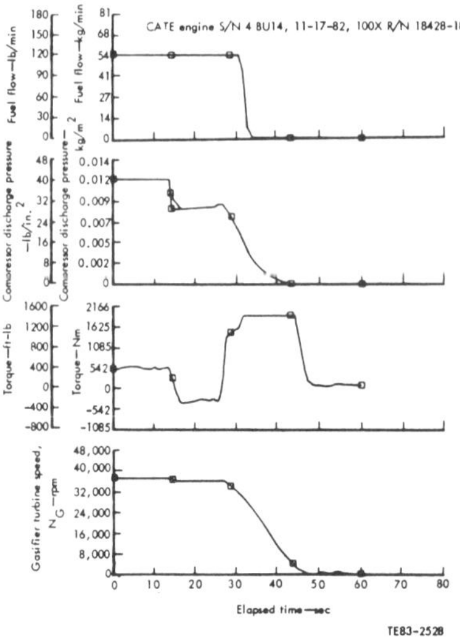

Figure 58. Data recorded at time of failure—engine C-4, BU14.

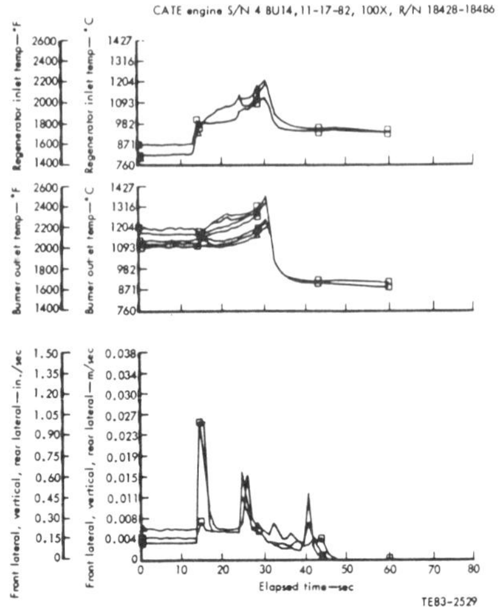

Figure 59. Data recorded at time of failure—engine C-4, BU14.

before auto shutdown due to T_6 (power turbine exit) overtemperature. Other parameters that indicated failure were loss of compressor discharge pressure, high negative torque, and high T_4. Inspection revealed that a failure did occur in the gasifier turbine. The engine was moved to the assembly floor and a failure investigation was conducted.

Discussion: Failure Investigation

Prior to engine removal from the test stand, the following failure investigation plan was formulated:

- review all data before engine removal
- identify potential failure modes
- prepare engine disassembly checklists addressing each potential failure mode
- direct engine disassembly in accordance with the checklists

Engine disassembly was conducted based on the previous plan. The data review resulted in the following conclusions:

- All three starts were similar and characterized by a slow temperature rise which did not exceed established limits.
- There was no abnormal engine vibration or noise other than at the moment of failure.
- All transients through the blade fundamental frequency response range were completed in less than 60 sec.
- All performance data were normal and engine operation was considered acceptable up to the moment of failure.

The potential failure modes identified were the following:

- gasifier turbine (blade failure)
- metal turbine tip shroud—cracked braze joint in gasifier shroud assembly
- foreign object damage (nonceramic source)
- gasifier turbine nozzle failure
- engine structural distortion causing mechanical loading
- plenum/inner annulus/combustor failure

The investigation team directed the engine disassembly and ensured that each item on the checklist was satisfied. The following is a summary of the gasifier turbine section damage:

- ceramic blades (airfoils missing—six failures below the platforms)
- outer vane support ring (multiple fractures—180 deg quadrant missing, left side as viewed from the rear; missing pieces)

- inner vane support ring (cracked through bottom lug; inside flange chipped at the crack; three lugs chipped)
- vane retaining ring (six fractures [7 o'clock position] clockwise; one segment missing)
- vanes (15 intact [right side of assembly as viewed from the rear])
- shroud (flow path missing; pieces of flange in place; some portion of each lug available)
- strut shells (one cracked, outside of the hot gas flow path)

Damage to the remainder of the engine was minimal considering the fact that the gasifier section failed. The power turbine nozzle and rotor were eroded by ceramic particles as were the inboard faces of both ceramic regenerator disks. The ceramic disks cannot be salvaged by remachining the inboard face. The regenerator seals are acceptable for continued use, and the Lamilloy combustor must have the small ceramic particles removed.

The remaining part of the Engine Test Section concentrates on each of the potential failure modes. One of the modes, plenum/inner annulus/combustor failure, was eliminated because disassembly revealed those parts to be in good condition. Each potential mode of failure was analyzed individually and each is discussed in the following paragraphs.

Discussion: Gasifier Turbine Blade Failure

Since the gasifier turbine blade was suggested as being a primary failure, the background of the blades was reviewed. The pertinent facts are outlined in the following:

1. rotor part number, EX 134536
2. rotor serial number, CX 42737
3. blades, 40 each, P/N EX 130180
   - source, CBO follow-on order No. 1
   - attachment finish, longitudinally ground, Crafts, Inc
   - treatment; oxidized, polished stalk
   - proof test, each blade individually to 117% speed
   - compliant layer, IN-600 material, P/N EX 129476 ±0.0025 mm (±0.0001 in.) thickness tolerance
4. rotor assembly—three spin proof test builds, 10 hr of hot engine simulator testing, and the final build (No. 3) was installed in engine C-4, BU14
   - Build 1—blade failure at rotor proof speed of 93%, traced to nonuniform thickness compliant layers, blades discarded
   - Build 2—40 blades per the previous description; four blades failed at the attachment at 107% speed in an attempt to reach the goal of 108%, three adjacent blades damaged:

	Rotor position			
4 failed	8	23	24	26
3 airfoil tip damage	7	22	—	25

     The blade in position 26 was primary. The blade dovetail bearing surface was nonuniform (not flat) due to local polishing to remove a flaw.
   - Build 3—the previous seven blades replaced; FPI check made on the remaining 33 blades—all cleared (The rotor was successfully proof spin tested to the following schedule: 5 minutes at 90%, 100%, 105%, and 108% speed. The rotor was exposed to the hot engine simulator rig [HESR] for a thermal proof test. The testing was for 10 hr total, including two starts [transients] and 30 minutes at 100% speed and an average temperature of 940°C [1725°F]. Table VII lists the blade serial number by blade position.)

Details of the overall engine build, test, and failure were described in the previous section. The following discussion will address the postfailure investigation of the turbine section with emphasis on the ceramic bladed gasifier rotor. The discussion begins with summary points.

The following points are the result of the blade failure investigation:

1. Of the six blades that failed below the platform, one (No. 31) is a suspect primary failure. Evidence of a large flaw and heat tint is seen at the attachment area failure origin, and the failure surface resembles a spin test tensile failure. The remaining five exhibit a ragged, contact-type failure surface.
2. Of the blades that failed at the airfoil root, the origins were suction side, crown, near the root. The failure surface is contact in nature.
3. Irregular thinning of the compliant layers was measured at teardown (TD). This substantiates the layers are performing the intended design function. Concurrent with thinning is a change in the surface bearing stress distribution and a change with the attachment internal stress distribution.
4. Primary failure due to rub or block distortion is concluded to be remote.

The condition of the gasifier rotor at engine teardown is illustrated in Figures 60, 61, 62, and 63.

Table VII.
Record of blade serial numbers, engine C-4/BU14 gasifier data.

Rotor: P/N EX 134536, S/N CX 42737
Roll pin at TDC
CW viewed from rear

Position	Blade Serial No.
1	FX 24932
	FX 24893
	FX 24900
	FX 24901
5	FX 24906
	FX 24907
	FX 24913
	FX 24897
	FX 24931
10	FX 24923
	FX 24924
	FX 24925
	FX 24926
	FX 24928
15	FX 24933
	FX 24940
	FX 24945
	FX 24948
	FX 24949
20	FX 24950
	FX 24951
	FX 24952
	FX 24959
	FX 24971
25	FX 24973
	FX 24961
	FX 24974
	FX 24965
	FX 24967
30	FX 24969
	FX 24972
	FX 24976
	FX 24977
	FX 24980
35	FX 24981
36	FX 24983
	FX 24989
	FX 24985
	FX 24986
40	FX 24987

Figure 60. Gasifier turbine rotor C-4, TD14, blades No. 1 through 10 of 40 total.

Figure 61. Gasifier turbine rotor C-4, TD14, blades No. 10 through 20 of 40 total.

Figure 62. Gasifier turbine rotor C-4, TD14, blades No. 20 through 30 of 40 total.

The six blades failed below the platform are No. 9, 10, 23, 31, 34, and 36. The failure surface is typically very ragged, except for No. 31. The blade remnants were easily removed from the wheel and the compliant layer contact pattern was uniform for every blade.

Clocking of the attachment failure blades suggested a pattern of failure at 120-deg intervals, as shown in Figure 64. However, a condition of blades selectively vibrating (such as at every 120 deg) has

Figure 63. Gasifier turbine rotor C-4, TD14, blades No. 30 through 40 of 40 total.

never been experienced in any Allison turbines or compressors. Standing waves have been observed in rotor assemblies, particularly for thin disks. The CATE turbine wheel is not in this category; it is a thick small diameter wheel and all frequencies are well in excess of the engine operating range. Wheel mode vibration was probably not the cause of blade failure.

A description of the laboratory investigation and results follows. The interesting attachment failure was identified as No. 31. The other failures were impact type ragged fracture surfaces. Figure 65 shows the attachment remnant and a sketch looking radially inward, and Figure 66 includes a magnified view of the failure surface. The features are the following:

- heat (color) tint—sufficient temperature during oxidation heat treatment flaw masked by glassy layer; operating temp in engine 649°C (1200°F) will not tint
- tensile fracture surface is similar to spin test fast fracture surface
- all other failures, airfoil root and attachment, appear contact in nature

The following conclusions were reached:

- Scenario suggests presence of flaw during all proof spin and HESR testing, plus 2.5 hr engine steady state at 100%.
- Actual "trigger" that caused release at 2 hr and 39 minutes cannot be identified, but compliant layer thinning and corresponding attachment stress redistribution is suspect.
- Blade No. 31 was concluded to be a probable primary failure.

In the conclusion of the gasifier turbine blade investigation it was noted that the compliant layers irregularly thinned during test. The following discussion addresses this.

Table VIII lists the measured posttest compliant layer thickness at the contact surface. Data are reported for three airfoil root failures and three failures below the platform, including the suspect primary failure, blade No. 31. Typically, the compliant layers irregularly thinned from the manufactured thickness of 0.1778 mm (0.007 in.). This is the design function of the layers—accommodation of local irregularities between the turbine wheel and the ceramic blade. Hot cyclic spin tests were conducted and reported in EDR 10156 (Ref 5) to specifically measure thinning for IN-600 and other materials. The engine time (2.55 hr at 100%) and measured thinning of approximately 0.05 mm (0.002 in.) generally agree with the hot cyclic test experience.

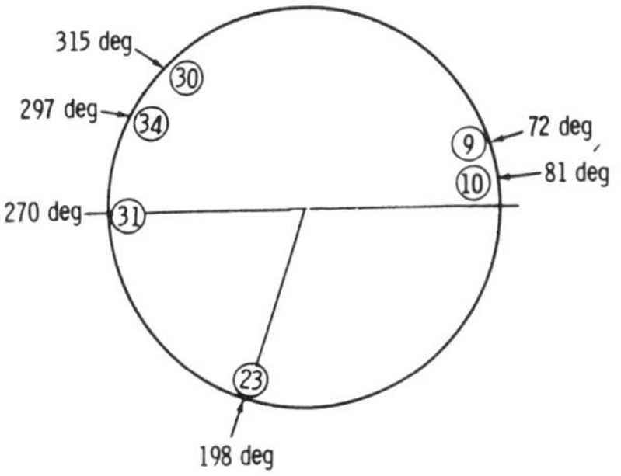

Failed blades—after approximately 2 hr and 39 min steady state

- Clocking
- Comments
  1. Never-observed blades selectively vibrating (as every 120 deg) in any DDA turbine or compressor.
  2. CATE wheel
     All frequencies well in excess of operating range

TE83-2534

Figure 64. Clocking the position of failed gasifier turbine blades, engine C-4, TD14.

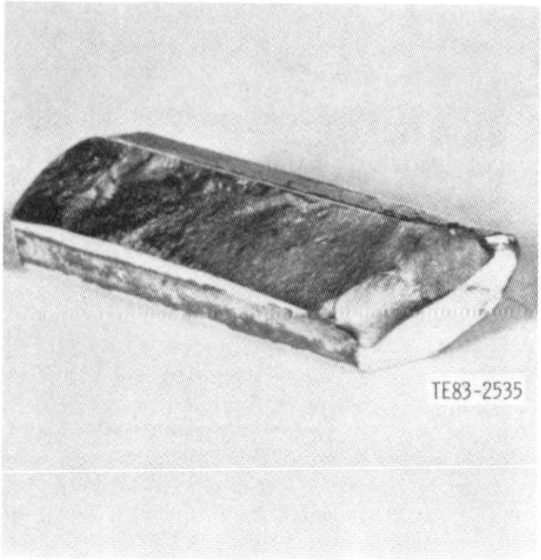

Figure 65. Blade No. 31, attachment remnant, engine C-4, TD14.

Stress redistribution results from local changes in the contact (bedding) pattern. The measured compliant layer thinning is suspected to have triggered the flaw in blade No. 31 with subsequent blade release. Note in Table VIII, blade No. 31, that the layer area adjacent to the failure location was thinned, indicating eventual increase in load at location 6.

A few identifiable pieces of the shroud ring were trapped in the power turbine inlet nozzle (see Figure 67). These pieces were sent to the materials lab for evaluation but a meaningful analysis was not possible. The abradable layer appeared to have been blasted by debris.

No obvious cause of failure could be determined from the pieces remaining. The radial clearance between the shroud and the blade tip was an average 1.07 mm (0.042 in.) cold and a calculated average of 0.838 mm (0.033 in.) hot. The calculated minimum hot radial clearance was 0.508 mm (0.020 in.). Previous test experience with metal blades and a ceramic shroud indicate that with an average hot radial clearance of 0.419 mm (0.0165 in.) the rotor would just rub the shroud. Therefore, with ceramic blades growing less than the metal blades and with twice the clearance (0.838 mm versus 0.419 mm), it is concluded that no rub was possible.

Figure 66. Failure surface, gasifier rotor blade No. 31, engine C-4, TD14.

Table VIII.
Post test compliant layer thickness engine C-4/TD 14.

Table D

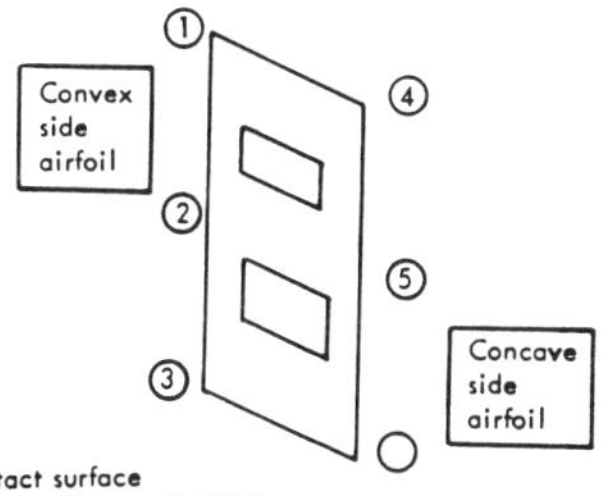

Drawing spec @ contact surface
0.1778 mm ± 0.0025(0.007 in. ± 0.0001)

Blade No.	Failure	Compliant layer thickness of six positions—mm (in.) 1	2	3	4	5	6
3	Airfoil	0.1524 (0.006)	0.1524 (0.006)	0.127 (0.005)	0.1524 (0.006)	0.1778 (0.007)	0.1778 (0.007)
12	Airfoil	0.1524 (0.006)	0.1524 (0.006)	0.1524 (0.006)	0.1524 (0.006)	0.1524 (0.006)	0.1524 (0.006)
18	Airfoil	0.1524 (0.006)	0.1524 (0.006)	0.1397 (0.0055)	0.1524 (0.006)	0.1524 (0.006)	0.1778 (0.007)
23	Stalk	0.1778 (0.007)	0.1524 (0.006)	0.1524 (0.006)	0.127 (0.005)	0.1524 (0.006)	0.1524 (0.006)
31	Attachment	0.1524 (0.006)	0.1397 (0.0055)	0.1397 (0.0055)	0.1524 (0.006)	0.1524 (0.006)	(0.1778) (0.007)
						Failure origin	
36	Stalk	0.1524 (0.006)	0.1524 (0.006)	0.127 (0.005)	0.1397 (0.0055)	0.1524 (0.006)	0.1778 (0.007)

Discussion: Metal Turbine Tip Shroud

Postfailure investigation did reveal a concern for potential failure of the tip shroud from adverse thermal gradients (thermal stress). At teardown, a crack was observed in the metal gasifier turbine shroud assembly as illustrated in Figure 68. The crack in a braze joint, as in Figure 65, extended over a 70 deg arc (10 to 2 o'clock) centered on the burner. The concern was for strut cooling air bleeding through the crack and impinging on the outer diameter of the ceramic shroud ring, thus creating local thermal stresses. Analysis revealed the following:

- visual evidence of a difference in flow deposits (debris) in the area of the crack versus adjacent areas
- minimal potential for cooling air to flow through the crack, indicated by secondary flow calculations
- a minimum of oxidation in the failed braze joint, determined by metallurgical analysis
- poor wetting and flowing of the braze in the joint

As a result of all the evidence, thermal stress failure of the ceramic tip shroud is considered possible but not likely.

Discussion: Foreign Object Damage

The possibility for a failure due to foreign object damage in engine C-4/BU14 is similar to the failure that occurred in C-4/BU13. That is, any airborne particles entering the engine must pass through the

Figure 67. Crack in braze joint of shroud assembly.

regenerator disks before they reach the combustor and gasifier section. Since the regenerators act as very good filters for debris (particles larger than 0.508 mm [0.020 in.] in diameter cannot pass) investigation of FOD was restricted to an area between the combustor and exhaust diffuser. This area includes the T_3 cavity, combustor, plenum, plenum inner annulus gasifier nozzle assembly, and T_6 cavity. Items such as nuts, bolts, washers, locktaps, shim stock, etc, were inspected to determine if any of these had dislodged and could possibly go through the turbine. The combustor was intact as was the igniter. All shim stock holding down instrumentation leads and instrumentation itself was also intact. However, pieces of H-cement (high temperature ceramic cement) around each of the strut shells and chips of zirconia plasma spray in the plenum gas flow path were missing. Figures 69, 70, and 71 show the area where the H-cement and zirconia plasma spray were missing. The sizes of these pieces can fit through the vane throat and could impact the gasifier turbine rotor. The pieces of H-cement located around the plenum strut shells ceramic rope packing seal could have either been ingested directly into the plenum or found their way into the combustor dilution holes (see Figure 72). It is also possible that the dislodging of these pieces could have occurred as a result of high vibration at the moment of failure.

Two possible sources for FOD were identified as the following:

- one or several pieces of H-cement dislodged from an area around the strut shell ceramic rope packing seal at the plenum and plenum inner annulus and ingested into the gasifier turbine
- chips of zirconia plasma spray in an area around No. 5 and 6 thermocouples in the plenum dislodged and ingested into the gasifier turbine

Discussion: Gasifier Turbine Nozzle

Upon removal of the gasifier nozzle assembly from the engine, some initial observations were made. Figures 73 and 74 show the gasifier section looking from the rear. Approximately half the nozzle section was missing and rotor airfoils were missing. No rub marks were seen on the outlet side of the gasifier turbine wheel.

Figure 75 shows the nozzle assembly after removal of the gasifier turbine rotor. The following were noted:

1. Approximately one half the outer vane support ring was in place.
2. The inner vane ring was fractured (180 deg from top).
3. The inner vane support ring lip seal is locally chipped (225 deg).
4. High temperature cement (H cement) was missing from the strut shell rope seals. This is further illustrated in Figure 70. Cement debris could pass through the primary gas path.
5. Some thermal barrier material was missing from the plenum flow-path surface.

Figure 76 is a view into the inlet of the power turbine nozzle. Large pieces of the vane outer support ring, vanes, and shroud are seen as indicative of the rapid 32-sec shutdown, and the gasifier tur-

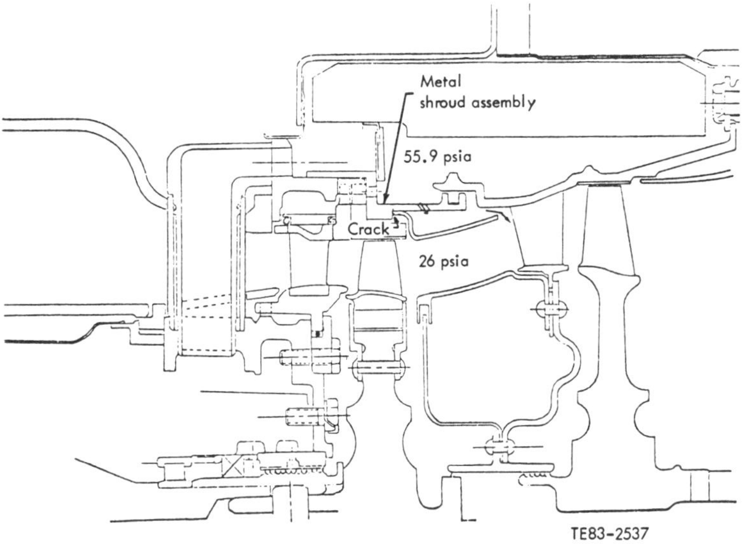

Figure 68. Cross section illustrating the gasifier turbine shroud assembly.

Figure 69. Missing H-cement around strut shells at the plenum inner annulus for engine C-4, TD14.

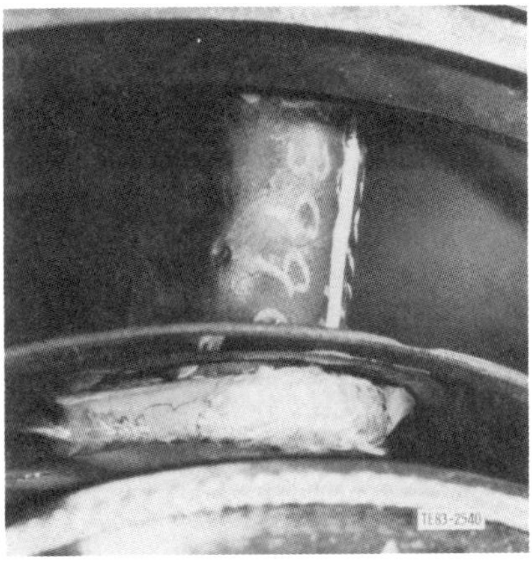

Figure 70. Detail illustration, missing cement, strut shell rope seal for engine C-4, TD14.

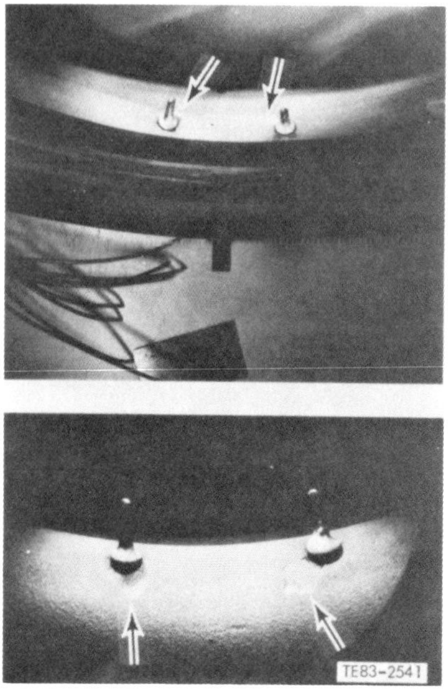

Figure 71. Missing zirconia chips in flow path of plenum for engine C-4, TD14.

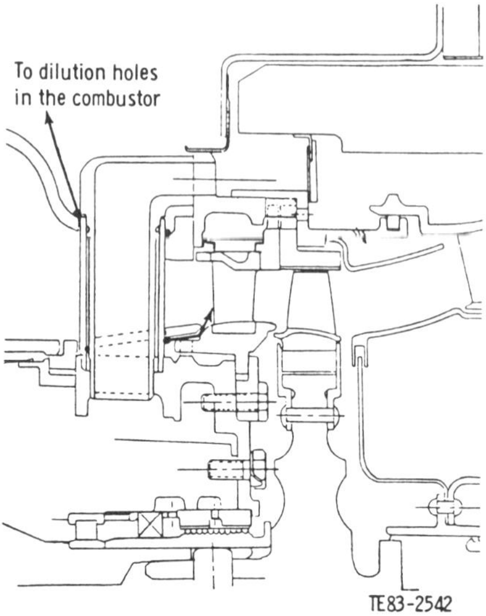

Figure 72. Possible path the H-cement could take to enter gasifier turbine for engine C-4, TD14.

bine blades were gone when these pieces passed through the gasifier rotor.

A nozzle failure could have been precipitated by a failure of any of the assembly components. Each component will be discussed separately.

Figures 77 and 78 further illustrate the failed inner vane support ring This ring is RBSiC material and had successfully accumulated 890 test hours in engine C-4 with a metal gasifier turbine rotor. Temperature was limited to 1038°C (1900°F). The chips in three of the ring retention lugs are consistent with prior test experience. A possible explanation is that fast fracture produces a recoil action which is reacted at the ring lugs and causes the chipping type failure. The ring fracture origin is near the maximum predicted stress location (see Figure 79). No heat tint was observed on the failed surface. The

Figure 73. Gasifier turbine assembly for engine C-4, TD14 (overall view).

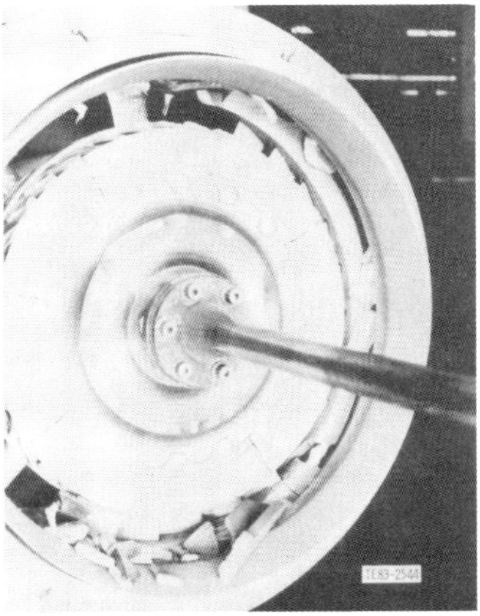

Figure 74. Gasifier turbine assembly, for engine C-4, TD14 (close-up view).

Figure 75. Gasifier nozzle assembly for engine C-4, TD14.

Figure 76. Inlet view, power turbine nozzle assembly for engine C-4, TD14.

magnitude of residual stresses which may have been stored in the ring prior to the failure was estimated by installing i.d. and o.d. strain gages and closing the fracture gap. The measured stresses were small (-6.2 MPa [-0.90 ksi] at the i.d. and 10.34 MPa [1.5 ksi] at the o.d.). No evidence of rotor rub was seen on the ring.

The locally chipped blade seal lip (see Figure 75) showed a contact origin at the i.d. of the lip; however, this could be the result of a blade release from an attachment failure.

Considering the impact nature of the failure origin with no evidence of heat tint, the IVSR failure was determined to be secondary.

Postfailure investigation revealed no evidence of a primary vane failure. This is based on the following:

1. steady-state engine operation at time of failure, predicted vane peak stress less than 69 MPa (10 ksi)
2. fractured vanes—impact type failure surfaces with no evidence of heat tint
3. the vanes that survived (approximately half)—free of any distress indications

Postfailure investigation of engine C-4/BU13 raised a concern for potential vibration of the slotted OVSR and subsequent failure. This concern was addressed in C-4/BU14 by configuring the engine with an unslotted ring. The rings are illustrated in Figure 80. The BU14 ring was a new, reaction-bonded SiC ring that had successfully passed the thermal shock

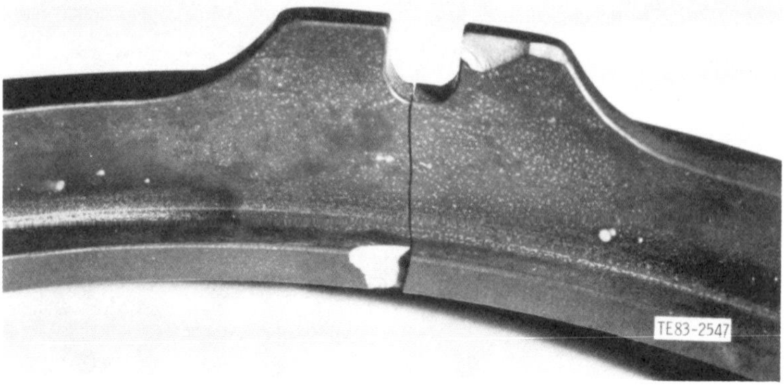

Figure 77. Cracked inner vane support ring for engine C-4, TD14.

rig proof test. Failure analysis on the remaining pieces found all fracture origins to be due to impact. Primary failure of the OVSR was determined to be unlikely.

The vane retaining ring is a simple hoop structure isolated from the primary gas path. Analytically, no failure stresses or dynamic modes are predicted. The majority of the ring pieces were located and tensile failure origins identified. Even in the event that the vane retaining rings were primary, it is unlikely that ring failure would precipitate a catastrophic engine failure.

Discussion: Engine Block Distortion

The concern of this portion of the investigation is for evidence of block distortion that could result in the power turbine nozzle imposing loads on the ceramic gasifier nozzle. The gasifier nozzle slips into the power turbine nozzle and seals a nominal radial gap of 0.686 mm (0.027 in.) with a piston ring. Measurements taken at teardown suggest block distortion was typical and did not result in line-to-line contact at the piston ring seal. The measurements were taken in two steps: (1) the diameter and runout of the gasifier and power turbine nozzle relative to the centerline of the gasifier rotor and (2) the diameter and runout of the power turbine nozzle relative to the centerline of the power turbine rotor. Each of the steps is described in the following paragraphs.

The diameter and runout relative to the centerline of the gasifier turbine at locations A, B, and C per Figure 81 were taken. These served to define the radial engagement of the gasifier nozzle into the power turbine nozzle. The results are shown in Figure 82. Note that the center of the power turbine nozzle is displaced 0.23 mm (0.009 in.) from the center of the gasifier nozzle. The piston ring seal accommodates this with no line-to-line contact.

The diameter and runout relative to the centerline of the power turbine rotor at seal 7 per Figure 81 were taken. The measurements are shown in Figure 83 and confirmed the build observation of a freely rotating power turbine. A complete picture of the centerline geometry of the gasifier/power turbine engagement is shown in Figure 84. The offsets are typical of the IGT engine. No witness marks were detected at teardown; thus it was concluded that forcible distortion of the ceramic nozzle by the power turbine nozzle/block was unlikely.

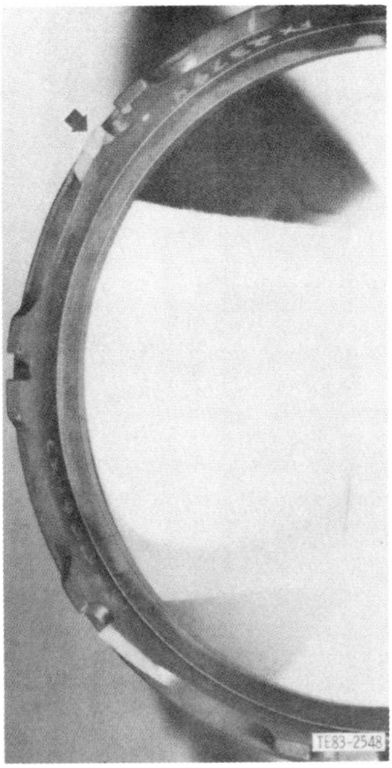

Figure 78. Chipped retention lugs inner vane support ring for engine C-4, TD14.

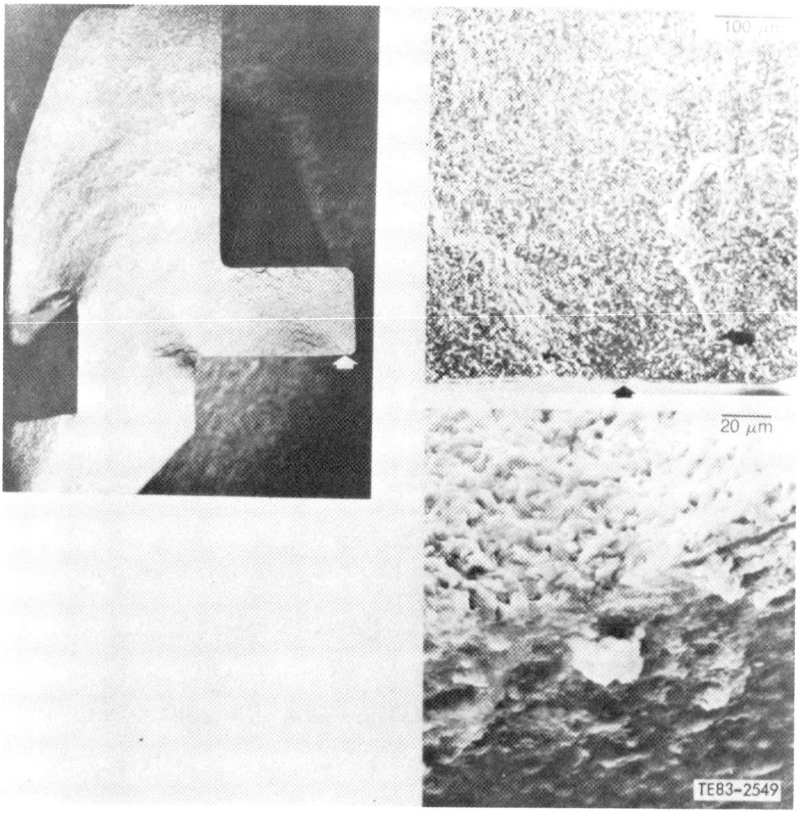

Figure 79. Fracture surface, inner vane support ring for engine C-4, TD14.

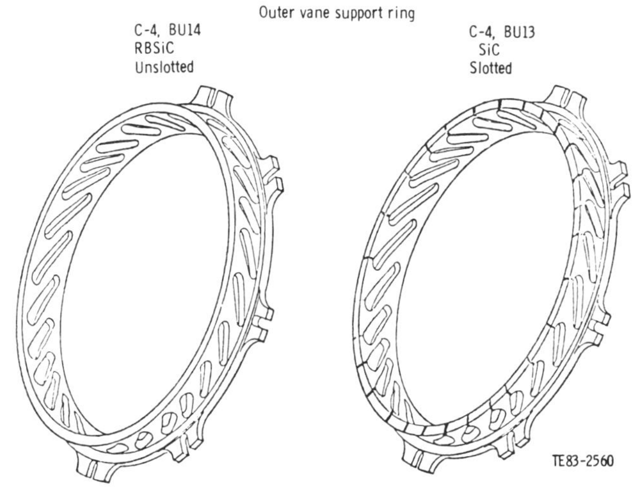

Figure 80. Outer vane support ring configuration for engine C-4, TD14.

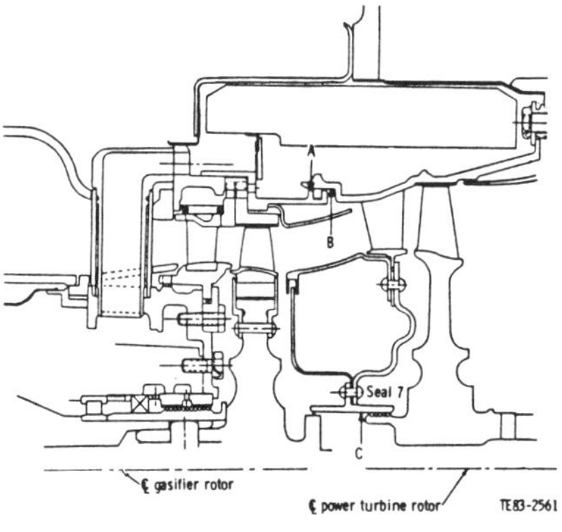

Figure 81. Measurement locations for checking engine block distortion for engine C-4, TD14.

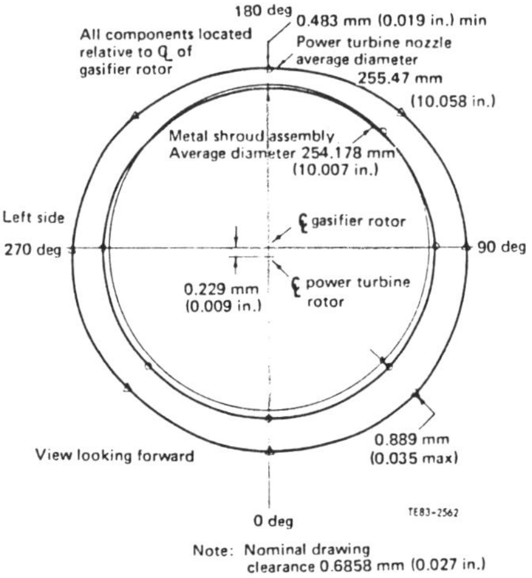

Figure 82. Engagement of the gasifier nozzle into the power turbine nozzle for engine C-4, TD14.

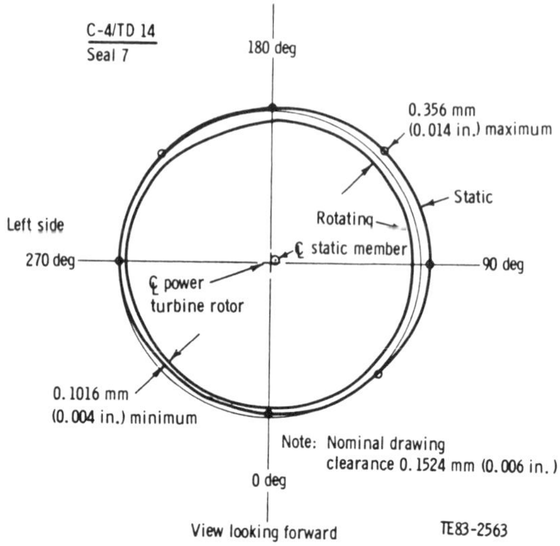

Figure 83. Engagement of labyrinth seal 7 for engine C-4, TD14.

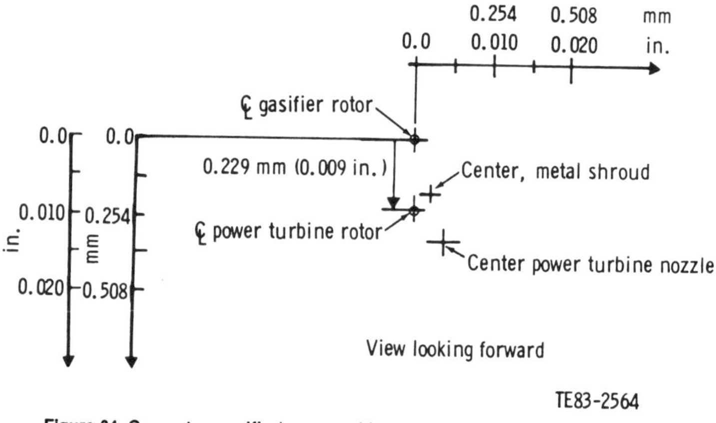

Figure 84. Geometry, gasifier/power turbine engagement for engine C-4, TD14.

III. Ceramic Materials Development

Summary

The ceramic materials development section documents the characterization and qualification of candidate materials and structural ceramic components, the development of nondestructive evaluation techniques, investigations of turbine tip shroud abradability, and development of machining technology for ceramic components.

The materials behavior and characterization subsection covers the testing and evaluation of candidate ceramic materials, including microstructure, density, and fast fracture strength.

The component characterization and qualification subsection details assessments of ceramic parts fabricated using established processing technology: vanes, support rings, retaining rings, and plenums.

The ceramic component process development subsection reviews the development efforts on components for which improved material characteristics and/or fabrication methodology were required: vanes, support rings, plenums, and blades.

The development of nondestructive evaluation techniques subsection covers the development of nondestructive evaluation methods for flaw detection in ceramic components. These techniques include high frequency ultrasonics, scanning laser acoustic microscopy, scanning photoacoustic microscopy, and ultrasonic velocity.

The turbine tip shroud abradability subsection documents the development of ceramic abradable seal materials systems compatible with both ceramic and metal turbine blades.

The ceramic component machining technology subsection discusses the development of tooling and machining parameters for the machining of ceramic components.

The ceramic blade process development subsection provides a brief summary of the ceramic blade process development experiment and the results obtained.

Materials Behavior and Characterization

SUMMARY

The materials behavior and characterization efforts focused on the testing and evaluation of candidate ceramic materials. These materials included sintered silicon carbide (SiC) from Carborundum (alpha SiC), Annawerk (CD-100), and NGK Spark Plug (EC-424); reaction-bonded silicon carbide from Carborundum (KX-02), Norton (NC-430 and NC-433), Pure Carbon (REFEL), Kyocera (SC-410), and NGK Spark Plug (EC-414); and NGK Spark Plug sintered silicon nitride (EC-129). Evaluation of the fast fracture strength (modulus of rupture) and microstructure of these materials was of primary concern. Particular attention was given to determining the nature of strength-controlling flaws through fracture surface analysis.

OBJECTIVE

The primary objective of this task was to establish appropriate material characteristics to support the design, development, and testing of hot-section ceramic components for the 1900°F- and 2070°F-configuration engines. A secondary objective was to evaluate new candidate ceramic materials and to assess which, if any, should be used in specific component development activities.

APPROACH

The material characterization activities were focused on microstructural, density, and fast fracture strength (MOR) examinations of candidate ceramic materials. Fracture surface analysis was also used to determine the nature and location of the strength-controlling defects. In addition, elevated temperature strength and oxidation resistance were evaluated for select materials.

DISCUSSION: SINTERED SILICON CARBIDE

Carborundum Sintered Alpha Silicon Carbide

Preliminary characterization of Carborundum (CBO) pressed and sintered alpha silicon carbide material was conducted in 1978-79, including strength/microstructure, oxidation, and elevated temperature strength.

The die-pressed sintered silicon carbide supplied by CBO was an equiaxed fine-grained material, with an average grain size of 6.5 μm, as shown in Figure 85. The average density measured 3.13 g/cm^3, which corresponds to 97.5% of theoretical density.

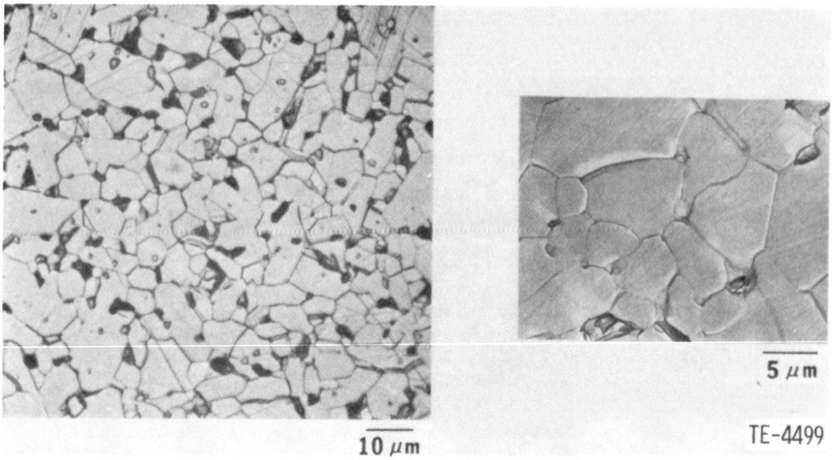

Figure 85. Microstructure of CBO sintered alpha SiC: (a) optical and (b) TEM carbon replica.

The room temperature MOR of longitudinally ground specimens averaged 335.72 MPa (48.69 ksi) with a standard deviation of 55.30 MPa (8.02 ksi). The primary strength-controlling defects were observed to be small pores and inclusions, shown in Figure 86. All fracture modes were observed to be transgranular cleavage.

CBO sintered alpha silicon carbide displayed the best oxidation resistance among the silicon carbide and silicon nitride candidate materials. The results of this testing are summarized in Table IX. No strength degradation was observed in this material after oxidation at temperatures up to 1371°C (2500°F).

Localized oxidation did occur even at low temperatures to form large, glassy spots, as shown in Figure 87a. In general, however, oxidation proceeded uniformly over the surface of the sample. Figure 87b shows the granular oxide surface obtained at 1050°C (1922°F) after 1000 hr. Most of the surface-connected pores were still open, with some sphere-shaped hillocks present on the surface. At a higher temperature, 1250°C (2282°F), these granules smeared out and became indistinguishable, as shown in Figure 87c. Pitting, crazing, and featherlike crystalline phases, shown in Figure 87d, also began to appear at this temperature. The round hillocks that formed were composed mainly of silicon with minor amounts of vanadium. (Small amounts of a vanadium secondary phase are generally found in alpha silicon carbide.)

Oxidation of alpha silicon carbide modulus of rupture (MOR) bars at temperatures up to 1371°C (2500°F) did not significantly alter either the strength or the critical flaw population controlling fast fracture strength. Strength-controlling flaws were still the pores and ball inclusions. Figure 88 shows the fractography of an MOR bar oxidized at 1050°C (1922°F) for 1000 hr. Failure originated at the large ball inclusion located at the surface, which is surrounded with oxide.

The fast fracture strength of CBO sintered SiC tested at a temperature of 1150°C (2102°F), 314.21 MPa (45.57 ksi), remained virtually unchanged from the room temperature strength value, 335.72 MPa (48.69 ksi). A second group of specimens oxidized for 500 hr at 1250°C (2282°F) and tested at 1150°C (2102°F) recorded an average strength of 317.72 MPa (46.08 ksi). In both cases, the same type of defects previously described controlled specimen failure.

Annawerk CD-100 Sintered Silicon Carbide

Annawerk CD-100 is an isopressed and sintered silicon carbide material that has a uniform fine-grained microstructure with an average grain size of 3.5 μm (see Figure 89). Clusters of irregularly shaped pores occur throughout the microstructure. The average density of this material measures 3.26 g/cm³.

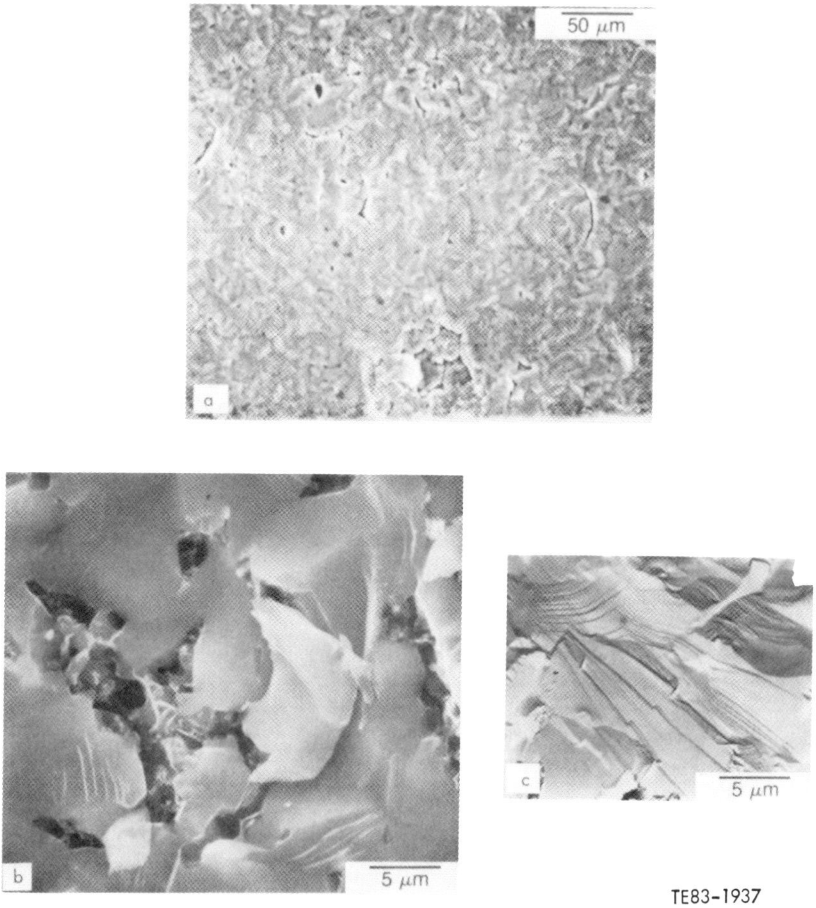

TE83-1937

Figure 86. Fracture surfaces of CBO sintered SiC test material: (a) surface pore and (b) and (c) transgranular cleavage.

The average MOR of longitudinally ground bars tested at room temperature measured 328.31 MPa (47.62 ksi) with a standard deviation of 31.21 MPa (4.53 ksi). The primary strength-controlling defects observed in this material were found to be surface and internal pores. Figure 90 shows typical examples of a surface pore and an internal pore.

NGK Spark Plug EC-424

NGK Spark Plug EC-424 is a fine-grained sintered silicon carbide with an average grain size of 4.5 μm, as shown in Figure 91. The average density of this material measures 3.12 g/cm^3. The room temperature fracture strength of test bars with a machined surface condition averaged 316.65 MPa

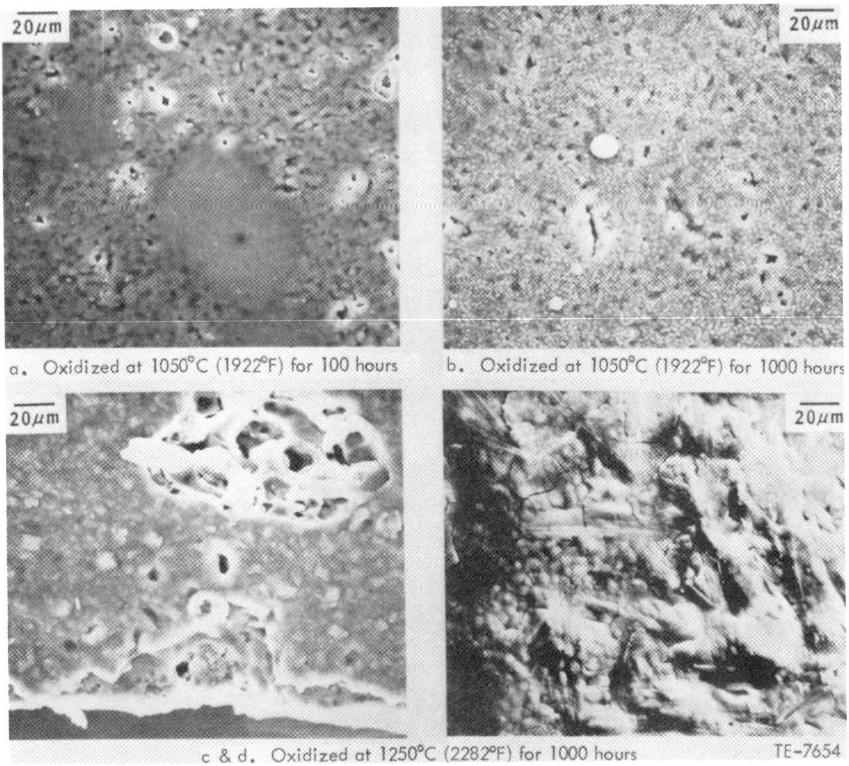

Figure 87. Oxide surface morphologies of CBO sintered SiC.

(45.93 ksi) with a standard deviation of 25.80 MPa (3.74 ksi). The primary strength-controlling flaws observed in these specimens, shown in Figure 92, were surface cracks (probably a result of the machining operation) with secondary flaws of surface and internal porosity.

DISCUSSION: REACTION-BONDED SILICON CARBIDE

Carborundum KX-02 Silicon Carbide

Injection-molded siliconized silicon carbide (KX-02) test bars from CBO were evaluated in August 1982. KX-02 is a dense material with very fine-grained silicon carbide particles dispersed in a matrix of free silicon. The microstructure of the KX-02 is shown in Figure 93. An average density of 2.83 g/cm^3 was recorded for this material, corresponding to a free silicon content of approximately 40% by volume.

Test bars with both as-fired and longitudinally ground tensile surface conditions were evaluated in four-point bend tests at room temperature. The average MOR of the as-fired test material measured 338.48 MPa (49.09 ksi) with a standard deviation of 49.51 MPa (7.18 ksi). All failures of these specimens originated at surface sites, probably as a result of surface roughness. A typical example of failure in

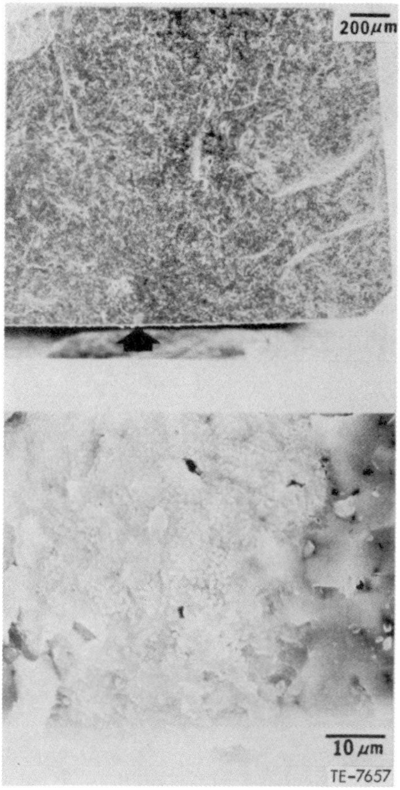

Figure 88. SEM fractographs of oxidized CBO alpha SiC (1050°C [1922°F], 1000 hr); failure origin, indicated by the arrow, is a commonly observed spherical inclusion, but it is oxidized as viewed at higher magnification.

as-fired KX-02 is shown in Figure 94. The room temperature MOR of the longitudinally ground test bars averaged 421.77 MPa (61.16 ksi) with a standard deviation of 113.77 MPa (16.50 ksi). All failures again initiated from surface sites. The strength-controlling flaws in most cases could not be identified; however, several test bars had unusual features located at the failure origin. The following critical defects were observed: (1) in bar No. 121.0069, a cluster of pores (see Figure 95); (2) in bar No. 121.0078, a deep surface scratch running at an angle to the normal machining direction (see Figure 96); and (3) in bar No. 121.0090, a preexisting crack (see Figure 97). Excluding bars No. 121.0078 and No. 121.0090, which failed as a result of machining or handling damage, the average MOR of the longitudinally ground material was increased to 462.03 MPa (67.01 ksi) with a standard deviation of 82.81 MPa (12.01 ksi).

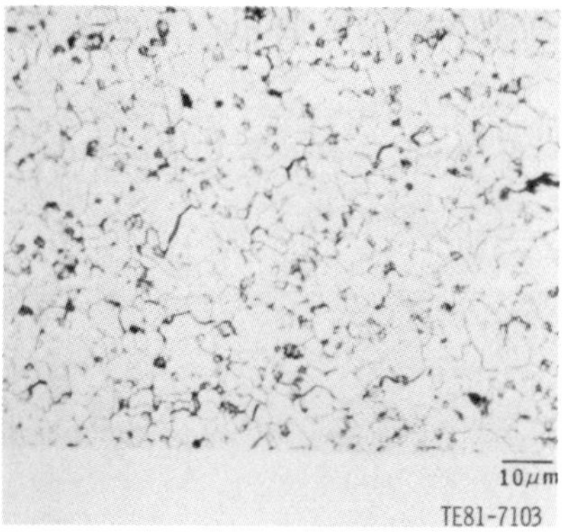

Figure 89. Micrograph of Annawerk CD-100 sintered SiC; the average grain size is 3.5 μm.

Norton NC-430 Reaction-Bonded Silicon Carbide

Characterization of Norton NC-430 material was conducted in 1978-79. This included microstructural and strength analysis as well as a study of the effects of oxidation for varying temperatures and exposure times. In addition, the elevated temperature strength was also evaluated.

Norton NC-430 is a slip-cast reaction-bonded silicon carbide material. The microstructure consists of a bimodal mixture of large (50-200 μm) and small (25 μm) silicon-carbide grains surrounded by a network of free silicon, as shown in Figure 98. The surface structure of the as-fired test material, shown in Figure 99, has a coarse granular appearance. The average density of the NC-430 material measures 3.10 g/cm^3.

The average room temperature fracture strength of bars with an as-fired surface condition measured 176.65 MPa (25.62 ksi) with a standard deviation of 28.61 MPa (4.15 ksi). The room temperature fracture strength of the machined test bars aver-

Table IX.
Effect of oxidation on Carborundum sintered silicon carbide.

	Room temperature	1050°C (1922°F) 100 hr	1050°C (1922°F) 1000 hr	1250° (2282°F) 100 hr	1250° (2282°F) 1000 hr	1371°C (2500°F)—100 hr
MOR—MPa (ksi)	335.72 (48.69)	345.16 (50.06)	366.47 (53.15)	340.20 (49.34)	357.23 (51.81)	341.37 (49.51)
Standard deviation—MPa (ksi)	55.30 (8.02)	19.17 (2.78)	30.20 (4.38)	37.23 (5.40)	15.93 (2.31)	27.37 (3.97)
Weight gain—mg/cm²	—	0.117	0.130	0.129	0.216	0.278
Oxide layer thickness—μm	—	1	1.7	1.9	2.6	2.8

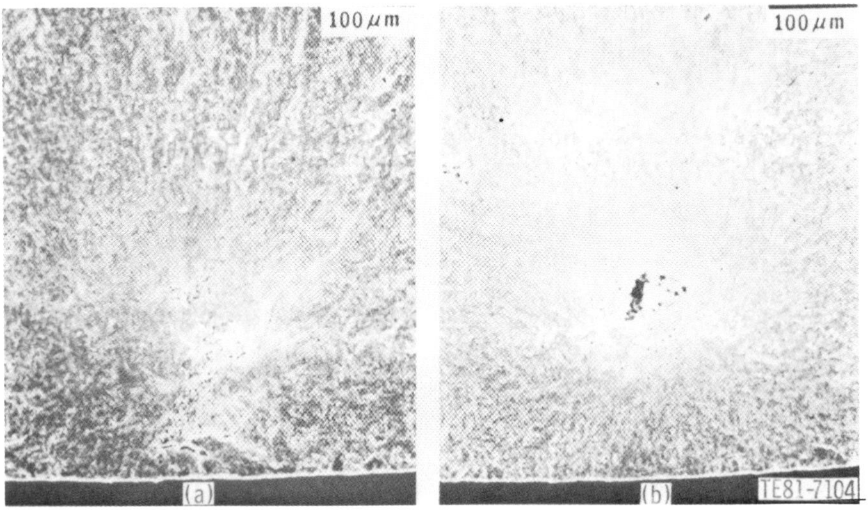

Figure 90. Typical fracture origins observed in Annawerk CD-100 sintered SiC: (a) surface pore and (b) internal pore.

aged 225.05 MPa (32.64 ksi) with a standard deviation of 14.07 MPa (2.04 ksi). Due to the coarse nature of the grain structure, establishing failure origins in this material with certainty is virtually impossible. Failure, however, appeared to originate from the surface, presumably from the large silicon carbide grains which form during firing.

The oxidation/strength behavior of NC-430 was evaluated for both as-fired and machined surface conditions. These results, summarized in Table X, indicate that both before and after oxidation at temperatures up to 1250°C (2282°F) the strength of the machined surface is greater than that of the as-fired surface. Further, oxidation increased fast fracture

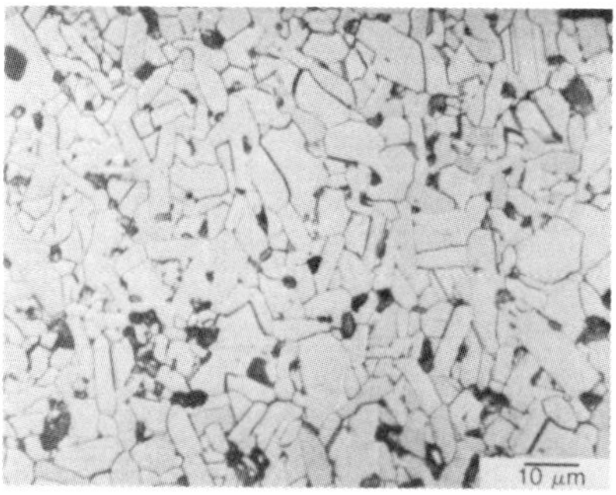

TE83-1927

Figure 91. Microstructure of NGK Spark Plug EC-424 sintered SiC test bars; the average grain size is 4.5 μm.

strength. The only exception occurred for the machined surface condition oxidized at 1371°C (2500°F) for 100 hr, where melting of the silicon phase was observed. More detailed results of this testing are reported in EDRs 9722 and 9951.

NC-430 exposed to 1050°C (1922°F) for 100 hr showed very slight oxidation. Most machining traces were still visible, with only scattered glassy spots (approximately 20 μm) detected. These areas had traces of aluminum present, as shown in Figure 100. Exposure at 1050°C (1922°F) for 1000 hr produced a uniform oxide layer and tripled the weight gain. Microscopically, the general appearance of the oxide surface shows a network of nonsealing granular oxide, as shown in Figure 101. The average strength of both the as-fired and machined surfaces increased to 194.92 MPa (28.27 ksi) and 251.32 MPa (36.45 ksi), respectively, after 100 hr and 190.44 MPa (27.62 ksi) and 273.66 MPa (39.69 ksi) after 1000 hr.

NC-430 oxidized at 1250°C (2282°F) exhibited similar but increased oxidation, with a completely sealed oxide layer (with numerous cracks) observed on the machined surface, as shown in Figure 102. Exposure for 100 hr resulted in an average strength of 222.16 MPa (32.22 ksi) for the as-fired surface condition and 257.53 MPa (37.35 ksi) for the machined surface condition. After 1000 hr, the as-fired strength decreased slightly to 194.03 MPa (28.14 ksi) while the machined strength increased to 284.63 MPa (41.28 ksi).

A smooth thick oxide layer was observed in NC-430 oxidized at 1371°C (2500°F) for 100 hr. After exposure, the strength of the as-fired material increased slightly to 201.89 MPa (29.28 ksi), while that of the machined specimens decreased to 152.38 MPa (22.10 ksi). Examination of the machined surface revealed the presence of occasional balls of silicon, indicating temperatures exceeding the melting point of the silicon phase were reached during the

Table X.
Effect of oxidation on Norton NC-430 reaction-bonded silicon carbide.

	Room temperature	1050°C (1922°F) 100 hr	1050°C (1922°F) 1000 hr	1250°C (2282°F) 100 hr	1250°C (2282°F) 1000 hr	1371°C (2500°F)—100 hr
MOR—MPa (ksi)						
As-fired	176.65 (25.62)	194.92 (28.27)	190.44 (27.62)	222.16 (32.22)	194.03 (28.14)	201.89 (29.28)
Machined	225.05 (32.64)	251.32 (36.45)	273.66 (39.69)	257.53 (37.35)	284.63 (41.28)	152.38 (22.10)
Standard deviation—MPa (ksi)						
As-fired	28.61 (4.15)	29.58 (4.29)	23.93 (3.47)	26.75 (3.88)	30.20 (4.38)	30.13 (4.37)
Machined	14.07 (2.04)	26.27 (3.81)	32.20 (4.67)	18.00 (2.61)	18.20 (2.64)	42.68 (6.19)
Weight gain—mg/cm^2						
As-fired	—	0.037	0.026	0.262	0.250	0.386
Machined	—	0.079	0.151	0.178	0.282	0.420
Oxide layer thickness—μm						
As-fired	—	1	2	6	3.5	
Machined	—	1	2	4	4	10

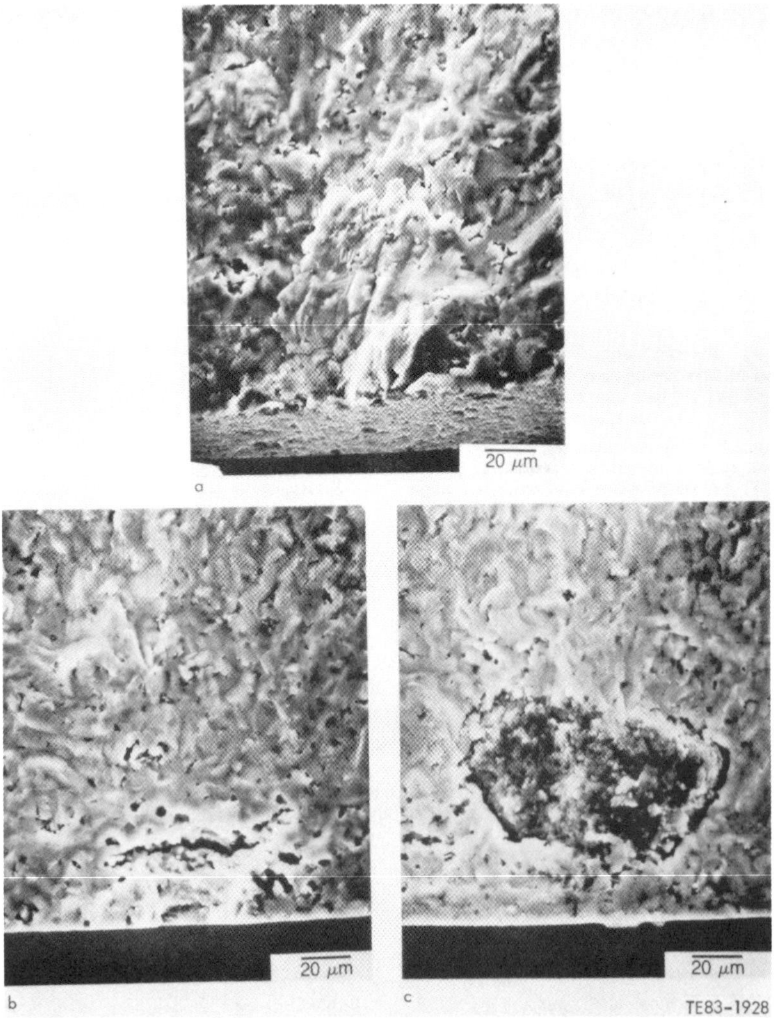

Figure 92. Typical strength-controlling flaws observed in NGK Spark Plug EC-424 sintered SiC test bars: (a) surface crack, (b) surface pore, and (c) internal pore.

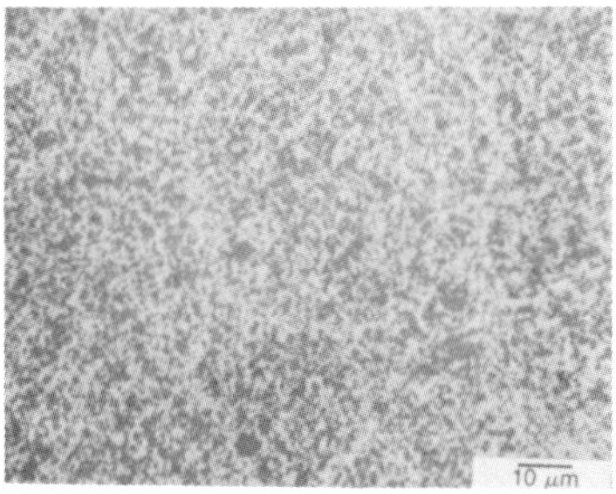

Figure 93. Microstructure of CBO KX-02 silicon carbide.

exposure run. Such losses of strength, if melting of the silicon phase occurs, are well known in NC-430. The as-fired surfaces, however, were not degraded.

Oxidation of the as-fired surface was observed to be different than that of the machined surface. During the early stages of oxidation—at 1050°C (1922°F), 100 hr—features on the as-fired surface began to smooth out with the formation of a thin, glassy layer. At 1250°C (2282°F) and 1000 hr, large areas of the as-fired surface were covered by a thick, glassy oxide, as shown in Figure 103a. At 1371°C (2500°F) and 100 hr, the formation of large, glassy puddles with bubbles underneath was observed, as shown in Figure 103b. The reason for obtaining different oxide surfaces may be rationalized by the impurities detected. The glassy oxides on the as-fired surface always contain alumina, which is a good glass former. Particles of alumina were detected on the as-fired surface of the MOR bars prior to oxidation. Traces of aluminum could also have been picked up from the substrate during oxidation.

These exposure tests show NC-430 (1) to have excellent oxidation resistance and (2) to not suffer a degradation in strength due to oxidation—at least not for short times at temperatures below 1371°C (2500°F).

Preliminary elevated temperature testing also indicated that the baseline strength level of NC-430 material is maintained to a temperature of 1250°C (2282°F). The fast fracture strength of NC-430 specimens with an as-fired surface condition was found to increase slightly with temperature from the value of 176.65 MPa (25.62 ksi) recorded at room temperature to 196.65 MPa (28.52 ksi) at 1150°C (2102°F) and 212.99 MPa (30.89 ksi) at a temperature of 1250°C (2282°F).

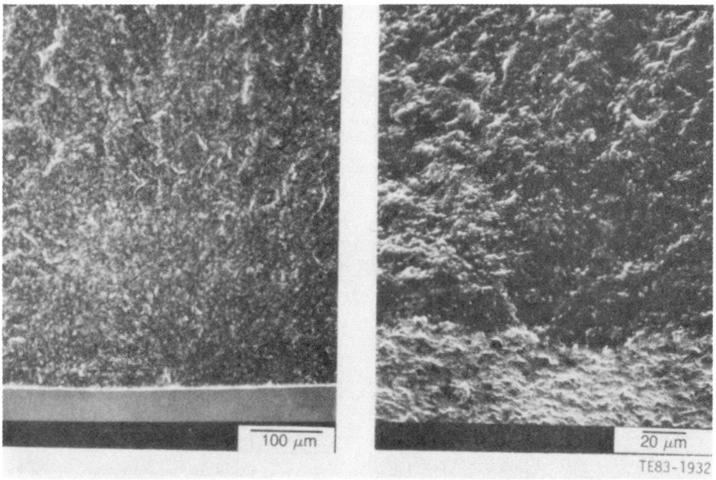

Figure 94. Typical failure origin (surface flaw) found in as-fired CBO KX-02 test bars.

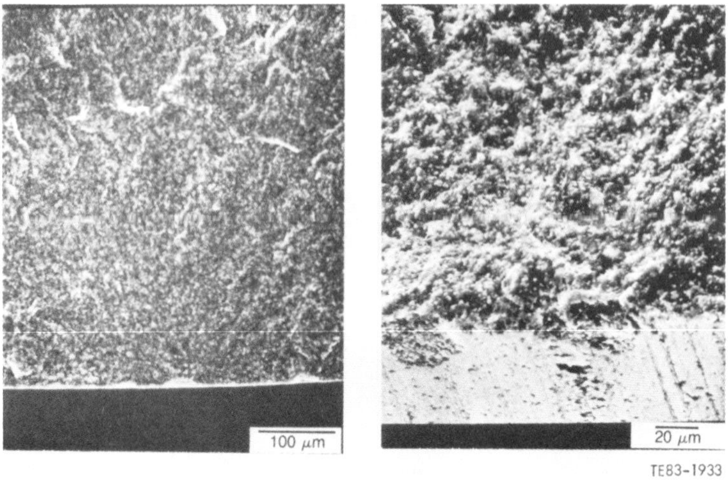

Figure 95. Failure origin (pore cluster) observed in longitudinally ground CBO KX-02 test bar No. 121.0069.

Figure 96. Failure origin (surface scratch) observed in longitudinally ground CBO KX-02 test bar No. 121.0078.

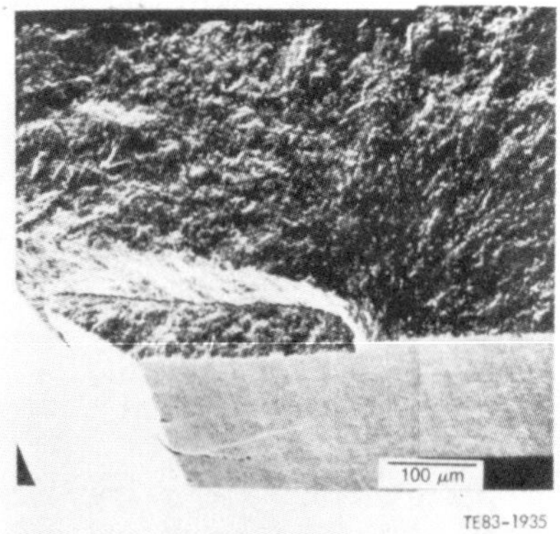

Figure 97. Failure origin (preexisting crack) observed in longitudinally ground CBO KX-02 test bar No. 121.0090.

Figure 98. Microstructure of Norton NC-430 RBSiC.

Figure 99. As-fired surface of Norton NC-430 test bar.

Norton NC-433 Reaction-Bonded Silicon Carbide

Norton NC-433 is a siliconized silicon carbide material featuring a bimodal distribution of silicon carbide grains, as shown in Figure 104. The average density was found to be 3.08 g/cm^3. NC-433 is the same as the Norton NC-430 material except that NC-433 has a much smaller grain size and a higher percentage of free silicon present.

The room temperature fracture strength for the NC-433 material averaged 166.03 MPa (24.08 ksi), with a standard deviation of 10.20 MPa (1.48 ksi), for the as-fired surface condition and 299.71 MPa (43.47 ksi), with a standard deviation of 22.82 MPa (3.31 ksi), for the machined surface condition. In all cases, surface sites were identified as the locations of the failure origins. A typical fracture surface is shown in Figure 105.

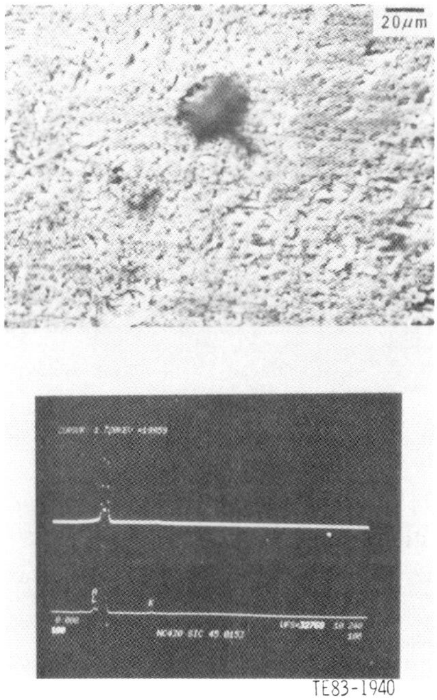

Figure 100. Light oxidation on machined NC-430 oxidized at 1050°C (1922°F) for 100 hr.

Pure Carbon Refel Reaction-Bonded Silicon Carbide

Characterization and evaluation of two Pure Carbon Refel RBSiC materials—one a slip cast product, the other an extruded product—were conducted in October 1979. Material microstructure, strength characteristics, and oxidation behavior were examined.

Pure Carbon's RBSiC materials have a microstructure consisting of a network of equiaxed SiC

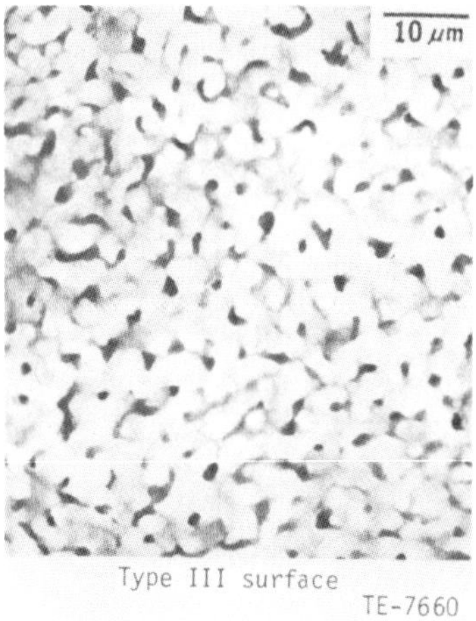

Figure 101. SEM micrograph of machined surface of NC-430 oxidized at 1050°C (1922°F) for 1000 hr.

grains surrounded by free silicon. The detail of the microstructure is different for each of the two fabrication processes. The slip-cast reaction-bonded SiC material has a very fine-grained structure, with an average grain size of 3.5 μm. The structure is generally uniform throughout. Occasionally, however, a thin lamination of free silicon is observed 400-500 μm below the as-fired surface (see Figure 106a). On the other hand, the extruded material has a much larger grain size, as shown in Figure 106b. In this case the average silicon-carbide grain diameter is approximately 11.6 μm or three times that found in the slip-cast material. The as-fired surface of the extruded material, however, has a silicon-rich layer approximately 300 μm in thickness, as shown in Figure 107.

The as-fired surface structure of both the slip-cast and extruded materials is similar. A typical example of the as-fired surface structure is shown in Figure 108. Microprobe analysis shows barium and sulfur are present in addition to silicon and small amounts of calcium and aluminum. The impurities tend to be concentrated in glassy puddles that are commonly found over the entire as-fired surface.

The four-point bend strengths of the Pure Carbon extruded and slip-cast RBSiC materials are summarized in Table XI. The strengths for the as-fired surface condition of the slip-cast and extruded materials were 265.87 MPa (38.56 ksi) and 306.62 MPa (44.47 ksi), respectively. In the slip-cast material, failure generally originated at the surface. The most common strength-controlling defects were glassy puddles. Occasionally failure also originated from silicon-filled subsurface laminations containing iron silicide particles. Such laminations become strength-controlling when microporosity is associated with the iron silicide. Failure also originated from surface sites for the extruded material. The strength-controlling defects in all cases were glassy puddles. The difference in strength between these two materials (significant at the 0.05 level) must be associated with differences in grain size and silicon content.

The average strength of the machined surface condition was similar for both materials (no difference exists at a 0.05 significance level). The slip cast material failed exclusively from surface sites, presumably from machining flaws, since no unusual microstructural feature could, in general, be associated with a given origin. However, in the case of the extruded material, failure originated at both surface and subsurface sites. Surface failures were associated with machining defects; the subsurface failures originated from carbon particles. A typical

Table XI.
Strength of Pure Carbon reaction-bonded silicon carbide.

		MOR		Standard deviation	
Material	**No. of specimens**	**MPa**	**(ksi)**	**MPa**	**(ksi)**
Slip cast					
As-fired	29	265.87	(38.56)	50.00	(7.25)
Machined	9	469.14	(68.04)	77.84	(11.29)
Extruded					
As-fired	29	306.62	(44.47)	49.02	(7.11)
Machined	10	430.59	(62.45)	35.79	(5.19)

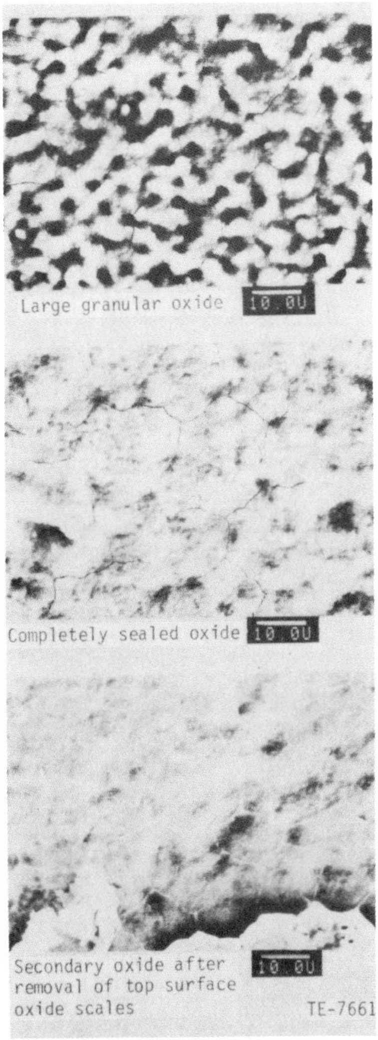

Figure 102. SEM micrograph of machined surface of NC-430 oxidized at 1250°C (2282°F) for 1000 hr.

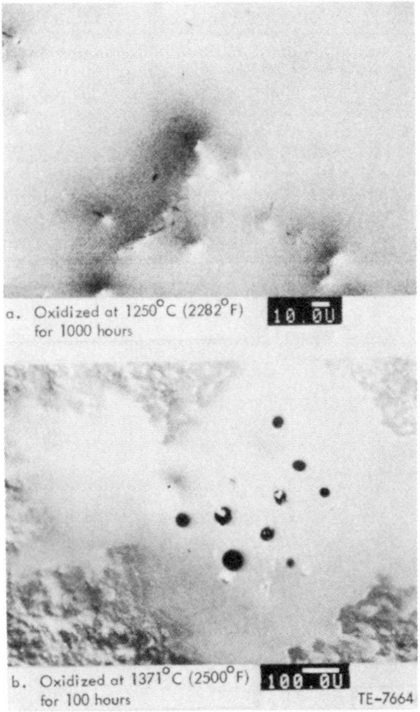

Figure 103. Oxide morphology of as-fired NC-430 SiC; aluminum was detected in the glassy oxide.

example of a subsurface origin is shown in Figure 109.

The effect of oxidation at 1250°C (2282°F) was evaluated in preliminary fashion. Samples of each surface finish were exposed for 100 hr in ambient air. After exposure, changes in surface structure and the resulting effect on material strengths were investigated.

After exposure, the surface of the as-fired slip-cast material was covered by a crazed glass. Very large circular spots, approximately 2 mm (0.078 in.) in diameter, were visible, as shown in Figure 110. The oxide layer, in addition to containing silicon and oxygen, contains barium, calcium, and aluminum. An acicular phase, which is free of barium, was distributed throughout the oxide. The large glassy puddles

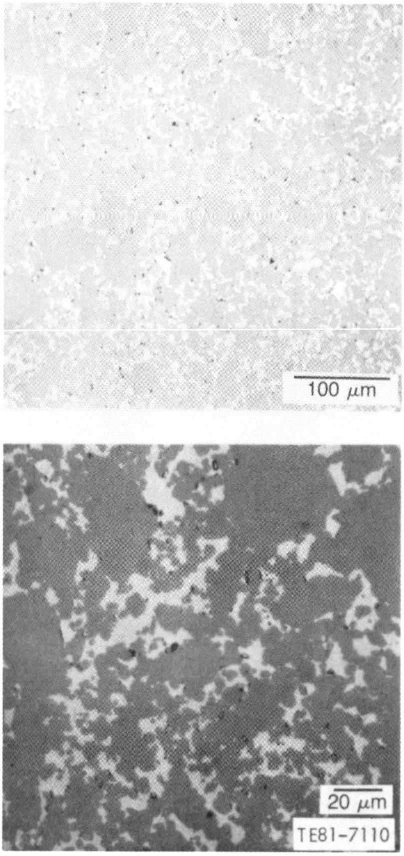

Figure 104. Typical microstructure of Norton NC-433 RBSiC.

generally contained high levels of calcium. Iron silicide was often present directly beneath the glassy bubbles.

The machined and oxidized slip-cast material was covered with a crazed oxide with numerous hills and bumps, as shown in Figure 111. These bumps

Figure 105. Typical fracture surface of Norton NC-433 RBSiC test bar.

contained high levels of calcium; however, no barium was found in the surface oxide.

The oxidized surfaces of both the as-fired and machined extruded material were similar to those found on the slip-cast material of Figure 112. The strength levels after exposure to 1250°C (2282°F) for 100 hr for both the as-fired and machined surface conditions are summarized in Table XII. No significant change in strength occurred in either material regardless of surface finish. The as-fired and machined slip-cast materials failed exclusively from surface sites—either glassy puddles, shown in Figure 113, or large grains, shown in Figure 114. The extruded material also failed from surface sites—either glassy puddles for the as-fired surface or oxidized bumps for the machined surface.

Kyocera SC-410 Reaction-Bonded Silicon Carbide

Kyocera International SC-410 is a dry-pressed RBSiC material with an average density of 3.11 g/cm³. The microstructure, shown in Figure 115, consists of an SiC network with a bimodal grain size distribution. The majority of grains are approximately 0.5 μm, with a few grains as large as 2.0 μm. The remainder of the microstructure consists of free silicon, with an estimated volume fraction of 10%.

The average room temperature fracture strength for material with an as-fired surface condition was found to be 381.20 MPa (55.29 ksi) with a standard deviation of 70.32 MPa (10.20 ksi). The average strength for the longitudinally ground surface condition measured 426.42 MPa (61.85 ksi) with a standard deviation of 112.90 MPa (16.37 ksi). The critical strength-controlling defects were usually pores

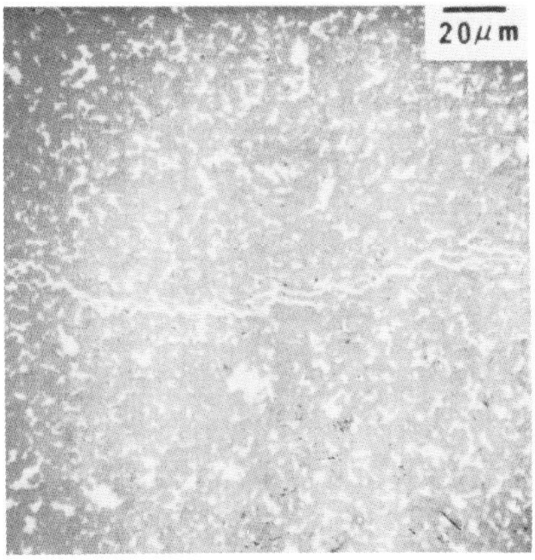

A

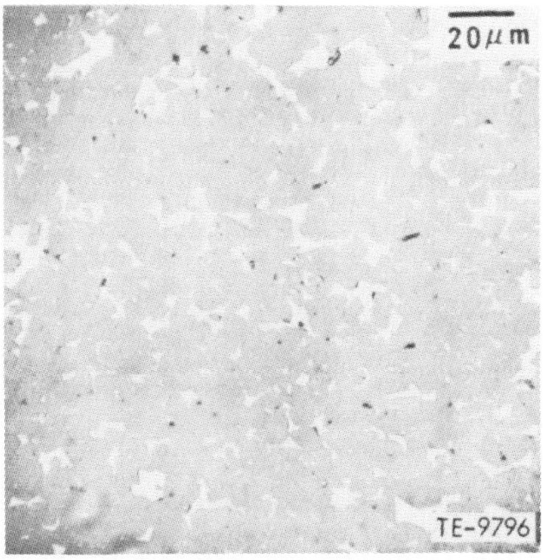

B

Figure 106. Microstructure of Pure Carbon Refel reaction-bonded SiC: (a) slip cast and (b) extruded.

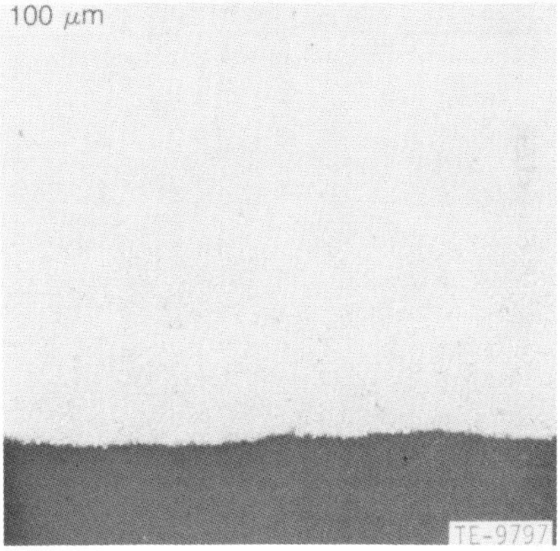

Figure 107. Micrograph of extruded Pure Carbon Refel reaction-bonded SiC showing silicon-rich layer on the as-fired surface.

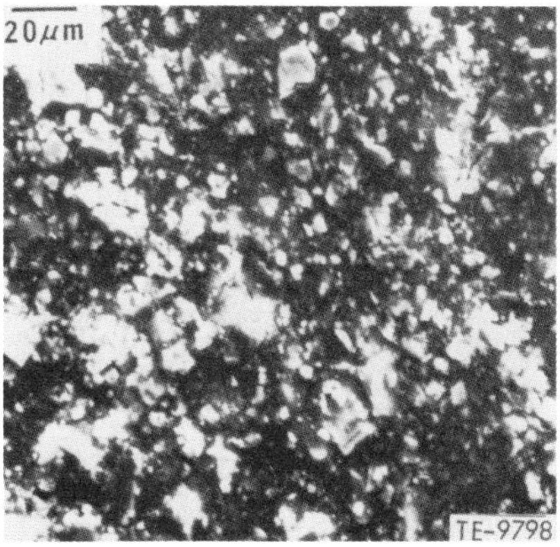

Figure 108. Surface structures of as-fired slip-cast Pure Carbon Refel reaction-bonded SiC.

and surface flaws. Large surface cracks, such as the one shown in Figure 116, were frequently observed in the machined test specimens. The bar in Figure 116 registered a strength of only 303.80 MPa (44.07 ksi).

NGK Spark Plug EC-414

NGK Spark Plug EC-414 is a reaction-bonded silicon carbide material, with a microstructure as shown in Figure 117. The average density measures 3.05 g/cm^3. The average room temperature fracture strength of machined test bars was 263.18 MPa

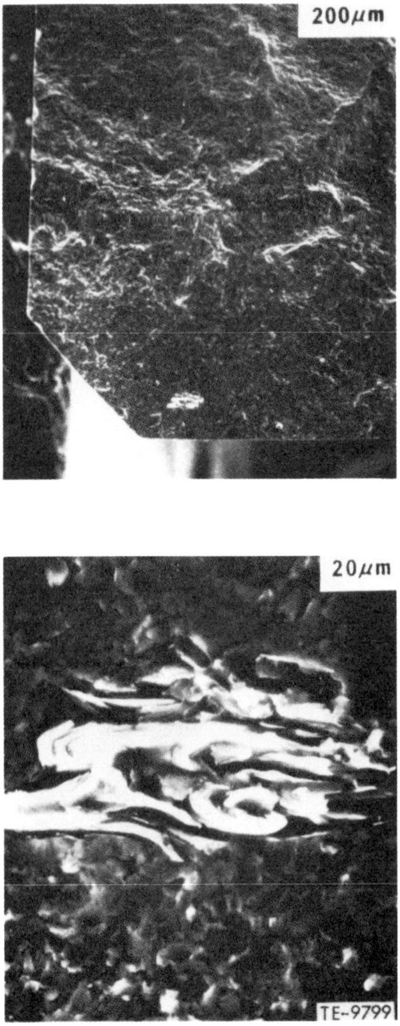

Figure 109. Fracture surface of machined, extruded Pure Carbon Refel RBSiC, showing a carbon particle at the failure origin.

(38.17 ksi) with a standard deviation of 30.86 MPa (4.48 ksi). This strength level is considerably lower than the 500.00 MPa (72.52 ksi) published by NGK Spark Plug. The primary strength-controlling flaws in EC-414 material were surface cracks and pores, shown in Figure 118.

DISCUSSION: SINTERED SILICON NITRIDE (NGK SPARK PLUG EC-129)

NGK Spark Plug EC-129 is a sintered silicon nitride with a fine grained microstructure, as shown in Figure 119, using MgO and Al_2O_3 as sintering additives. The average density of this material is 3.22 g/cm^3. Test bars with a machined surface condition registered an average room temperature fracture strength of 576.26 MPa (83.56 ksi) with a standard deviation of 124.84 MPa (18.11 ksi). The strength-controlling flaws observed in these specimens were surface pores, internal pores, and surface cracks, examples of which are shown in Figure 120. The fracture strength of machined bars tested at a temperature of 1100°C (2012°F) averaged 434.41 MPa (63.01 ksi) with a standard deviation of 46.96 MPa (6.81 ksi). Failures of these test bars originated from internal low density pockets and pores, shown in Figure 121. A thin oxidized glassy layer was observed on the exterior surface of the samples and, in some instances, on the fracture surface near the failure origin.

Component Characterization and Qualification

SUMMARY

The component characterization and qualification activities were divided into two classifications: ceramic parts fabricated using established processing technology and components for which either improved material characteristics and/or further development of fabrication methodology were required. The development component activities are presented in the ceramic component process development subsection of this report. Components in the established class included 1900°F-configuration vanes from Annawerk; 2070°F-configuration vanes from CBO and Pure Carbon; inner vane support rings from Pure Carbon, NGK-Locke, and Kyocera; outer vane support rings from CBO, Pure Carbon, and NGK-Locke; vane retaining rings from Pure Carbon and NGK-Locke; and plenums from Norton.

OBJECTIVE

The primary objective of this task was the evaluation and characterization of ceramic components

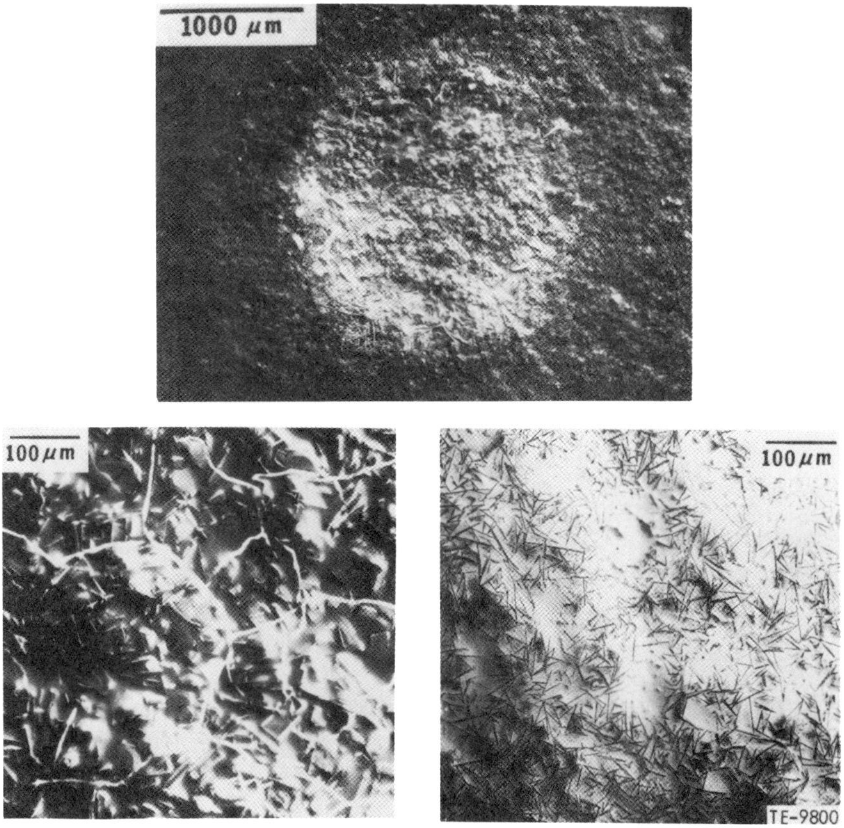

Figure 110. As-fired surface structure of slip-cast Pure Carbon Refel RBSiC after exposure at 1250°C (2282°F) for 100 hr.

with appropriate material and dimensional properties for successful use in rig and engine testing. Included in this objective was furnishing information back to the ceramic suppliers for iterative component improvements.

APPROACH

The component characterization activities were directed toward nondestructive evaluations of ceramic components. These evaluations included vis-

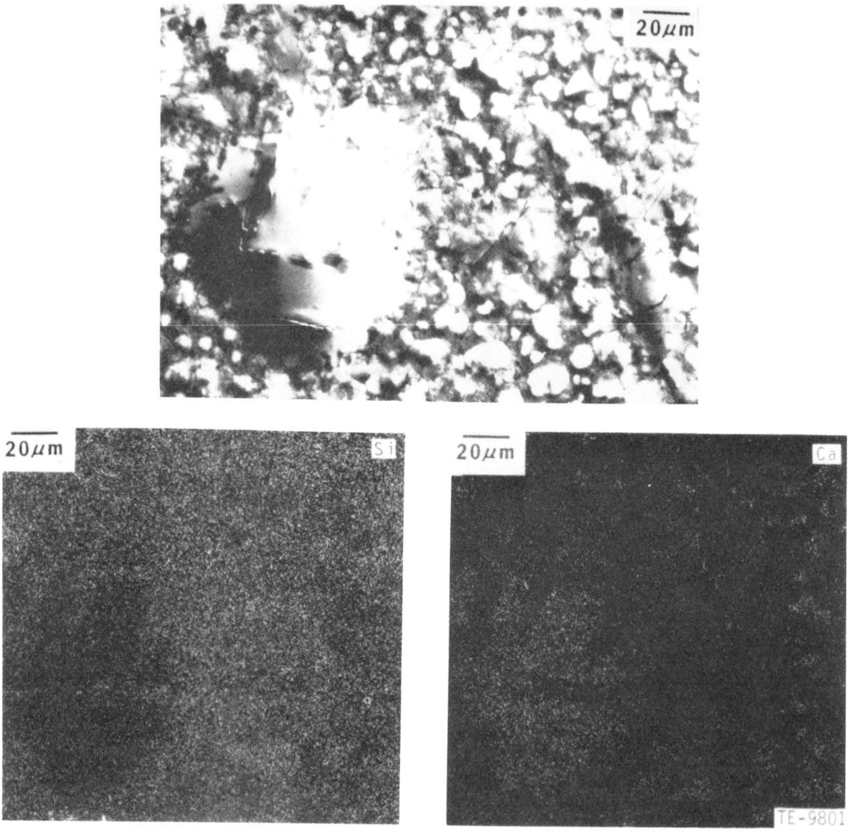

Figure 111. Machined surface structure of slip-cast Pure Carbon Refel RBSiC after exposure at 1250°C (2282°F) for 100 hr.

ual, fluorescent penetrant, and radiographic inspections, as well as microstructural determination. In addition, qualification test bars processed with the components were evaluated for both MOR strength and to determine the nature of the strength-controlling defects.

DISCUSSION: 1900°F-CONFIGURATION VANES (ANNAWERK CS-600 RBSiC)

The 1900°F-configuration vanes from Annawerk were fabricated from slip-cast CS-600, an RBSiC material. The microstructure was coarse,

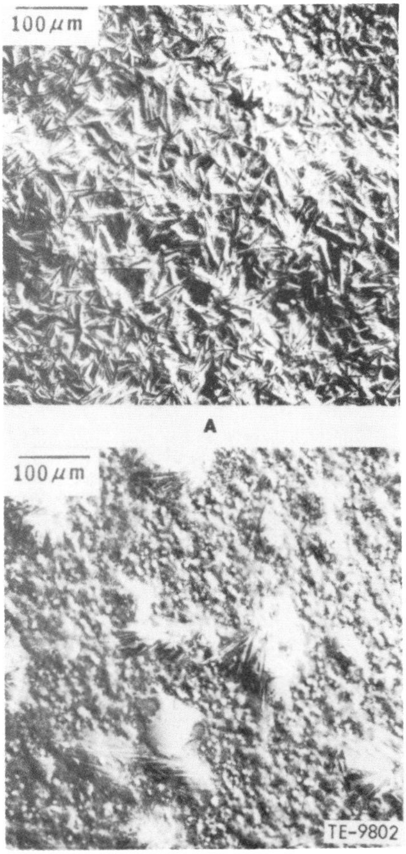

Figure 112. Surface structure of (a) as-fired and (b) machined, extruded Pure Carbon Refel RBSiC after exposure at 1250°C (2282°F) for 100 hr.

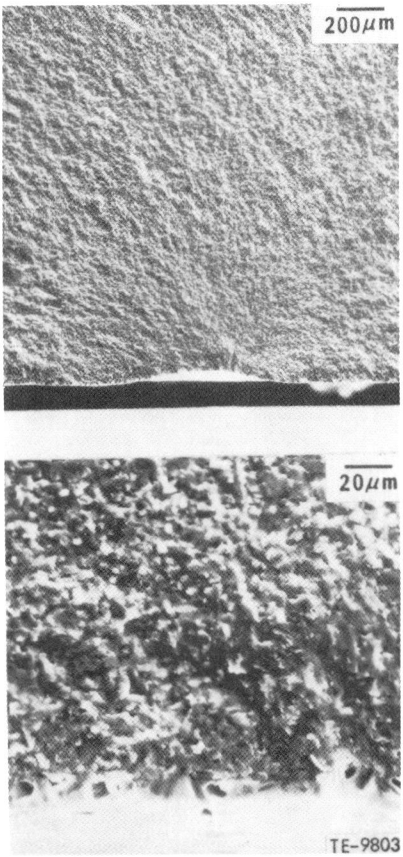

Figure 113. Fracture surface of slip-cast, Pure Carbon Refel RBSiC after oxidation showing glassy puddle at failure origin.

consisting of very large (20-50 µm diameter) and small (1-3 µm) SiC grains with free silicon occupying the remaining porosity, as is shown in Figure 122. A small quantity of large graphite particles (10-30 µm) were also present. NDE inspections of these vanes revealed no objectionable defects present. The average density of these components measured 3.06 g/cm^3.

The average room temperature MOR strength for qualification bars with an as-fired surface condition measured 182.10 MPa (26.41 ksi) with a standard deviation of 12.41 MPa (1.80 ksi). The MOR strength of machined bars was 291.95 MPa (42.34 ksi) with a

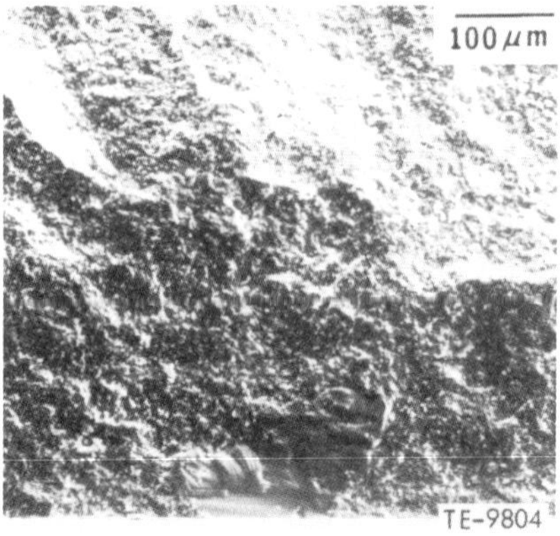

Figure 114. Photomicrograph of fracture surface of oxidized slip-cast Pure Carbon Refel RBSiC showing large grain at failure origin.

Figure 115. Micrograph of Kyocera SC-410 RBSiC; in the dark area are SiC grains and in the light area is free silicon.

standard deviation of 36.78 MPa (5.33 ksi). All specimens failed from surface sites in both the as-fired and machined materials. A typical fractograph of a failure origin is shown in Figure 123.

DISCUSSION: 2070°F-CONFIGURATION VANES

Carborundum Sintered Alpha Silicon Carbide

Development efforts for the fabrication of 2070°F-configuration vanes from injection-molded sintered alpha silicon carbide were conducted at CBO in 1979 and 1980. In the 2265°F engine, which was subsequently deleted due to funding limitations, some vanes would have been exposed to temperatures above the limit of the reaction-bonded SiC material. For this reason, experience was needed with the fabrication of 2070°F alpha SiC vanes.

An injection-molded plastic replica from the vane and test bar tool appears in Figure 124. NDI evaluations of these sintered SiC components were generally excellent, with only small surface pits measuring 50-75 μm (0.002-0.003 in.) observed in several of the vanes. The average density of both the vanes and test bars was 3.15 g/cm^3 (98% theoretical density).

The average room temperature fracture strength of test bars molded with the vanes measured 326.16 MPa (47.31 ksi) with a standard deviation of 66.66 MPa (9.67 ksi) for an as-fired surface condition. The primary strength-controlling flaws observed in these specimens were surface cracks. The MOR of bars with a machined surface condition

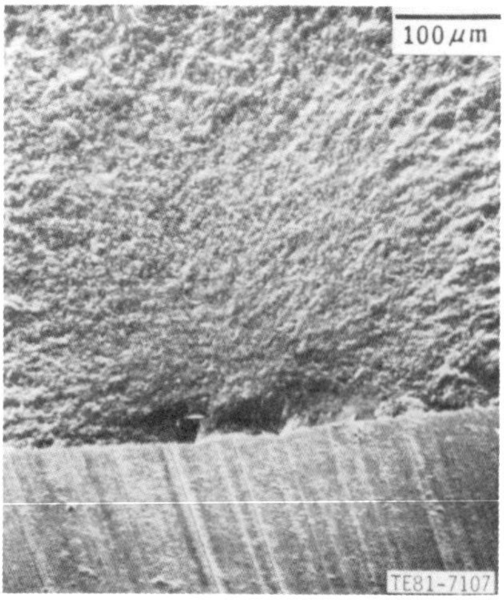

Figure 116. Typical example of failure origin (surface defect) observed in machined Kyocera SC-410 RBSiC test bars.

Table XII.
Oxidation effect on strength of Pure Carbon RBSiC materials.*

Material	No. of specimens	MOR		Standard deviation	
		MPa	(ksi)	MPa	(ksi)
Slip cast					
As-fired	5	276.49	(40.10)	87.43	(12.68)
Machined	5	483.82	(70.17)	87.29	(12.66)
Extruded					
As-fired	5	355.92	(51.62)	88.46	(12.54)
Machined	5	421.15	(61.08)	30.20	(4.38)

*100 hr exposure in ambient air at 1250°C (2282°F)

averaged 316.92 MPa (45.97 ksi) with a standard deviation of 61.91 MPa (8.98 ksi). Internal pores, such as the one in Figure 125, were identified as the fracture origins in all instances.

Pure Carbon Refel Reaction-Bonded Silicon Carbide

The 2070°F-configuration vanes received from Pure Carbon were fabricated of injection-molded Refel reaction-bonded silicon carbide. Visual, FPI, and X-ray inspections of these components indicated no objectionable defects, although the unmachined surfaces of the vanes were rough. The average density of these parts measured 3.07 g/cm^3.

The qualification test material accompanying these components had an average density of 3.09 g/cm^3. Microstructural evaluations of these test bars revealed the presence of laminations in the free silicon and large internal and surface agglomerates (300 μm), composed primarily of silicon with iron impurities and traces of titanium, sulfur, and calcium (see Figure 126). Several samples also exhibited fine particles of silicon and iron interspersed in the network of free silicon, as shown in Figure 127. The average room temperature fracture strength of test material with an as-fired surface condition measured 267.03 MPa (38.73 ksi) with a standard deviation of 47.29 MPa (6.86 ksi). Surface flaws, as shown in Figure 128, were identified as the primary strength-controlling defects in all of these specimens. The average MOR of test bars with a longitudinally ground surface measured 332.12 MPa (48.17 ksi) with a standard deviation of 69.48 MPa (10.09 ksi). The primary strength-controlling defects in these test bars were internal and surface inclusions of silicon and iron; the secondary defects were surface flaws and pores, shown in Figure 129.

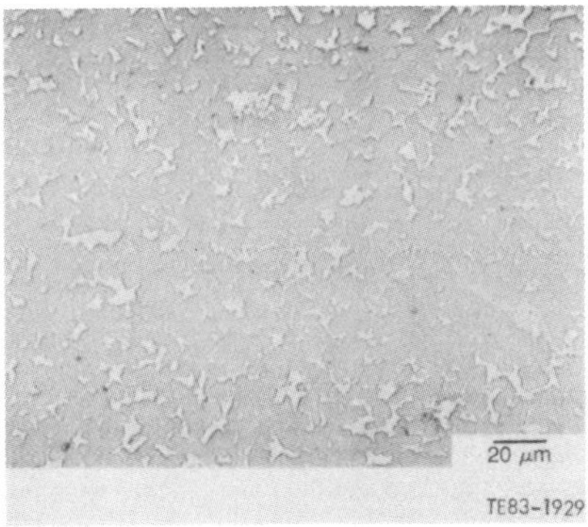

Figure 117. Microstructure of NGK Spark Plug EC-414 RBSiC test bars.

DISCUSSION: INNER VANE SUPPORT RINGS

Pure Carbon Refel Reaction-Bonded Silicon Carbide

Three ceramic inner vane support rings from Pure Carbon were received and evaluated. These rings were fabricated of isopressed and green-machined Refel RBSiC. The typical microstructure of this material is shown in Figure 130. Visual, FPI, and X-ray examinations of these components revealed a circular crack on the lip of the aft surface of one of the rings, which resulted in a "C" quality classification. The other two rings were "A" quality with no objectionable defects. The average density of these components measured 3.11 g/cm^3.

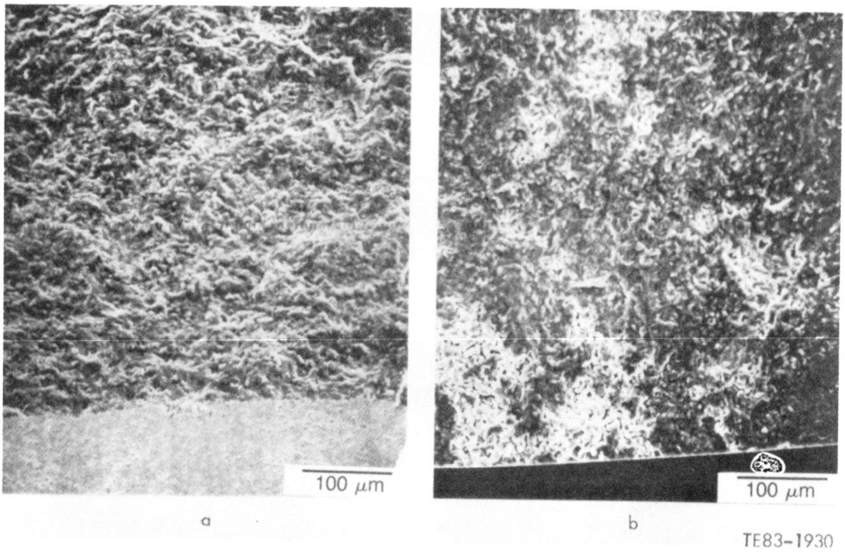

Figure 118. Typical strength-controlling defects observed in NGK Spark Plug EC-414 RBSiC test bars: (a) surface crack and (b) surface pore.

The average density of the accompanying qualification test material measured 3.12 g/cm^3. The average room temperature fracture strength of material

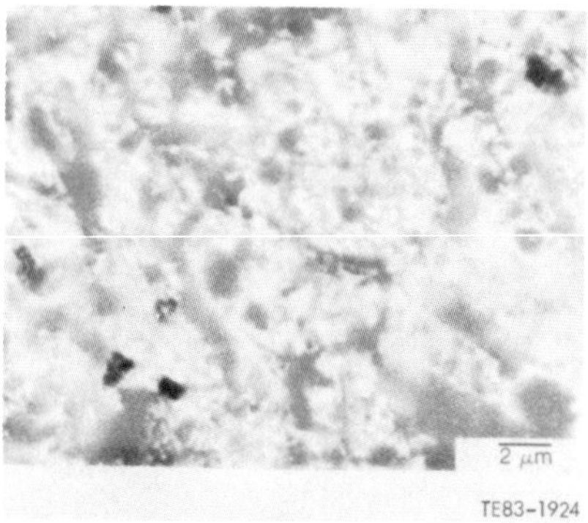

Figure 119. Microstructure of NGK Spark Plug EC-129 sintered Si_3N_4 test bars.

with an as-fired surface condition measured 294.10 MPa (42.66 ksi) with a standard deviation of 36.12 MPa (5.24 ksi). All failures of these specimens initiated from surface flaws, shown in Figure 131. Test material with a machined surface condition registered an average MOR of 337.12 MPa (48.90 ksi) with a standard deviation of 49.50 MPa (7.18 ksi). Graphite inclusions, both external and internal, were identified as the fracture origins in all of these samples (see Figure 132).

NGK-Locke SN-50 Sintered Silicon Nitride

The NGK-Locke SN-50 Si_3N_4 inner vane support ring had a component density of 3.11 g/cm$_3$. The microstructure of the SN-50 silicon nitride material, shown in Figure 133, is fairly uniform with an average grain size of 2 μm. Energy dispersive analytical X-ray (EDAX) analysis revealed the presence of magnesium (Mg), cerium (Ce), and strontium (Sr)—presumably sintering additives. Standard nondestructive evaluation (NDE) inspections of this component (visual, FPI, and X-ray) indicated a relatively porous surface, with numerous small surface pits.

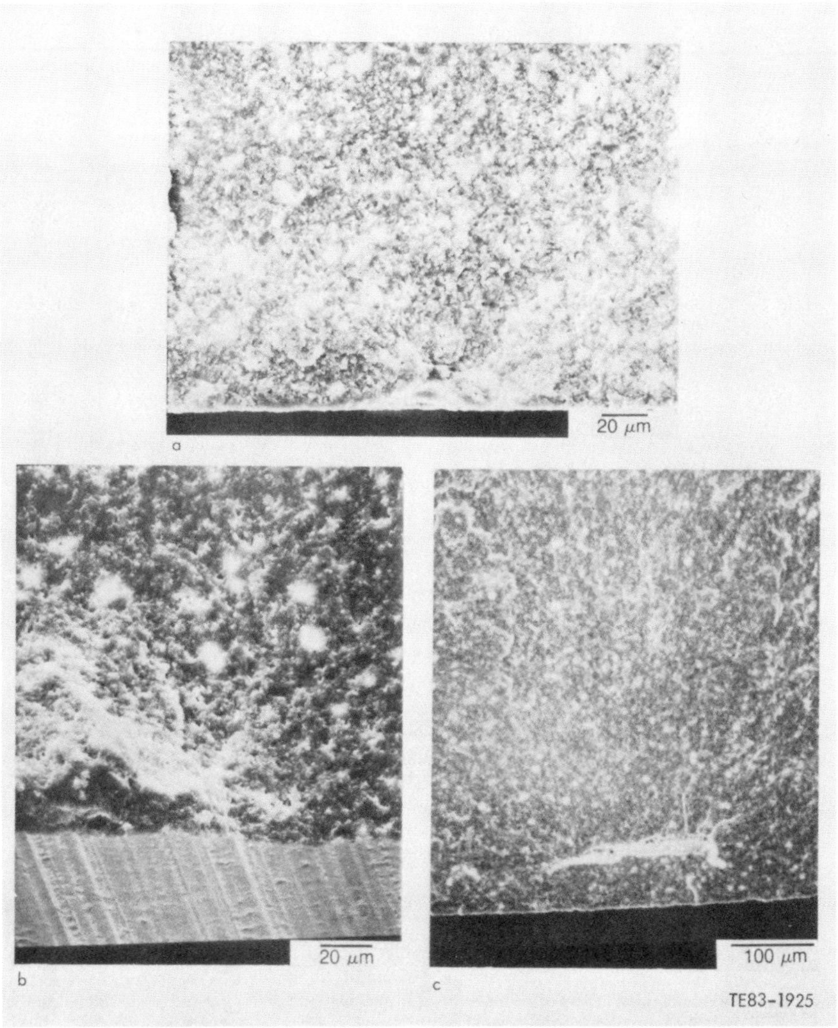

Figure 120. Typical strength-controlling defects observed in NGK Spark Plug EC-129 sintered Si_3N_4 test bars at room temperature: (a) surface pore, (b) surface crack, and (c) internal pore.

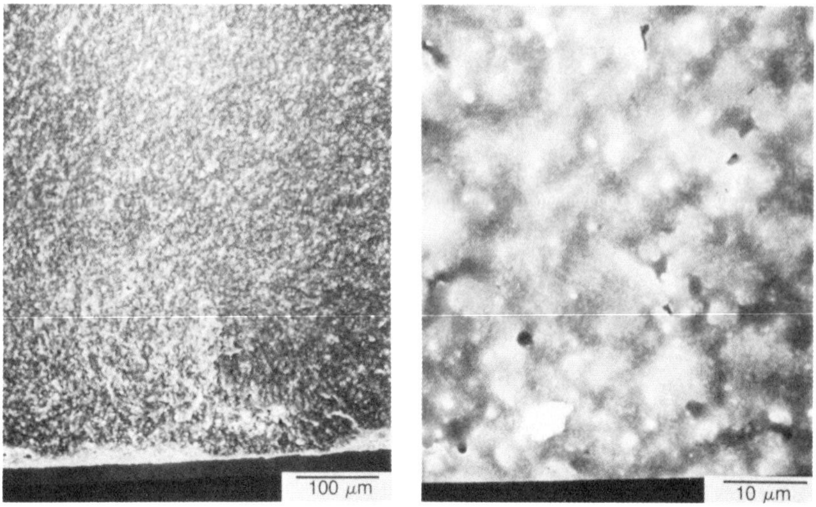

Figure 121. Fracture surface of NGK Spark Plug EC-129 sintered Si_3N_4 test bar at 1100°C (2012°F).

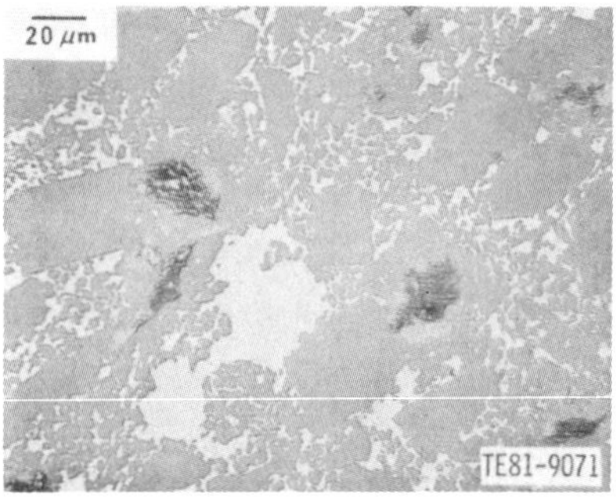

Figure 122. Microstructure of Annawerk CS-600 RBSiC 1900°F vane qualification bar: the gray phase is SiC; the white phase is free silicon; and the dark phase is graphite.

Figure 123. SEM fractograph of an Annawerk RBSiC test bar; arrow indicates failure initiated at surface site.

Qualification test material had an average density of 3.20 g/cm^3, which was slightly higher than the component density. The average room temperature fracture strength of material with a machined surface condition measured 422.44 MPa (61.27 ksi) with a standard deviation of 23.99 MPa (3.48 ksi). The

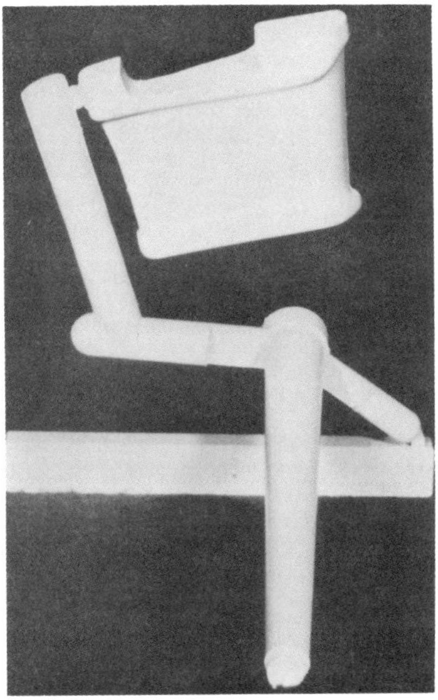

Figure 124. Plastic replica of injection-molded CBO alpha SiC vane and test bar.

Figure 125. Typical failure origin (internal pore) observed in machined CBO sintered alpha SiC vane test bars.

strength-controlling defects associated with these specimens were large, low-density pockets, a typical example of which is shown in Figure 134.

Kyocera SC-201 Sintered Silicon Carbide

Six inner vane support rings, received from Kyocera International, were made from SC-201. SC-201 is a single-phase (α), fine-grain silicon-carbide material, with an average grain size of 3.2 μm, as shown in Figure 135. These components were fabricated from isopressed and green machined rings, which were subsequently diamond-ground on all surfaces after sintering to achieve finished dimensions. Visual, FPI, and X-ray examinations of the rings revealed intermittent white spots and linear discoloration on the exterior surfaces; however, no

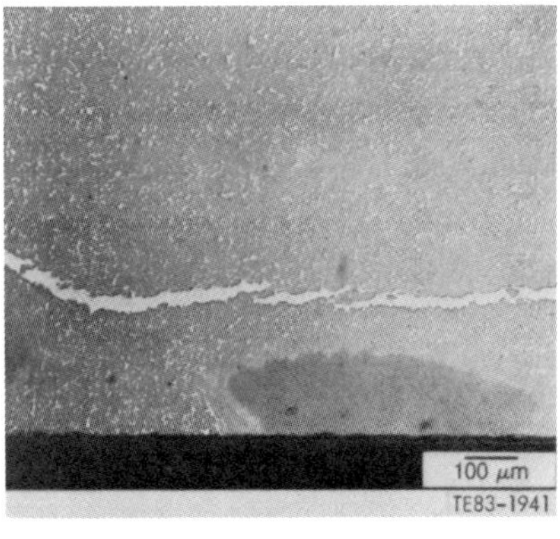

Figure 126. Microstructure of Pure Carbon Refel RBSiC 2070°F vane qualification test bar with silicon lamination and large surface agglomerate.

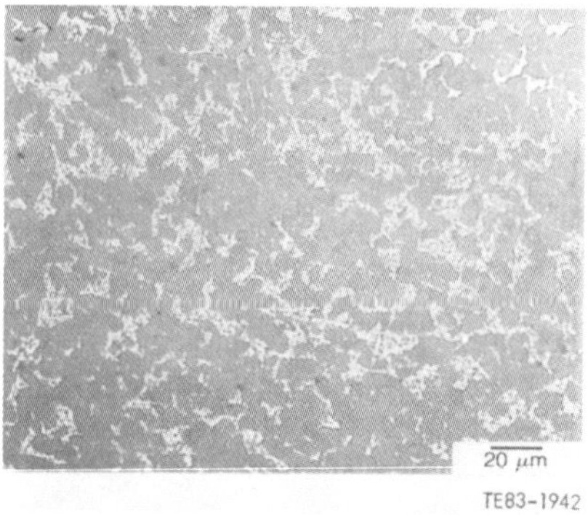

Figure 127. Microstructure of Pure Carbon RBSiC with fine particles of silicon and iron in the free silicon.

objectionable defects were observed. The average density of both the components and the accompanying qualification test bars measured 3.13 g/cm^3.

The average room temperature MOR of qualification test material with an as-fired surface condition measured 382.47 MPa (55.47 ksi) with a standard deviation of 28.34 MPa (4.11 ksi). The primary strength-controlling defects were found to consist of surface cracks and pores. A typical example of a surface crack representing the critical flaw is shown in Figure 136. Material with a machined tensile surface condition registered an average strength of 359.85 MPa (52.19 ksi) with a standard deviation of 54.13 MPa (7.85 ksi). The strength-controlling defects were again observed to be surface flaws and pores, with both types of defects affecting the strength of the material to approximately the same degree.

To further characterize these components, 17 test bars measuring 25.4 mm x 6.35 mm x 3.30 mm (1.0 in. x 0.25 in. x 0.125 in.) were cut from one of the rings and tested in three-point bending. The average three-point MOR of 438.73 MPa (63.63 ksi), which converts to an equivalent four-point MOR of approximately 329.00 MPa (47.72 ksi), correlated well with the average MOR measured for machined qualification material. In addition, similar types of strength-controlling flaws were observed in both sets of material.

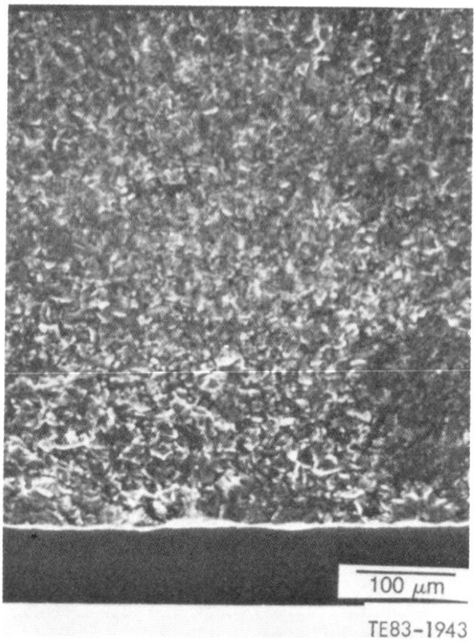

Figure 128. Typical strength-controlling defect (surface flaw) observed in Pure Carbon vane test bars.

DISCUSSION: OUTER VANE SUPPORT RINGS

Carborundum Sintered Alpha Silicon Carbide

Outer vane support ring blanks produced by Carborundum from sintered alpha SiC were received and evaluated in 1981. These components were fabricated by isopressing thick-walled cylindrical rings, which were then green machined to L-shaped cross sections. Following sintering, finished dimensions were attained by diamond grinding. Evaluation of these rings indicated an average density of 3.19 g/cm^3, while the density of the associated qualification test material measured 3.10 g/cm^3. The microstructure, shown in Figure 137, is composed of an equiaxed grain structure with a distribution of grain sizes ranging from 2 to 20 μm (the average grain size is 4.5 μm). Although visual examination revealed small chips on the aft portion of the flange in several of the rings, the overall quality of these components was good.

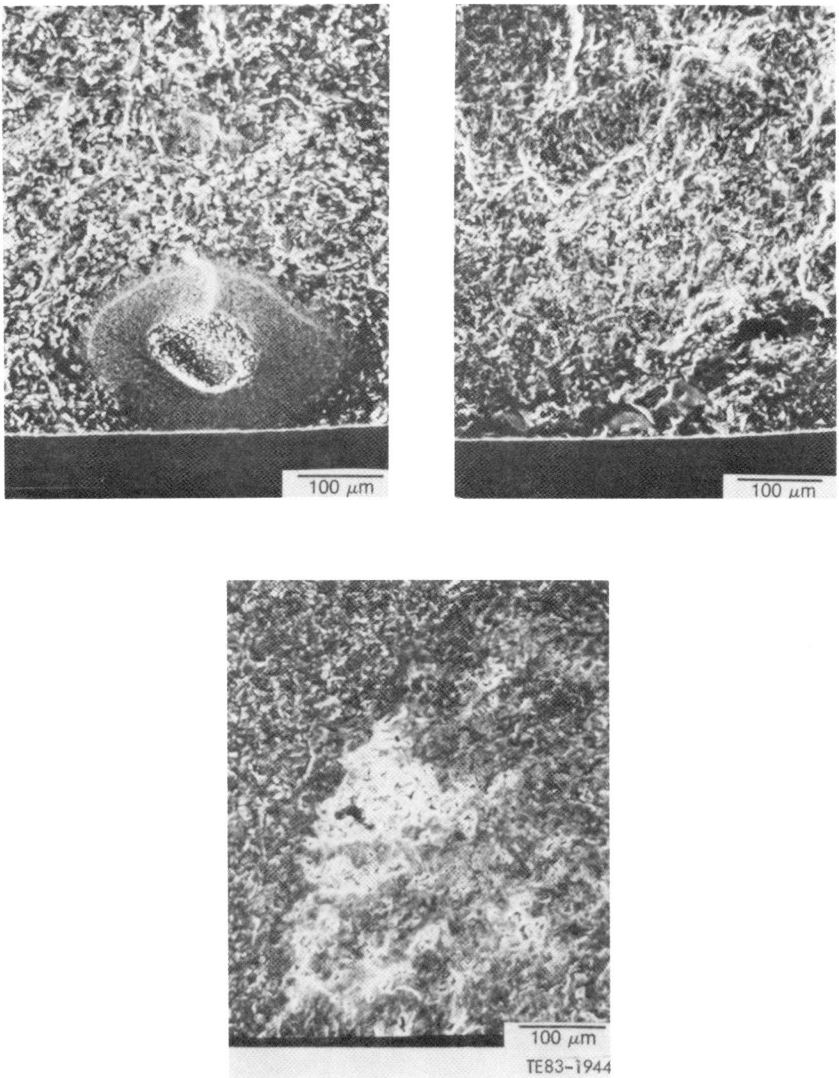

(c)

Figure 129. Examples of typical strength-controlling defects observed in machined Pure Carbon vane test bars: (a) inclusion with pore, (b) surface flaw, and (c) internal pore.

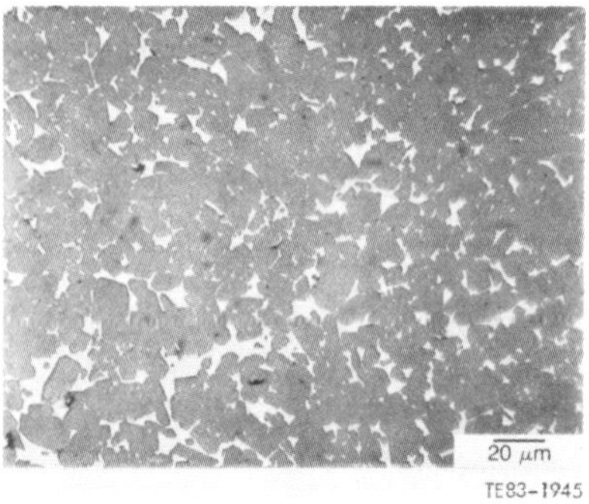

Figure 130. Typical microstructure of Pure Carbon RBSiC inner vane support ring test bars.

The average room temperature fracture strength of qualification test material with an as-

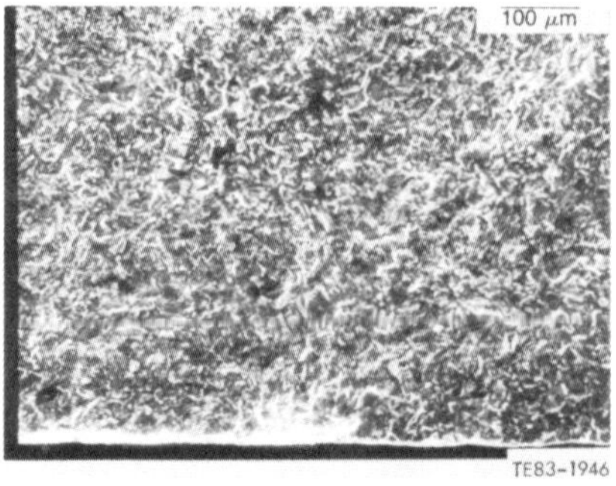

Figure 131. Failure origin (surface flaw) of as-fired inner vane support ring test bars.

fired surface condition measured 355.42 MPa (51.55 ksi) with a standard deviation of 70.81 MPa (10.27 ksi). The dominant strength-controlling defects in

Figure 132. Failure origin (internal graphite inclusion) of Pure Carbon RBSiC test bars with machined surface condition.

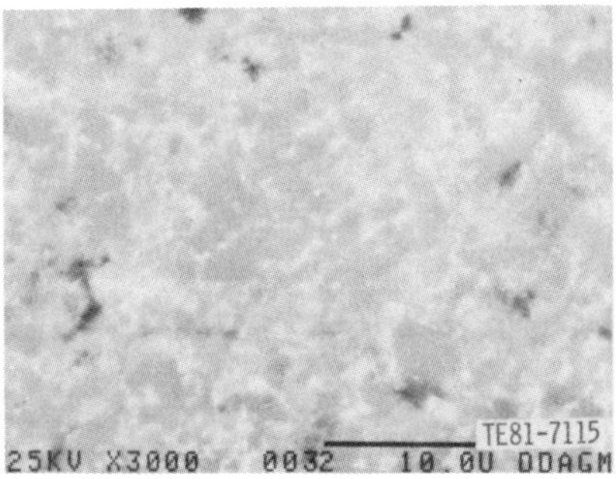

Figure 133. Microstructure of NGK-Locke SN-50 sintered Si_3N_4 inner vane support ring.

these bars were the surface and internal pores typically observed in this material. However, a number of test bars failed as a result of flaws (see Figure 138), which appeared to be surface cracks.

To determine the relationship between the strengths of qualification test bars and the actual components, standard-size four-point bend test bars, 50.8 mm x 6.35 mm x 3.20 mm (2.0 in. x 0.25 in. x 0.125 in.), were cut from the flange of one of the rings. The average density of these test bars measured 3.16 g/cm^3. The average fracture strength of material with an as-fired surface condition was 325.11 MPa (47.16 ksi), which is slightly lower than the strength measured for qualification material with a similar surface condition. The primary strength-controlling defects were surface flaws and cracks in all specimens. The fracture strength of the component test bars with a longitudinally ground surface condition averaged 394.69 MPa (57.25 ksi). The primary strength-controlling defects for the bars were surface flaws and pores; the secondary defects were internal pores.

Further component characterization was conducted using the internal pressure test (bladder test) described later in this subsection. All six CBO sintered SiC outer vane support rings tested successfully reached the design pressure level of 5170 kPa (750 psi), which generated a peak calculated stress of 73.1 MPa (10.54 ksi), with no apparent distress to the parts. One additional "C"-quality ring was pressure tested to failure. Two-dimensional finite element analysis predicted a 0.50 probability of failure (P_f) at 12.4 MPa (1800 psi) using Weibull characteristics corresponding to an MOR of 366.8 MPa (53.2 ksi) and a Weibull modulus of 8. This ring failed at a

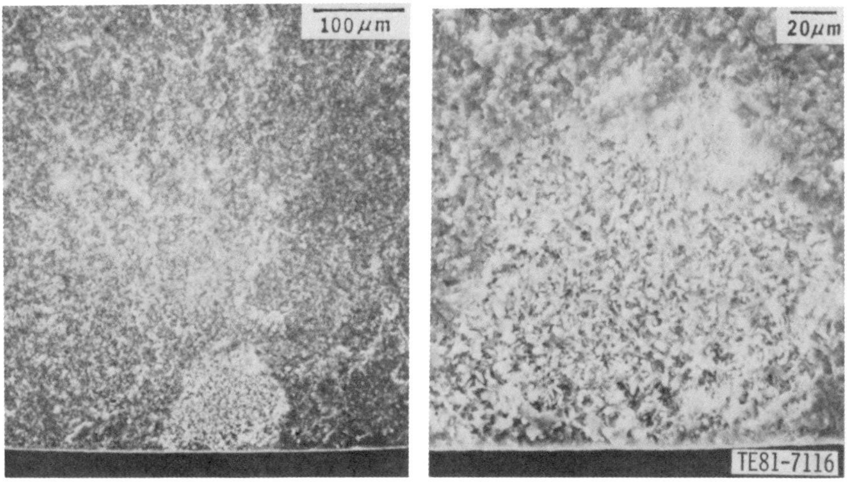

Figure 134. Fracture surface of NGK-Locke Si_3N_4 inner vane support ring test bar, showing large, low-density pocket.

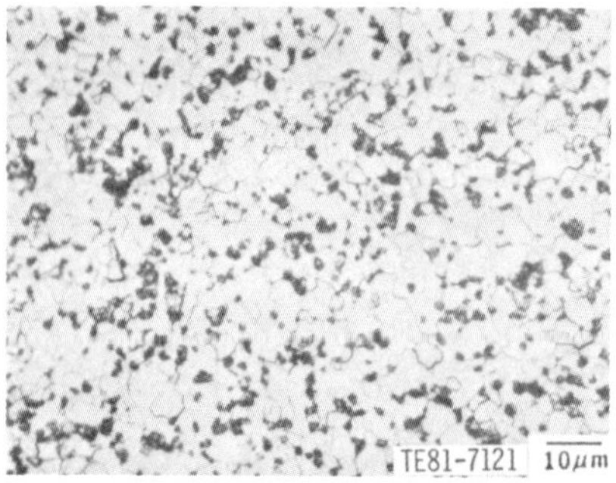

Figure 135. Microstructure of Kyocera SC-201 sintered SiC inner vane support ring test bar; average grain size is 3.2 μm.

pressure of 15.2 MPa (2200 psi), which corresponds to a peak calculated stress of 213.0 MPa (30.9 ksi). Fracture initiated from an internal pore located on the aft portion of the inside diameter of the ring, as illustrated in Figure 139.

Pure Carbon Refel Reaction-Bonded Silicon Carbide

Four ceramic outer vane support rings, fabricated by Pure Carbon from isopressed and green-machined Refel RBSiC, were evaluated. This is a dense material, with the components averaging a density of 3.11 g/cm^3 and the accompanying qualification bars averaging 3.14 g/cm^3 density. SEM micrographs of this material indicate that there is a relatively high concentration of metallic silicon in a 0.76 mm (0.03 in.) layer at the as-fired surface, as shown in Figure 140. Fairly large equiaxed SiC grains are also clearly shown in Figure 140. Visual, FPI, and X-ray examinations revealed indications of scattered porosity on the surface of these components, with several small cracks on the lugs.

The measured room temperature MOR of this material was relatively low. The average value of the as-fired surface was found to be 235.53 MPa (34.16 ksi) with a standard deviation of 42.95 MPa (6.23 ksi), and the machined surface condition was 289.30 MPa (41.96 ksi) with a standard deviation of 17.17 MPa (2.49 ksi). All failures of the test bars with an as-fired surface condition had origins at surface sites with no discernible defects present. However, in the

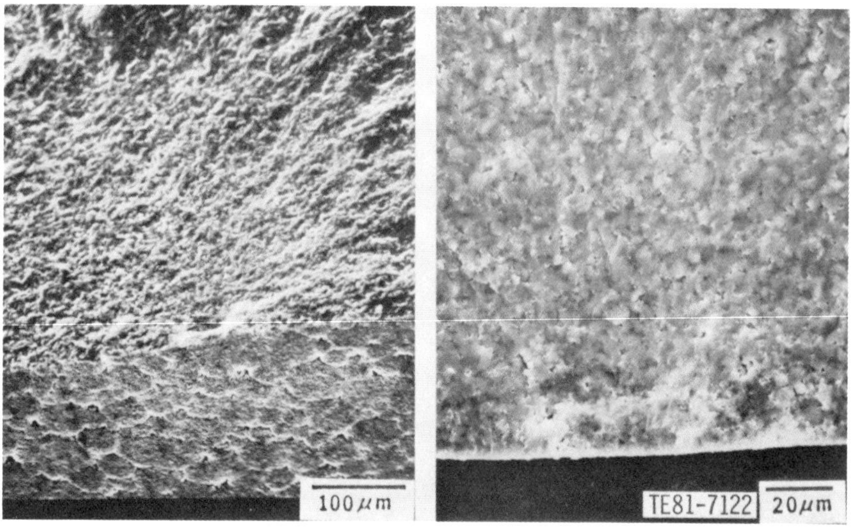

Figure 136. Typical strength-controlling flaw (surface crack) in Kyocera SC-201 test bars.

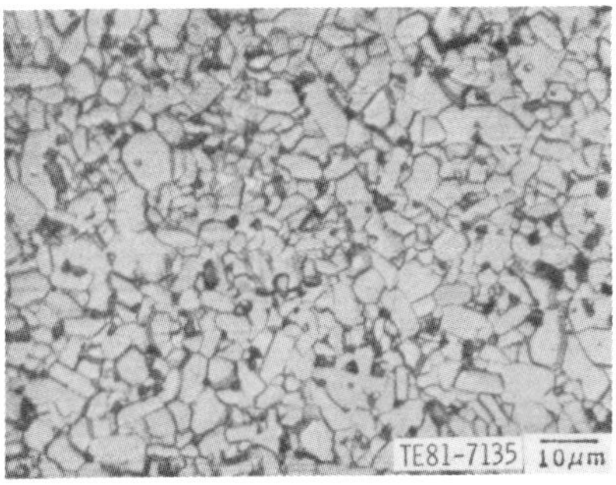

Figure 137. Microstructure of CBO sintered alpha SiC outer vane support ring test bars; average grain size is 4.5 μm.

test bars with a machined surface condition, several had fracture origins associated with either surface porosity or surface inclusions of graphite.

Additional characterizations of these components were conducted following the failure of a Pure Carbon RBSiC outer vane support ring in the ther-

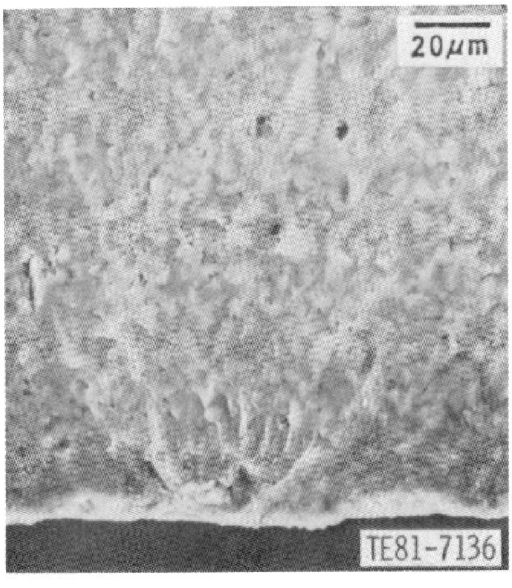

Figure 138. Strength-controlling critical flaw (surface crack) in as-fired CBO alpha SiC outer vane support ring bars.

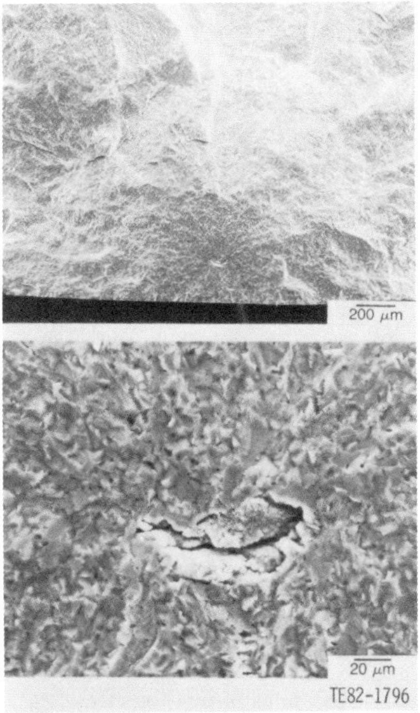

Figure 139. SEM fractograph of failure origin (internal pore) in CBO sintered alpha SiC outer vane support ring.

mal shock rig BU13. Test material was cut from the flange of the failed ring and tested in four-point bending. The tensile surface of these specimens was identical to that of the aft surface of the ring, i.e., machined and oxidized (24 hr at 1250°C [2282°F]). The strength level realized in this material compared favorably with that of the machined qualification test material. The average room temperature fracture strength of the component material measured 304.76 MPa (44.20 ksi) with a standard deviation of 51.23 MPa (7.43 ksi). The primary strength-controlling defects observed in these bars were surface and internal pores, shown in Figure 141. These defects were similar to the primary failure origin identified in the outer vane support ring, although somewhat smaller in size.

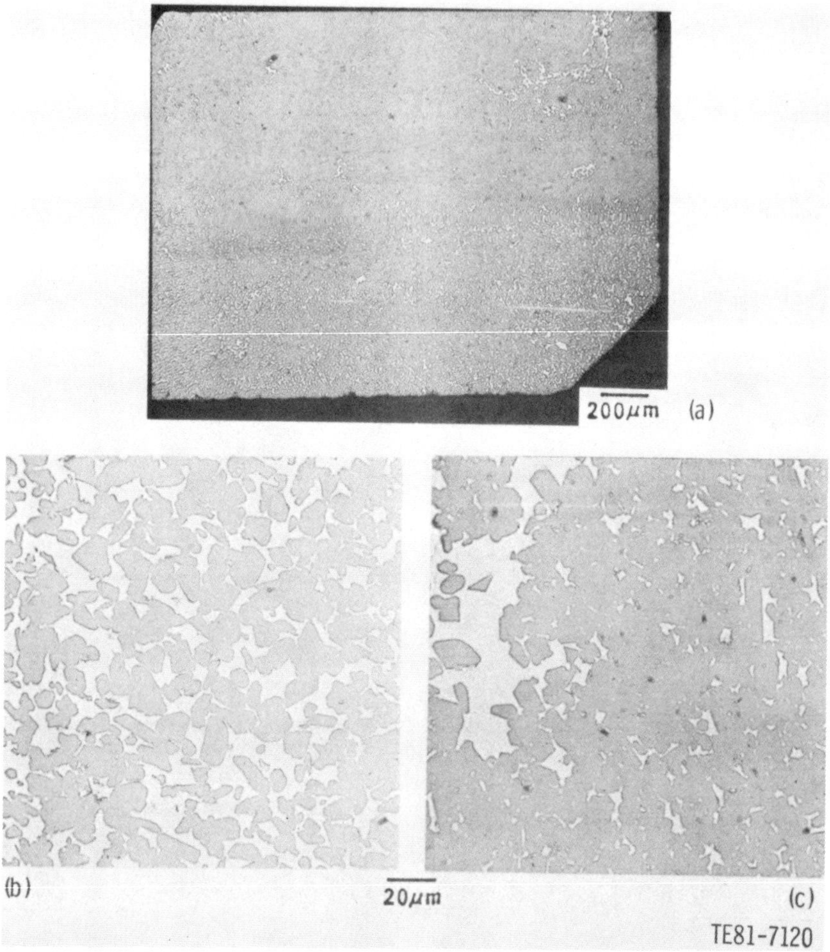

Figure 140. As-fired Pure Carbon RBSiC outer vane support ring test bars: (a) microstructure, (b) surface layer, and (c) interior.

NGK-Locke SN-50 Sintered Silicon Nitride

The outer vane support ring received from NGK-Locke was fabricated from SN-50 sintered silicon nitride that was isopressed and green machined to near net shape. The component density measured 3.15 g/cm^3, compared with the average density of 3.09 g/cm^3 measured for the qualification bars. NDE inspection of the ring structure revealed scat-

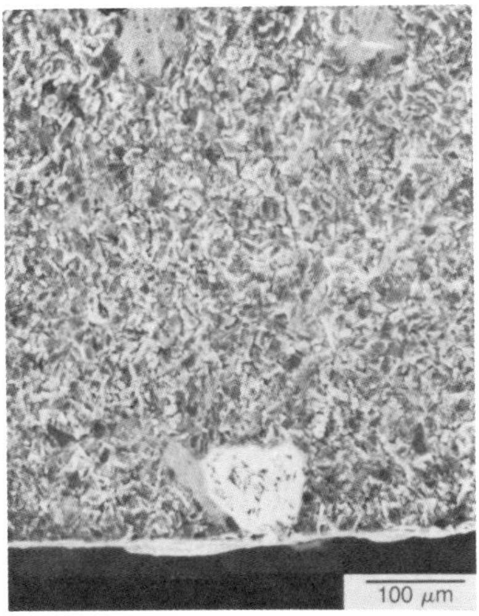

Figure 141. Typical failure origin (surface pore) observed in test bars cut from Pure Carbon RBSiC outer vane support ring.

tered surface porosity, although the porosity was not as severe as that observed for the inner vane support ring.

The qualification test bars appeared to have a finer grain size, as shown in Figure 142, and more porosity than was observed in the inner vane support ring bars. The average MOR of test bars with a machined surface condition measured 535.10 MPa (77.61 ksi) with a standard deviation of 48.06 MPa (6.97 ksi). The strength-controlling critical flaws were surface low-density pockets and pores.

DISCUSSION: VANE RETAINING RINGS

Pure Carbon Refel Reaction-Bonded Silicon Carbide

The vane retaining rings from Pure Carbon were fabricated of isopressed and green-machined Refel RBSiC. Standard NDE inspections of these components generally indicated no objectionable defects, although X-ray examination of one of the rings detected a large internal low-density region measuring 25.4 mm x 6.4 mm (1.0 in. x 0.3 in.). The av-

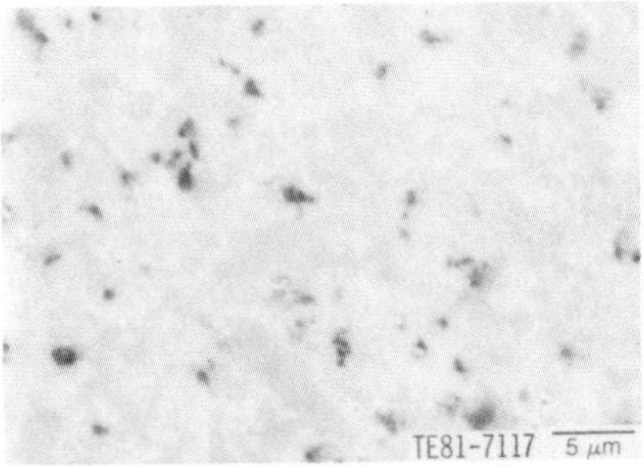

Figure 142. Microstructure of NGK-Locke SN-50 Si_3N_4 outer vane support ring.

erage density of these components measured 3.08 g/cm³.

Qualification test bars, processed concurrently with each component, were then evaluated to provide an estimate of the material strength and to determine the nature of the strength-controlling defects inherent in this material. The average density of these test bars measured 3.09 g/cm³. The surface topography of test bars with an as-fired surface condition was rough, as shown in Figure 143. Approximately 50% of the as-fired specimens had foreign materials loosely bonded to the surface, as in Figure 144. X-ray analysis indicated that these particles are composed primarily of barium and sulfur, with traces of calcium and aluminum. The average

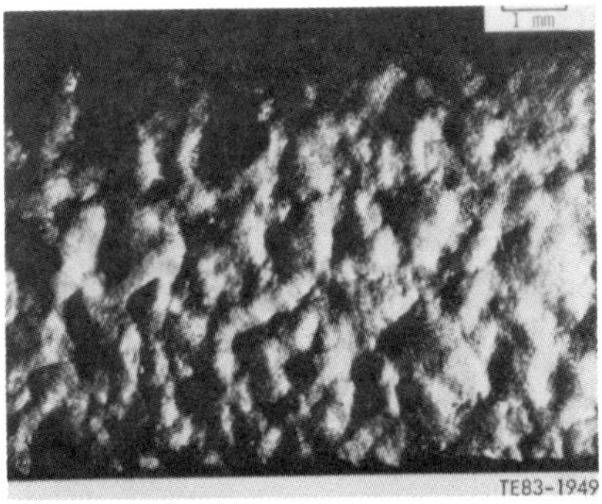

Figure 143. As-fired surface of Pure Carbon RBSiC vane retaining ring test bars.

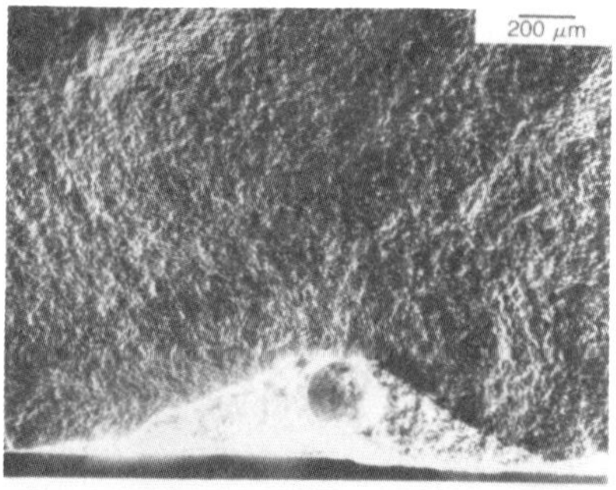

Figure 144. Pure Carbon RBSiC test bar showing foreign material (barium and sulfur) adjacent to the as-fired surface.

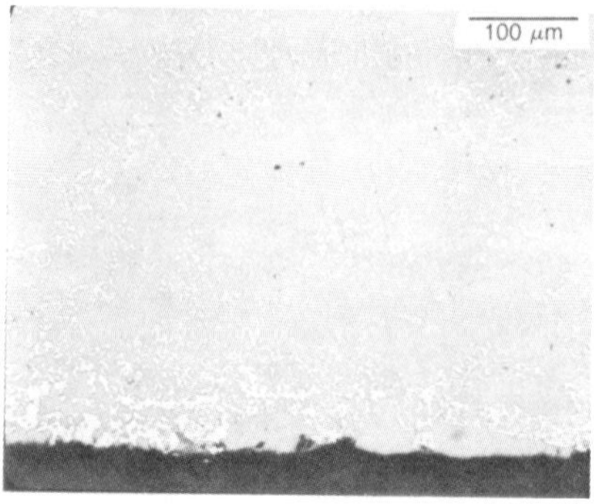

Figure 145. Typical microstructure of Norton NC-430 RBSiC plenum test bars.

MOR of these test bars measured 180.97 MPa (26.25 ksi) with a standard deviation of 29.44 MPa (4.27 ksi). All failures initiated from surface flaws. Test bars with a longitudinally ground surface condition registered an average strength of 311.88 MPa (45.24 ksi) with a standard deviation of 37.64 MPa (5.46 ksi). The primary strength-controlling defects observed in these specimens were internal graphite inclusions and surface flaws.

NGK-Locke SN-50 Sintered Silicon Nitride

The vane retaining ring received from NGK-Locke was fabricated from isopressed SN-50 sintered silicon nitride that was green machined to near net shape. The density of this component measured 3.07 g/cm^3. Inspection of the ring indicated scattered surface porosity.

The average room temperature MOR of machined qualification bars measured 635.15 MPa (92.12 ksi) with a standard deviation of 37.65 MPa (5.46 ksi). The primary strength-controlling flaws observed in these test bars were internal pores and low-density pockets.

DISCUSSION: PLENUMS (NORTON NC-430 RBSiC)

Seven 2070°F-configuration plenums were received from Norton and evaluated at Allison. These components were fabricated from slip-cast NC-430, an RBSiC material with a bimodal grain size distribution (see Figure 145). Visual, FPI, and X-ray examinations of these plenums indicated no gross defects, although the unmachined surfaces were relatively rough and several small surface pits were visible. The average component density was determined to be 3.11 g/cm^3.

The qualification test material accompanying these parts was evaluated to provide an estimate of the component strength and to determine the nature of the strength-controlling defects observed in this material. The average room temperature fracture strength of qualification test bars with a machined tensile surface condition measured 216.09 MPa (31.34 ksi) with a standard deviation of 11.45 MPa (1.66 ksi). Surface flaws were identified as the primary strength-controlling defects. The average MOR of qualification material with an as-fired surface condition measured 191.68 MPa (27.80 ksi) with a standard deviation of 23.86 MPa (3.46 ksi). All fractures of these specimens again initiated from surface flaws. As shown in Figure 146, several of the as-fired test bars possessed a thin layer of SiC grains on the surface that was not present in the other samples. This resulted in a slightly higher average strength of 225.88 MPa (32.76 ksi). Traces of zirconium (Zr) were identified on this surface in addition to the aluminum (Al) present in all specimens.

Test bars measuring 25.4 mm x 6.35 mm x 3.30 mm (1.0 in. x 0.25 in. x 0.125 in.) were cut from a Norton NC-430 plenum to provide an additional determination of component properties. Both the average strength of these plenum bars, tested with an as-fired surface condition (234.71 MPa [34.04 ksi] measured in three-point bending—equivalent to an average four-point bend strength of 176.03 MPa [25.53 ksi]), and the nature of the strength-controlling defects (all surface flaws) correlate well

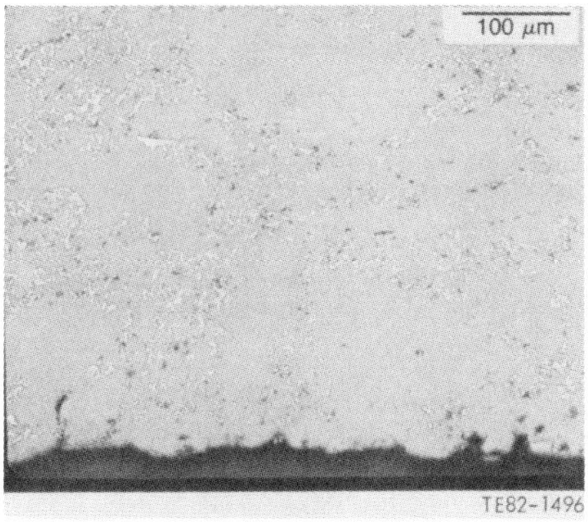

Figure 146. Microstructure of Norton NC-430 plenum test bars with a layer of SiC on the as-fired surface.

with results obtained for qualification material with a similar tensile surface condition.

Ceramic Component Process Development

SUMMARY

Extensive process development studies were conducted on those ceramic components for which either improved material characteristics and/or further development of fabrication methodology were required. These studies involved two ceramic suppliers: GTE Laboratories (blades) and Carborundum (blades, stationary ring components, plenums, and 2070°F-configuration vanes).

At GTE Laboratories a detailed process routing (AY6-404) was established for the fabrication of injection-molded sintered silicon nitride turbine blades. These blades demonstrated an average failure speed of 52,300 rpm (141.7% design speed). Recent processing modifications (AY6-405) have resulted in an average MOR strength of 617.03 MPa (89.49 ksi), indicating that substantial increases in failure speed can be expected.

At Carborundum, injection-molding parameters were developed for sintered alpha silicon carbide turbine blades. Spin test results indicate that current state-of-the-art material properties (as determined from companion test bars molded with the blades) are being realized in full-scale hardware. A gasifier turbine rotor utilizing 40 CBO sintered alpha SiC blades accumulated approximately 9 hr of operating time in the CATE C-4 engine, including 2 hr and 39 minutes at 100% speed and temperature (36,905 rpm and 1132°C [2070°F] rotor inlet temperature). A maximum temperature of 1371°C (2500°F) was observed during transient operation of the engine.

The definition of additional characterization activities (internal pressure test, determination of the effects of cross grinding on material strength, and evaluations of material cut from the components) was addressed for the CBO RBSiC compression-molded stationary ring components. These components were the inner and outer vane support rings and the turbine tip shroud. Initial development studies addressing the use of coarse-grained sintered alpha SiC for slip casting of the plenum were conducted. Also included was a material and process variation experiment involving the CBO transfer-molded RBSiC 2070°F-configuration vanes.

OBJECTIVE

The primary objective of this task was to develop ceramic components with appropriate material characteristics for successful use in rig and engine testing. Concurrent with these activities was the development of component fabrication techniques with the capability of producing dimensionally sound parts with a minimum of finish machining.

APPROACH

A two-path approach was utilized to develop near net-shape ceramic components which met both structural and reliability requirements. The first involved a series of iterative studies to translate state-of-the-art material properties into full-scale hardware. The second approach incorporated studies of material and processing variations to increase material strength characteristics and processing yields.

Component evaluations were conducted using both standard nondestructive inspections (visual, FPI, and radiography) and advanced NDE methods (high-frequency ultrasonics and photoacoustic spectroscopy). Qualification test bars processed with the parts were evaluated for microstructure, fracture strength (MOR), and failure origin type and location. In addition, test material was cut from the components to ascertain the relationship existing between the qualification test bars and the full-scale parts.

DISCUSSION

GTE Laboratories Sintered Silicon Nitride—Blades

During 1980-81, an initial development program with GTE Laboratories was conducted to establish processing parameters for the fabrication of injection-molded 2070°F-configuration turbine blades from sintered silicon nitride. The primary objective of this program was to produce the highest yield of defect-free blades that would meet dimensional requirements. This effort focused on the injection-molding segment of the blade processing sequence, with particular emphasis placed on the mixing, molding, and binder removal steps. The production of blades with a net shape capability was emphasized to minimize the finish machining required.

The material selected for blade fabrication was SN 502, a commercially available silicon nitride powder produced by GTE. The starting ceramic powder included 6% Y_2O_3 and 2% Al_2O_3 as sintering additives. Preliminary studies identified a suitable binder system, consisting of a mixture of low-density polyethylene and paraffin waxes, for injection molding. Upon completion of the development study, a detailed process routing (designated AY6-404) was established, as shown in Figure 147, and 32 blades were fabricated for laboratory and spin test evaluation. In addition, 100 qualification test bars were molded and processed with the blades for microstructural, fracture strength, and fracture surface evaluation.

Qualification test bars, processed simultaneously with the blades, were evaluated to provide an estimate of the material strength and to determine the nature of the strength-controlling defects associated with this material. The average density of these qualification test bars measured 3.22 g/cm^3 (98.7% theoretical density). The typical microstructure, shown in Figure 148, is very fine grained, with an average grain size of 0.9 μm. However, several of the test bars had small dark-colored pores and inclusions present, as shown in Figure 149. These pores and inclusions were composed primarily of iron silicide, introduced into the part from wear and scoring of the injection-molding barrel. A new barrel and plunger with borided surfaces were obtained. The boride greatly reduced the wear problem and eliminated the metallic inclusions.

The average room-temperature fracture strength of bars tested with an as-fired tensile surface condition measured 408.25 MPa (59.21 ksi) with a standard deviation of 54.63 MPa (7.92 ksi). Low-density pockets and pores, shown in Figure 150, were identified as the primary strength-controlling defects. Bars tested with a machined surface condition recorded an average strength of 438.12 MPa (63.54 ksi) with a standard deviation of 67.44 MPa (9.78 ksi). All fractures of these specimens initiated from surface low-density pockets and pores.

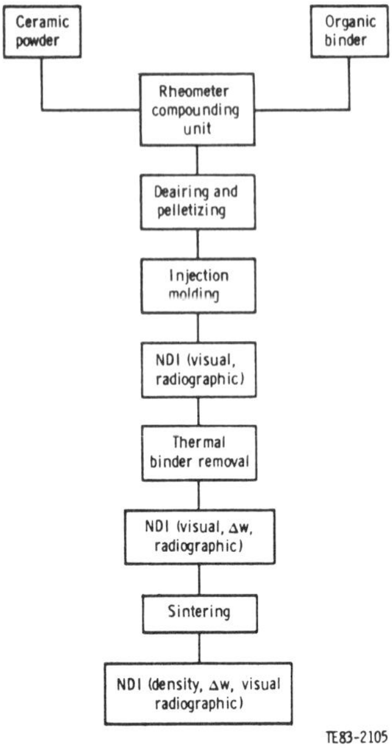

Figure 147. Process routing (AY6-404) for GTE injection-molded sintered silicon nitride turbine blades.

Nondestructive evaluations (visual, FPI, X-ray, and dimensional) of the 32 Si_3N_4 blades produced from this initial process development study were then conducted. These inspection results are summarized in Table XIII. Visual and fluorescent penetrant examinations revealed that eight blades (25%) had cracks at the platform near the trailing edge of the airfoil, and an additional six exhibited cracks or line indications in random locations. Radiographic analysis revealed the presence of high-density inclu-

Table XIII.
Inspection summary of GTE AY6-404 Si_3N_4 blades.

Inspection	Yield—%	Primary indications
Visual and FPI	81	Surface cracks
X-ray	100*	High-density inclusions
Dimensional	66	Positive airfoil twist
Overall yield	63	

*Although 18 of the 32 blades (56%) had high-density inclusions, these were not judged severe enough for rejection.

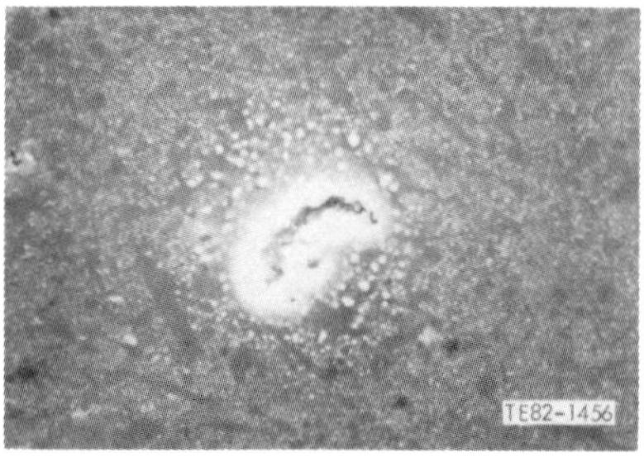

Figure 149. Subsurface defect (iron silicide inclusion) identified in GTE Si_3N_4 blade test bars.

sions, measuring 50 to 200 μm in diameter, in 18 of the blades, primarily in the airfoil region. These high-density indications were found to consist of iron (Fe) and molybdenum (Mo) inclusions. The iron inclusions, similar to those observed in the qualification test material, were found to be a result of spalling of the injection-molding plunger. As mentioned earlier, boriding the plunger and cylinder has resolved this problem. The molybdenum was traced to contaminations in the starting silicon nitride powder, which contained 0.036% Mo. Powder with the Mo content reduced to 0.006% has since been available and was used for all subsequent blade and bar fabrication. Dimensional inspections revealed that 11 blades (34%) had excessive positive airfoil twist.

On the basis of these inspections, 20 blades (63%) were judged to be acceptable for finish machining and subsequent overspeed-to-failure spin testing. Ten of these blades were supplied to Crafts for multiple-pass grinding of the attachment, with the remaining 10 shipped to Midwest Precision Services (MPS) for creep-feed grinding. The results of the overspeed-to-failure spin tests conducted on these blades are summarized in Table XIV. Weibull plots

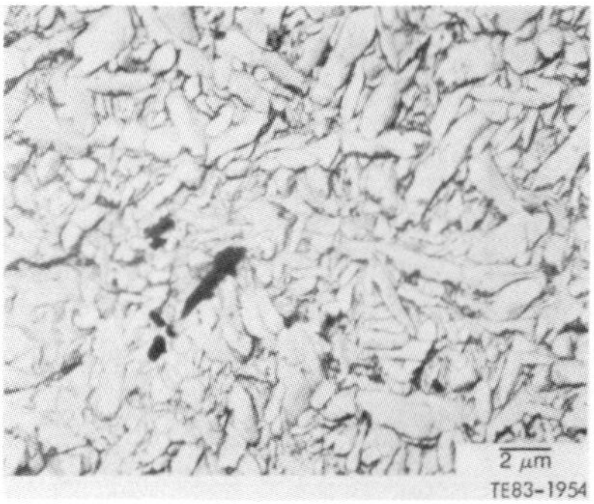

Figure 148. Typical microstructure of GTE sintered silicon nitride material; the average grain size is 0.9 μm.

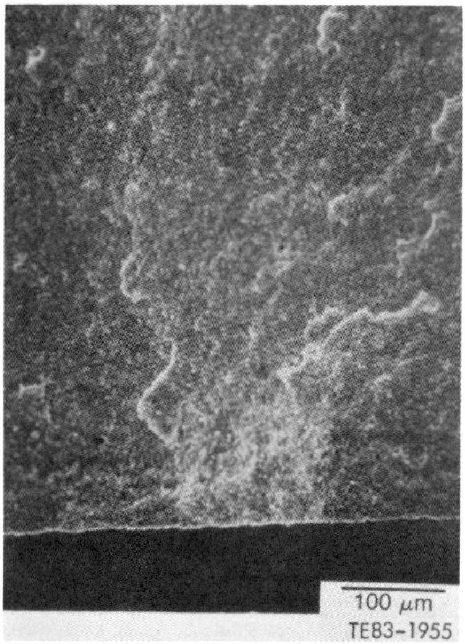

Figure 150. Typical strength-controlling flaw (low-density pocket) observed in GTE AY6-404 Si_3N_4 blade test bars.

Table XIV.
Overspeed-to-failure spin tests, GTE Si_3N_4 blades.

Grinder	No.	Blade Serial No.	Failure speed—rpm (%)	Failure origin
Crafts	1	430	48,500 (131.4%)	Convex side, stalk (shell)
	2	432	49,570 (134.8%)	Convex side, platform (shell)
	3	433	46,700 (126.5%)	Convex side, stalk (shell)
	4	437	48,600 (131.7%)	Convex side, stalk (shell)
	5	441	49,850 (135.1%)	Convex side, platform (shell)
	6	445	45,200 (122.5%)	Convex side, attachment (surface low density)
	7	485	49,450 (134.0%)	Convex side, stalk (shell)
	8	513	47,800 (129.5%)	Concave side, attachment (surface low density)
	9	554	49,750 (134.8%)	Airfoil root (shell)
	10	381	51,700 (140.0%)	Convex side, stalk (shell)
	Average	48,730 (132.0%)		
Midwest Precision Services	1	403	52,250 (141.6%)	Convex side, attachment (missing)
	2	404	50,600 (137.1%)	Convex side, attachment (pore)
	3	405	52,000 (140.9%)	Convex side, attachment (pore)
	4	412	53,500 (145.0%)	Concave side, attachment (crack)
	5	456	47,400 (128.4%)	Convex side, attachment (pore)
	6	459	52,900 (143.3%)	Convex side, stalk (pore)
	7	460	52,800 (143.1%)	Convex side, airfoil (shell)
	8	484	57,000 (154.5%)	Convex side, stalk (shell)
	Average	52,300 (141.7%)		

are shown in Figures 151 and 152. The average failure speed of the 10 blades machined at Crafts was 48,730 rpm—equivalent to 132.0% of design speed. The erratic nature of the failure locations (2 attachment, 5 stalk, 2 platform, and 1 airfoil) was attributed to a high frequency (80%) of atypical fracture origins, i.e., shell-like surface cracks, as in Figure 153. These artifacts were the result of an abrasive cleaning operation (grit blasting) performed on the blades after sintering to enhance the surface finish. This operation has been omitted in all subsequent blade fabrication. The average failure speed of the eight blades machined at MPS (two blades were broken during machining) was 52,300 rpm (141.7% design speed). Pores were identified as the primary strength-controlling defects, with only two blades failing from the shell-type surface cracks mentioned earlier in the section.

To further evaluate actual blade strength characteristics, test bars measuring 25.4 mm x 3.20 mm x 1.60 mm (1.0 in. x 0.125 in. x 0.063 in.) were machined from the remaining dovetails of the failed blades. Three-point MOR testing of bars cut from the base of the dovetail, where the surface condition was identical to that of the most highly stressed region of the blade attachment, revealed an average

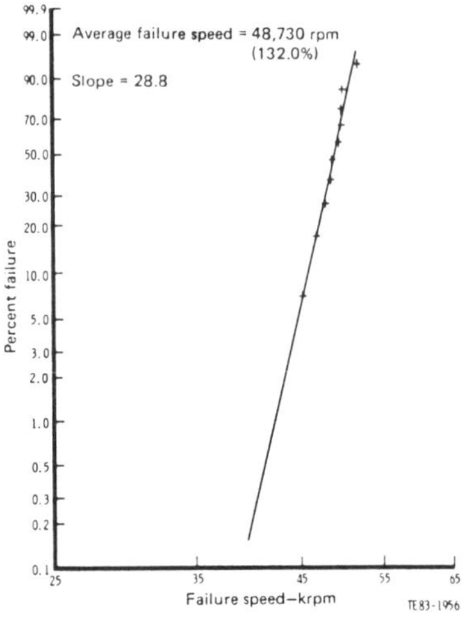

Figure 151. Weibull plot of spin-test results for GTE Si_3N_4 blades (machined at Crafts).

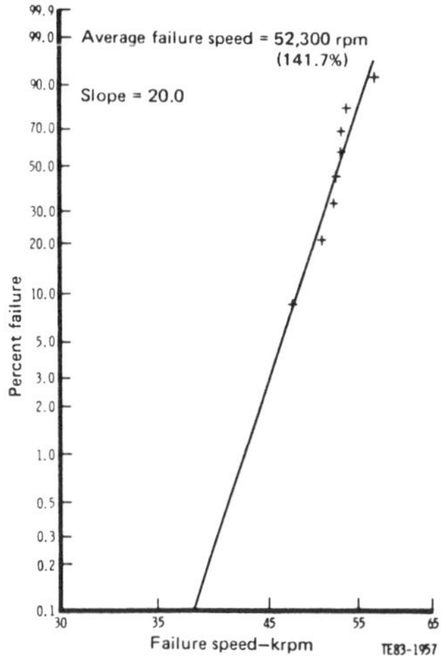

Figure 152. Weibull plot of spin-test results for GTE Si_3N_4 blades (machined at MPS).

fracture strength of 687.43 MPa (99.70 ksi). Surface pores and flaws were found to be the primary strength-controlling defects. Additional test bars were cut from the midsection (interior) of the dovetail. These bars were tested with a longitudinally ground surface condition, and an average strength

Figure 153. Atypical failure origins (shell-like surface cracks) observed in GTE AY6-404 Si_3N_4 blades.

of 485.06 MPa (70.35 ksi) was measured. Surface low-density regions and pores, shown in Figure 154, were identified as the primary strength-controlling flaws. Thus, while spin-test performance to date is considered excellent, the significant decrease in strength noted in the interior of the blades compared with that obtained from the surface indicated that additional development activities were required to enhance blade uniformity.

To address the difficulties encountered with internal low-density regions previously observed in the initial shipment of GTE blades, a follow-on order for an additional 180 blades and 100 injection-molded Si_3N_4 test bars was placed. An updated process specification, AY6-405, which incorporates a modified binder system with improved binder burnout characteristics, was developed. A thermosetting material, which was found to impart a beneficial rigidizing effect to the molded parts, was added.

Preliminary visual examinations of these blades revealed good overall quality. Initial dimen-

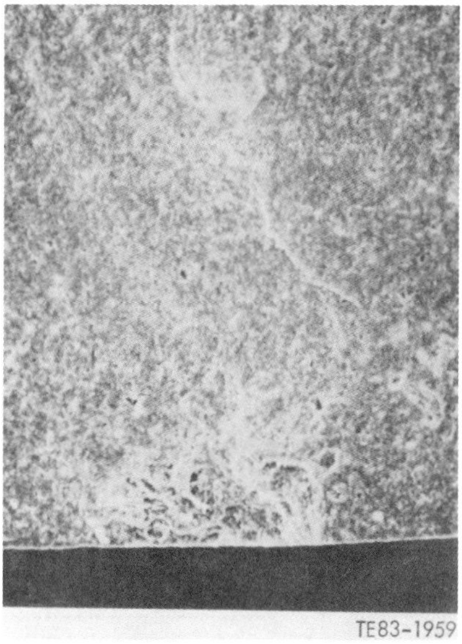

Figure 154. Surface low-density regions and pores, identified as the failure origins in test material cut from the interior of the dovetail of GTE AY6-404 Si_3N_4 blades.

sional characterizations indicated the dimensional stability of these blades was excellent, with all dimensions found to be within design specifications.

The strength characteristics of the AY6-405 Si_3N_4 test bars were examined at both room temperature and elevated temperatures. In addition, preliminary efforts to determine the effects of oxidation on this material were conducted at various temperature levels.

The average room-temperature fracture strength of test material with an as-fired surface condition measured 542.77 MPa (78.72 ksi) with a standard deviation of 65.23 MPa (9.46 ksi). Bars with a machined surface condition exhibited an average strength of 617.03 MPa (89.49 ksi) with a standard deviation of 65.78 MPa (9.54 ksi). These values represent the highest strengths obtained to date for any GTE injection-molded Si_3N_4 material. The primary strength-controlling flaws observed in these specimens were surface and internal pores, shown in Figure 155.

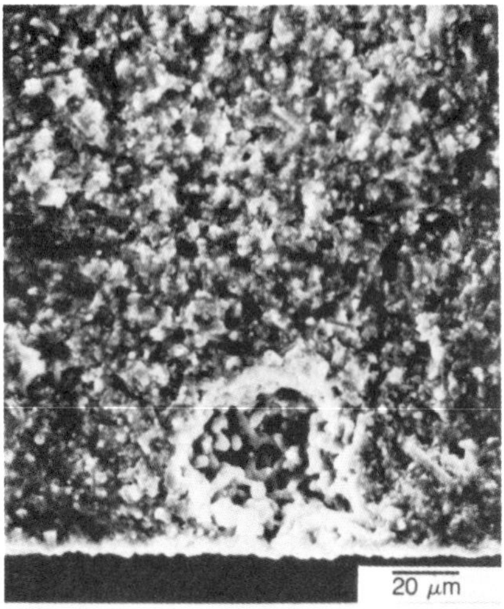

TE83-1960

Figure 155. Typical strength-controlling flaw (internal pore) observed in GTE AY6-405 sintered Si_3N_4 blade test bars.

The results of the elevated temperature strength tests are summarized in Table XV. These tests, all conducted on bars with machined surface conditions, showed the average fracture strength of this material was 496.65 MPa (72.03 ksi) at a temperature of 800°C (1472°F); 517.75 MPa (75.09 ksi) at 1000°C (1832°F); and 388.81 MPa (56.39 ksi) at 1200°C (2192°F).

Note that while a 37% decrease in strength was observed from room temperature to 1200°C (2192°F), the high temperature strength of this Si_3N_4 material is still comparable to state-of-the-art injection-molded SiC material strengths and is very close to the strength necessary to meet blade reliability requirements.

The oxidation/strength behavior of this material was evaluated for both as-fired and machined surface conditions. All oxidation was conducted in an air atmosphere with exposure times of 100 hr. The results of these tests are summarized in Table XVI.

Table XV.
Elevated temperature strength of GTE AY6-405 Si_3N_4 machined test bars.

Temperature—°C (°F)	Fracture strength MPa	(ksi)
Room temperature	617.03	(89.49)
800 (1472)	496.65	(72.03)
1000 (1832)	517.75	(75.09)
1200 (2192)	388.81	(56.39)

After oxidation for 100 hr at 1000°C (1832°F), the strength of the as-fired test bars dropped to 439.07 MPa (63.68 ksi), while the strength of the machined bars decreased to 474.72 MPa (68.85 ksi). Further decreases in strength after exposure at 1150°C (2102°F) were observed in both the as-fired and machined bars, which measured 405.98 MPa (58.88 ksi) and 388.33 MPa (56.32 ksi), respectively. The increases in strength observed after oxidation at 1250°C (2282°F)—450.17 MPa (65.29 ksi) as-fired and 473.96 MPa (68.74 ksi) machined—were attributed to the formation of a thin, glassy layer on the surface of the bars, shown in Figure 156. This layer was not seen after exposure at either 1000°C (1832°F) or 1150°C (2102°F).

These results demonstrate that the GTE injection-molded AY6-405 sintered silicon nitride material exhibits excellent room temperature strength and can be fabricated into complex geometries (such as blades) that meet both structural and

Table XVI.
Oxidation behavior of GTE AY6-405 Si_3N_4.*

Temperature	Surface condition	Fracture strength MPa	(ksi)
Room temperature	As-fired	542.77	(78.72)
	Machined	617.03	(89.49)
1000°C (1832°F)	As-fired	439.07	(63.68)
	Machined	474.72	(68.85)
1150°C (2102°F)	As-fired	405.98	(58.88)
	Machined	388.33	(56.32)
1250°C (2282°F)	As-fired	450.17	(65.29)
	Machined	473.96	(68.74)

*Specimens exposed for 100 hr in air

dimensional requirements. In addition, this material possesses good oxidation resistance and strength up to temperatures of 1250°C (2282°F).

Carborundum Sintered Alpha SiC—Blades

During 1979-80, an initial process development program for fabrication of injection-molded 2070°F-configuration turbine blades from sintered alpha silicon carbide was conducted at Carborundum. This program was focused on the injection molding segment of the blade processing sequence. The primary goal was to establish those injection-molding parameters conducive to the highest yields of defect-free as-fired blades (i.e., free of defects discernible with standard NDE techniques) that could meet dimensional requirements. Emphasis was placed on achieving net-shape capability so that finish machining could be limited to the attachment, platform edges, and blade tip. Upon completion of the study in late 1980, a detailed process routing was established for blade fabrication, and two large "production" runs (designated follow-on order No. 1 and follow-on order No. 2) were undertaken to quantitatively assess blade-fabrication capability and to provide blades for evaluation and rig/engine testing. This subsection will concentrate on characterization and qualification activities conducted on components from the most recent blade run—follow-on order No. 2. (Details of work performed on follow-on order No. 1 blades may be found in EDRs 10156, 10383, and 10672 [Ref 4, 5, and 6].) Process yield, structural/dimensional quality, and strength characteristics were evaluated for the current blades. In addition, the results of overspeed-to-failure blade spin testing are summarized.

The yields realized in each processing step for the large run of blades are summarized in Table XVII. The total process yield of fully machined blades that met both NDE and dimensional requirements was 28%. Yields for molding and baking were above 90% and, therefore, were judged to be satisfactory. However, the yield after firing as determined by visual, FPI, and X-ray assessments was low (43%). Finally, the yield of defect-free sintered blades meeting dimensional requirements of airfoil contour, thickness, and setting angle was 78%, which was thought to be acceptable.

Qualification test bars molded simultaneously with the blades were evaluated to assess blade strength characteristics indirectly. The results of this testing are summarized in Table XVIII. Note that all blade test-bar MOR strengths are corrected by appropriate Weibull surface and volume effects resulting from a 0.51 mm (0.020 in.) bevel on the edges of the bars. These bevels, illustrated in Figure 157, were incorporated to minimize machining damage to the test bar edges.

The average density of these test bars was found to be 3.15 g/cm^3. Microstructural evaluation re-

Table XVII.
Process yields of Carborundum sintered alpha SiC blades.

Processing step	Yield — %
Molding	92
Baking	91
Sintering	43
Dimensional	78
Total	28

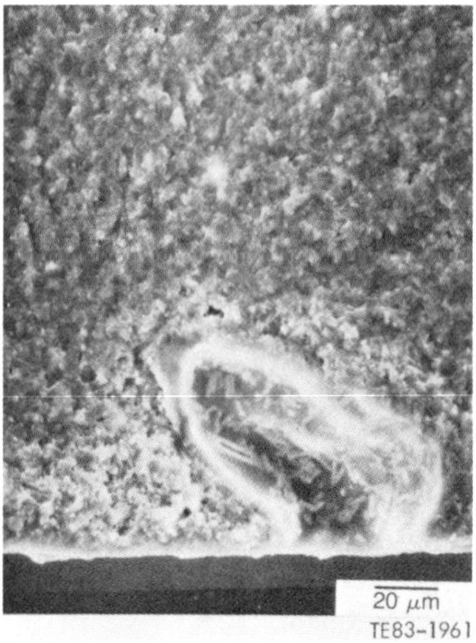

Figure 156. Typical failure origin (surface pore) observed in GTE AY6-405 Si_3N_4 test bars after exposure at 1250°C (2282°F) for 100 hr; a thin, glassy layer developed on the surface of the pore.

vealed that a majority of the bars possess an equiaxed fine-grained microstructure, as shown in Figure 158. The average grain size ranged from 4.2 μm to 5.8 μm with no preferred crystalline orientation. Several of the test bars were observed to have

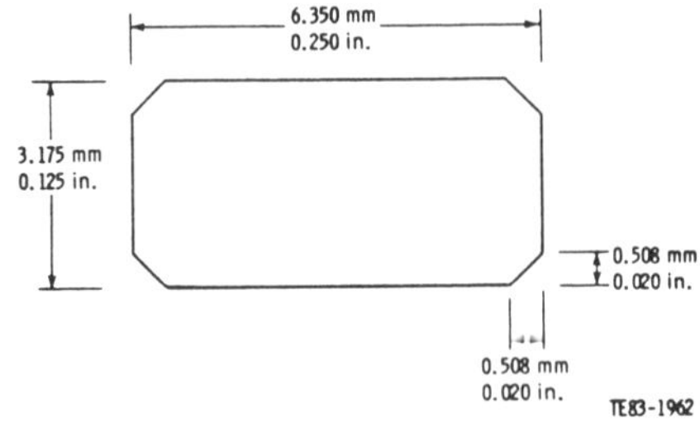

Figure 157. Cross section of CBO sintered alpha SiC blade test bar with 0.508 mm (0.020 in.) corner bevels.

a long, tabular grain structure, as in Figure 159. No relationship was found to exist between grain size, grain shape, and room-temperature fracture strength.

The average room-temperature fracture strength of 100 bars tested with a longitudinally ground surface condition was 418.87 MPa (60.75 ksi) with a standard deviation of 70.19 MPa (10.18 ksi). The primary strength-controlling defects observed in these specimens were internal pores (54%), with the secondary defects consisting of surface pores (34%) and surface flaws (12%). Typical examples of these defects are shown in Figure 160. A two-parameter Weibull analysis of the strength data provided a characteristic strength of 447.83 MPa (64.95 ksi) with a Weibull modulus (m) of 7.1, as shown in Figure 161.

Test bars with an as-fired surface condition exhibited an average MOR of 417.92 MPa (60.61 ksi) with a standard deviation of 46.38 MPa (6.73 ksi).

Table XVIII.
Room temperature fracture strengths of Carborundum sintered alpha SiC blade test bars.*

Surface condition	MOR		Std deviation	
	MPa	(ksi)	MPa	(ksi)
Longitudinally ground	418.87	(60.75)	70.19	(10.18)
As-fired	417.92	(60.61)	46.38	(6.73)
Transversely ground	317.60	(46.06)	40.87	(5.93)
As-fired and oxidized**	396.60	(57.52)	48.88	(7.09)
Transversely ground and oxidized	396.03	(57.44)	74.34	(10.78)

*All strengths corrected for 0.51 mm (0.020 in.) bevels
**Oxidized in air for 24 hr at 1250°C (2282°F)

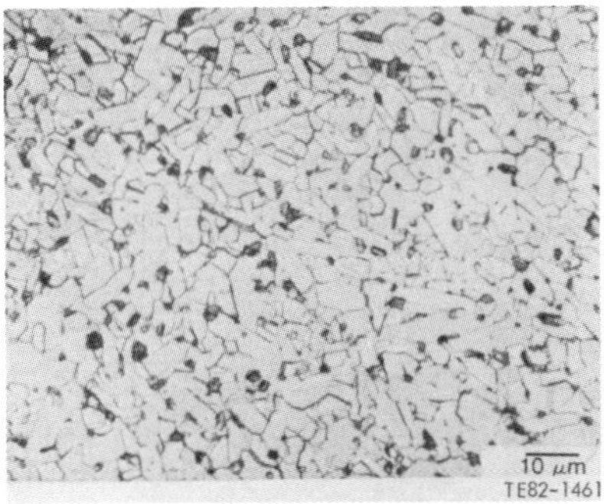

Figure 158. Typical microstructure of CBO sintered alpha SiC blade test material; the average grain size is 4.2 μm.

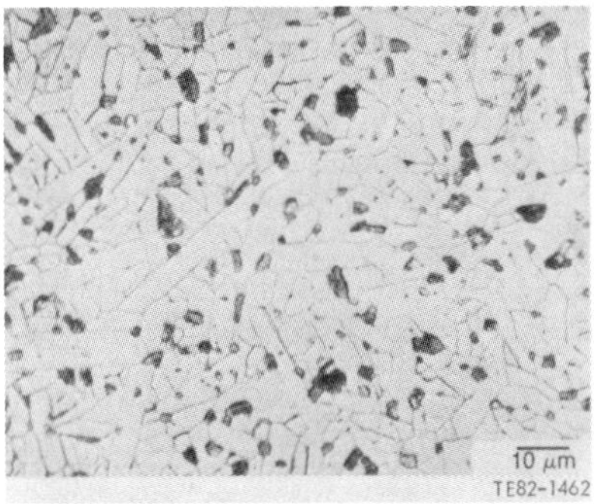

Figure 159. Microstructure of CBO sintered alpha SiC blade test material with tabular grains; the average grain size is 5.1 μm.

Surface pores (55%) were identified as the primary flaws, with internal pores (25%) and surface flaws (20%) constituting the secondary defects.

The average room-temperature fracture strength of blade test bars with a transversely ground surface condition measured 317.60 MPa (46.06 ksi) with a standard deviation of 40.87 MPa (5.93 ksi). The strength-controlling defects were found to be surface flaws (89%) and surface pores (11%).

Investigations to determine the effects of heat treatment on fracture strengths were conducted on this material. Blade test bars with tensile surface conditions identical to those found on the actual blades (as-fired and transversely ground) were heat treated for 24 hr at 1250°C (2282°F) in an air atmosphere to increase strength and heal possible surface damage. Test material with an oxidized as-fired surface condition registered an average room temperature MOR of 396.60 MPa (57.52 ksi) with a standard deviation of 48.88 MPa (7.09 ksi) for a 49 bar sample size. This represents a statistically insignificant decrease in strength (α = 0.01) when compared with the strength of 417.92 MPa (60.61 ksi) observed in the untreated bars with a similar tensile surface condition. The primary strength-controlling defects were found to be surface pores (72%), with internal pores (26%) and surface flaws (2%) observed as the secondary defects. A two-parameter Weibull function fitted to the strength distribution of these oxidized as-fired test bars yielded a characteristic strength of 417.54 MPa (60.56 ksi) and a Weibull modulus of 9.4, as shown in Figure 162.

A group of 47 test bars subjected to the same heat-treat procedure with a transversely ground surface condition recorded an average room-temperature fracture strength of 396.03 MPa (57.44 ksi) with a standard deviation of 74.34 MPa (10.78 ksi), a statistically significant increase (α = 0.01) of 78.43 MPa (11.38 ksi) over the virgin transversely ground material. The primary strength-controlling defects observed in these specimens were surface pores (66%), with internal pores (28%) and surface flaws (4%) also present. One test bar (2%), shown in Figure 163, failed from a large surface crystal. A two-parameter Weibull fit of these strengths provided a characteristic strength of 425.79 MPa (61.75 ksi) and a Weibull modulus of 6.2, as illustrated by Figure 164.

Room-temperature overspeed-to-failure blade spin testing was used to directly establish blade strength characteristics. These tests were conducted in a single slotted wheel using uniform thickness—0.203 mm (0.008 in.) ±0.0025 mm (0.0001 in.)—L605 compliant layers. All blade attachments were coated with a thin layer of boron nitride. Since the blade attachment is transversely ground, all blades were oxidized for 24 hr at 1250°C (2282°F). Previous experience has shown that such a treatment will restore strengths to a level comparable with that of the longitudinally ground surface condition. The results of the overspeed-to-failure spin tests are plotted in Figure 165. The average failure speed measured 51,100 rpm, equivalent to 138.5% design speed. The straight line in Figure 165 represents the predicted blade performance generated

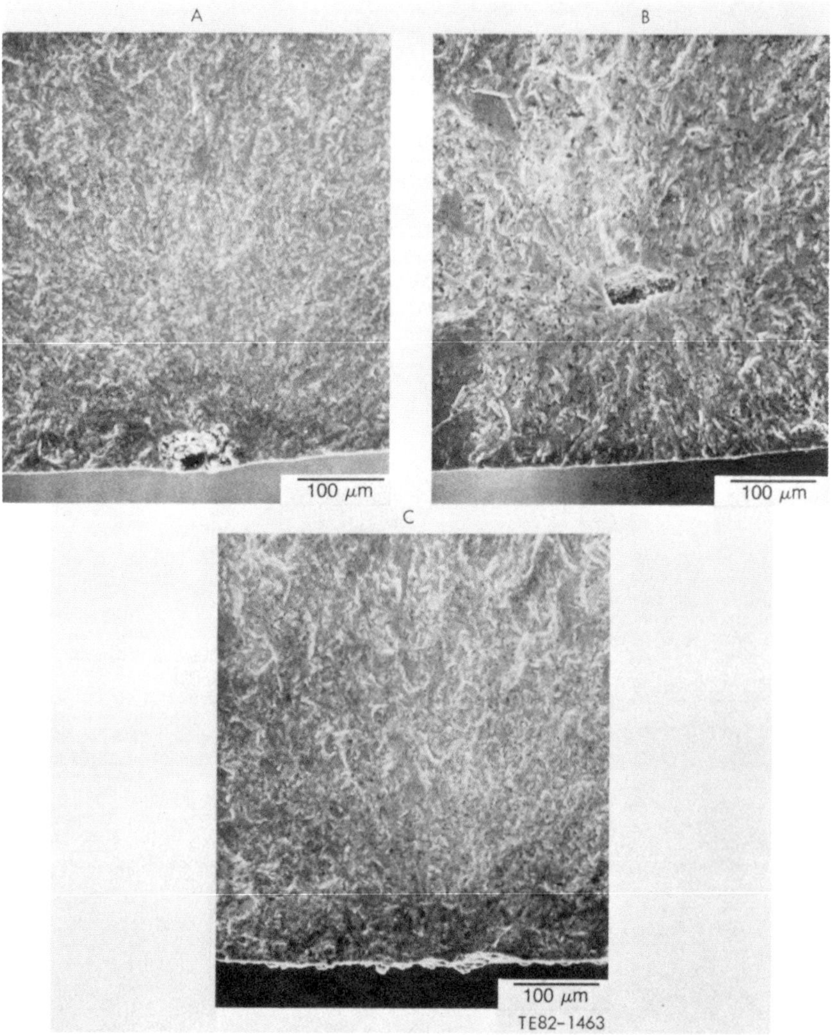

Figure 160. Examples of typical strength-controlling defects in CBO sintered SiC blade test bars: (a) surface pore, (b) internal pore, and (c) surface flaw.

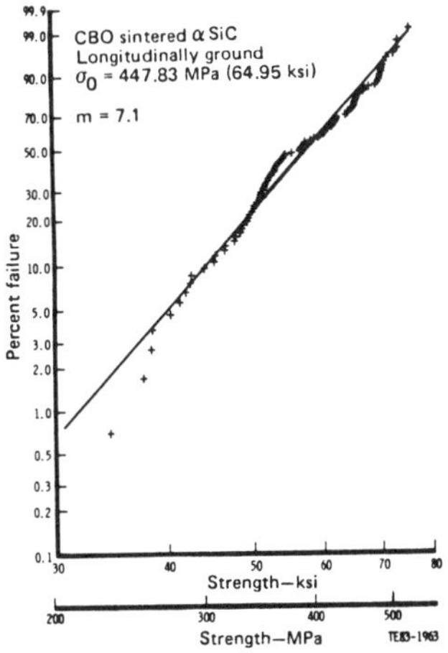

Figure 161. Strength distribution of CBO sintered alpha SiC blade test bars with an as-fired surface condition (heat treated for 24 hr at 1250°C [2282°F]).

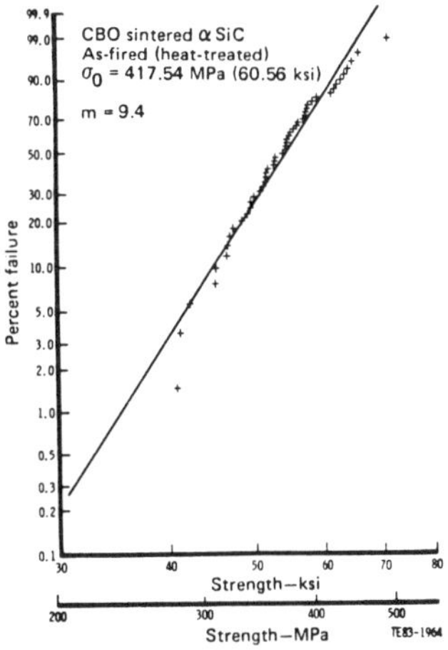

Figure 162. Strength distribution of CBO sintered alpha SiC blade test bars with an as-fired surface condition (heat treated for 24 hr at 1250°C [2282°F]).

from a three-dimensional finite element analysis using the material strength characteristics of the qualification test bars. These strengths, divided into both surface and volume characteristic strengths for each type of surface finish on the blades, are summarized in Table XIX. The correlation between predicted performance and actual performance is excellent. Furthermore, the types of defects which precipitated failure in the blades were identical to those observed in the companion test material, i.e., surface and internal pores.

These results indicate that the companion qualification test material is indeed representative of the material present in the actual blades and that present-day state-of-the-art silicon carbide material characteristics are being realized in actual blades.

Table XIX.
Weibull parameters generated from CBO alpha SiC blade test bars.

	Surface	
	σ_{os}—MPa (ksi)	**Surface Weibull modulus, m_s**
As-fired and oxidized*	661.096 (53.004)	10.915
Transversely ground and oxidized*	976.555 (48.535)	6.041
	Volume	
	σ_{ov}—MPa (ksi)	**Volume Weibull modulus, m_v**
All conditions	686.481 (32.293)	8.619

* Oxidized in air for 24 hr at 1250°C (2282°F)

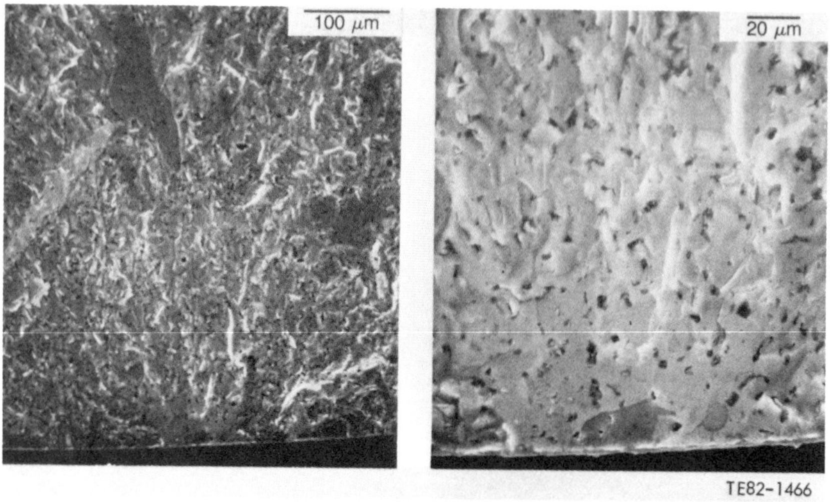

Figure 163. Strength-controlling defect (large surface crystal) observed in oxidized, transversely ground CBO sintered alpha SiC blade test bar.

Carborundum RBSiC—Stationary Ring Components

The three 2070°F-configuration stationary ring components—inner vane support ring, outer vane support ring, and turbine tip shroud—were all fabricated by Carborundum from the same material (RB-SiC) with identical process routings. Therefore, characterization and evaluation of these rings are presented together in this subsection. These parts were produced by warm compression molding simple cylindrical rings, which were subsequently green machined to near net shape. Following siliconization, finished dimensions were obtained by diamond grinding. Both the vane pockets and lug slots in the inner and outer vane support rings were then ultrasonically machined into the sintered parts.

Characterization of these components was conducted in a four step process: (1) nondestructive inspections (visual, FPI, X-ray, and density), (2) evaluation of qualification test bars processed with the rings, (3) evaluation of test material cut from the actual hardware, and (4) an internal pressure proof test. In addition, the effects of heat treatment on material strength were examined.

The overall quality of the rings was generally good, with standard nondestructive inspections of the latest shipments indicating no objectionable defects. The average density of the parts measured 2.93 g/cm^3.

The room-temperature fracture strengths for the as-fired, longitudinally ground, and transversely ground surface conditions were evaluated for the qualification test bars accompanying these components. The fracture strength for the as-fired surface condition was quite low, with an average MOR of only 137.27 MPa (19.91 ksi) and a standard deviation of 33.58 MPa (4.87 ksi). Several of these test bars possessed a relatively thick (80-100 μm) layer of SiC crystals on the as-fired surface. These specimens exhibited a particularly low strength—100.00 MPa (14.51 ksi). This SiC surface layer is shown in Figure 166, along with an example of the microstructure typically observed in this material. The longitudinally ground specimens had an average fracture strength of 350.39 MPa (50.82 ksi) with a standard deviation of 53.50 MPa (7.76 ksi). Qualification bars tested with a transversely ground surface condition averaged 183.19 MPa (26.57 ksi) with a standard devi-

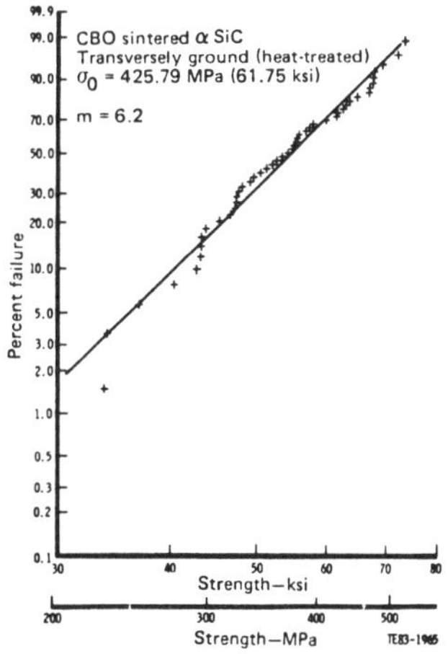

Figure 164. Strength distribution of CBO sintered alpha SiC blade test bars with a transversely ground surface condition (heat treated for 24 hr at 1250°C [2282°F]).

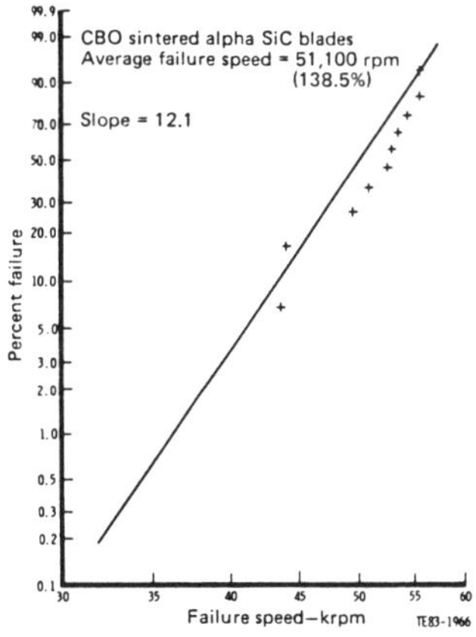

Figure 165. CBO sintered alpha SiC blade overspeed-to-failure blade spin test results. The straight line represents predicted blade performance.

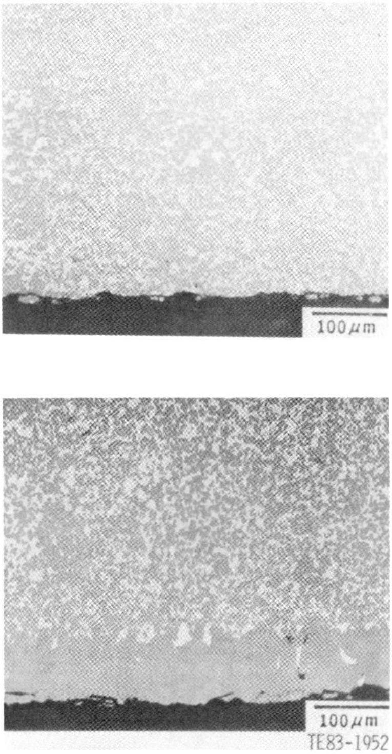

Figure 166. Microstructure of CBO RBSiC test bars: (a) typical microstructure and (b) as-fired surface with a thick layer of SiC.

ation of 19.10 MPa (2.77 ksi). In all cases, surface sites were identified as the locations of the fracture origins.

The relationship between the strengths of qualification test bars and the actual components was examined. Standard-size four-point bend test bars, measuring 50.8 mm x 6.35 mm x 3.30 mm (2.0 in. x 0.25 in. x 0.125 in.), were cut with a diamond saw from the flange of three outer vane support rings, as shown in Figure 167. The average room-temperature fracture strength of these test bars was determined

Table XX.
Fracture strength of CBO RBSiC.

Origin	Surface condition	No. of samples	Fracture strength MPa	(ksi)
Qualification bars	As-fired	29	137.27	(19.91)
	Longitudinally ground	29	350.39	(50.82)
Cut from rings	As-fired	12	193.54	(28.07)
	Longitudinally ground	15	349.98	(50.76)

and compared with that of the qualification material. These results are summarized in Table XX.

The average MOR of component test bars with an as-fired surface condition measured 193.54 MPa (28.07 ksi), somewhat higher than the strength of 137.27 MPa (19.91 ksi) recorded for the qualification bars with a similar surface condition. The fracture strength of the longitudinally ground test bars measured 349.98 MPa (50.76 ksi), virtually identical to the strength of 350.39 MPa (50.82 ksi) shown by the longitudinally ground qualification bars.

Preliminary efforts to determine the effects of heat treatment on CBO RBSiC material were conducted. The surface of RBSiC material was subjected to a grit blasting operation after reaction sintering to remove impurities and excess free silicon. In this investigation, both as-fired and transversely ground qualification test bars were heat treated for 24 hr at 1250°C (2282°F) in an air atmosphere to heal the surface damage resulting from the grit blasting operation and restore strength to that exhibited by longitudinally ground material. The room-temperature fracture strengths were then measured, and visual examinations were done on the tensile surfaces. A summary of the preliminary results is shown in Table XXI.

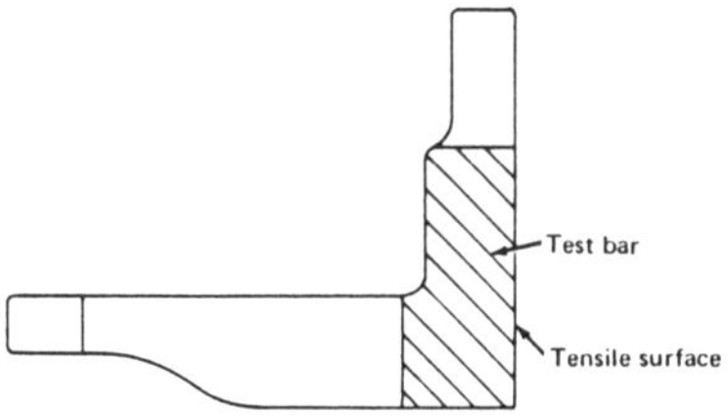

Figure 167. Cross section of CBO RBSiC outer vane support ring showing location of test bars.

The average room-temperature fracture strength of heat-treated qualification bars with an as-fired surface condition measured 244.00 MPa (35.39 ksi) with a standard deviation of 7.32 MPa (1.06 ksi). This represents a 78% increase in strength over untreated test bars with as-fired surfaces. A thin, glassy layer of silica (SiO_2) was observed on the oxidized as-fired tensile surface, shown in Figure 168.

The average MOR of heat-treated test bars with a transversely ground tensile surface measured 387.90 MPa (56.26 ksi) with a standard deviation of 49.16 MPa (7.13 ksi). This strength is 112% higher than that of similarly machined untreated bars. A

Table XXI.
Fracture strength of CBO RBSiC ring qualification bars.

Surface condition	Fracture strength MPa	(ksi)	% increase
As-fired	137.27	(19.91)	—
As-fired and oxidized*	244.00	(35.39)	78
Transversely ground	183.19	(26.57)	—
Transversely ground and oxidized*	387.90	(56.26)	112

*Heat treated for 24 hr at 1250°C (2282°F)

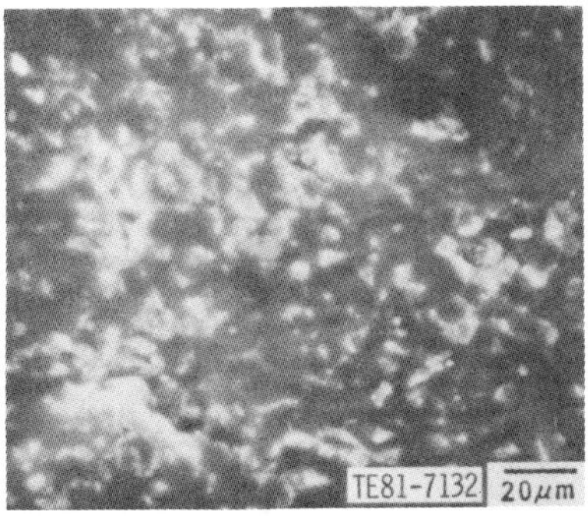

Figure 168. Photomicrograph of the surface of CBO RBSiC test bar with an as-fired surface condition heat treated at 1250°C (2282°F) for 24 hr.

thin, glassy layer was again observed on the oxidized tensile surface, as in Figure 169.

To further characterize component performance, a bladder proof test was designed and conducted on the Carborundum RBSiC rings. The bladder testing involved loading the inside diameter of the rings with internal pressure, shown in Figure 170, applied through a thick-walled rubber bladder pressurized with hydraulic fluid. The equipment used for this testing is shown in Figure 171. A two-dimensional finite element analysis was conducted on these rings to simulate various loading conditions and determine the stress levels in the parts. Based on this analysis, a pressure load of 5170 kPa (750 psi) was selected for the proof test. This internal pressure generated an average stress on the outer flange of the rings of 56.39 MPa (8.18 ksi), which was equivalent to the peak stress level developed in the rings under engine operating conditions.

Ten Carborundum RBSiC rings were subjected to the bladder proof test: nine outer vane support rings and one shroud ring. A summary of this testing is presented in Table XXII. A more detailed description of the bladder proof test may be found in EDR 11006 (Ref 8).

The results of initial bladder proof tests indicated that the as-fired surface strength was the controlling factor in ring performance. Since maximum stresses, both in bladder testing and in actual engine operation, are developed in areas with an as-fired surface condition, a decision was made to subject all further rings to heat treatment at 1250°C (2282°F) for 24 hr prior to testing. This heat treatment has resulted in a substantial increase in the

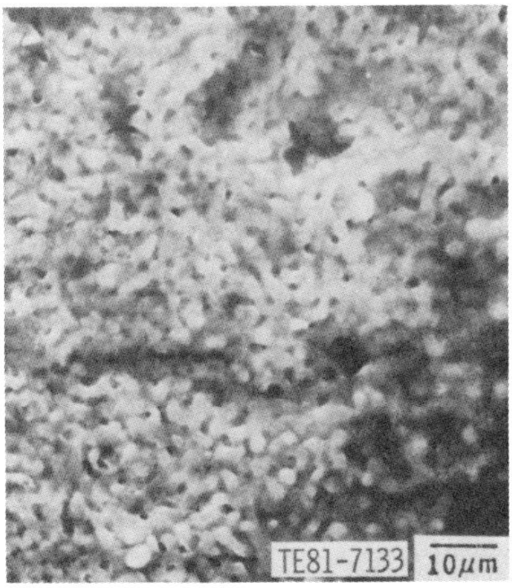

Figure 169. Photomicrograph of the surface of CBO RBSiC test bar with a transversely ground surface condition heat treated at 1250°C (2282°F) for 24 hr.

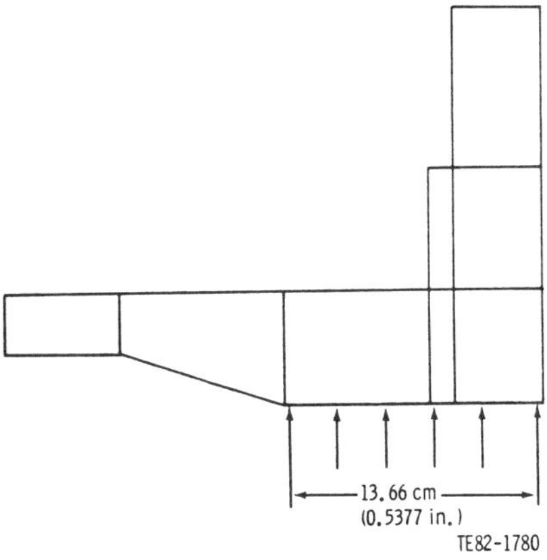

Figure 170. Area of rings loaded (internal pressure) in bladder test.

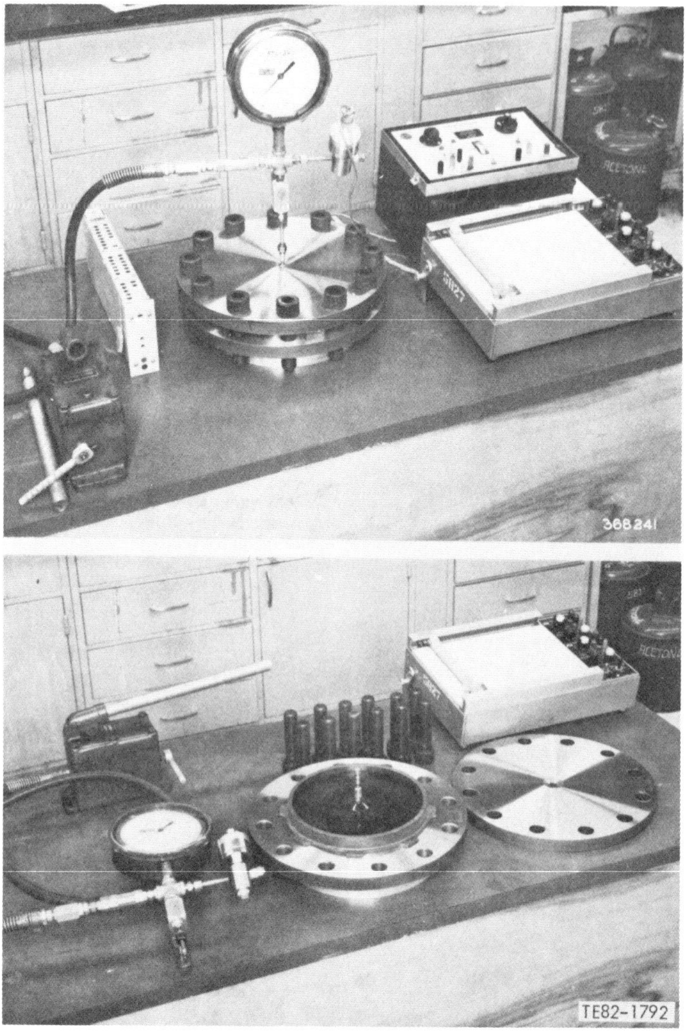

Figure 171. Bladder test fixture.

Table XXII.
Summary of CBO RBSiC ring bladder proof tests.

Component	Serial No.	Quality	Pressure — kPa (lb/in.²)
Outer vane support ring	FX 25837	A	3240 (470)
Outer vane support ring	FX 25838*	B+	5170 (750) passed
Outer vane support ring	FX 25839*	A	5170 (750) passed
Outer vane support ring	FX 25840*	A	5170 (750) passed
Outer vane support ring	FX 25842*	A	3895 (565)
Outer vane support ring	FX 25843*	C	2827 (410)
Outer vane support ring	FX 25844*	C	3344 (485)
Outer vane support ring	FX 25847*	C	3792 (550)
Outer vane support ring	FX 25848*	C	4585 (665)
Shroud ring	FX 29266*	A	5170 (750) passed

*Heat treated at 1250°C (2282°F) for 24 hr

as-fired strength of RBSiC test bars—from 137.27 MPa (19.91 ksi) to 244.00 MPa (35.39 ksi).

Of the seven rings that failed at pressures below the design level of 5170 kPa (750 lb/in.²), three were observed to have fracture origins located at relatively thick (150-200 μm) layers of silicon-carbide crystals on the ring outside diameter, as shown in Figure 172. These crystals were similar to those previously observed in the as-fired qualification test bars, which exhibited strengths of only 100.00 MPa (14.51 ksi). Carborundum identified the source of this phenomenon as a crack in the graphite tube of the vacuum induction furnace used to fire these components. This resulted in localized hot spots in the sintering furnace which promoted exaggerated grain growth on the surface of these rings. This problem has since been resolved by the replacement of the defective tube and the incorporation of additional inspections after firing.

Additional component characterization was performed on the latest shipment of CBO RBSiC shroud rings. These rings incorporate an extended flange, as in Figure 173, from which test material was cut into four-point MOR test bars for laboratory evaluation. Nondestructive inspections of the nine shroud rings were excellent, with no objectionable defects observed in any of the components.

The average density of the test bars cut from these rings was 2.87 g/cm³. The average room-temperature MOR of specimens with a longitudinally ground surface condition measured 402.67 MPa (58.40 ksi) with a standard deviation of 61.23 MPa (8.88 ksi), an increase of 15% over the strength of 349.98 MPa (50.76 ksi) observed for machined bars cut from earlier CBO RBSiC rings. Surface cracks were identified as the failure origins in all specimens.

Carborundum Sintered Alpha Silicon Carbide—Plenums

Efforts were initiated in 1980 at Carborundum to develop a sintered alpha SiC material suitable for slip-casting large complex components (such as the plenum). A total of four slip-cast sintered alpha SiC plenums were received from CBO as a result of this effort.

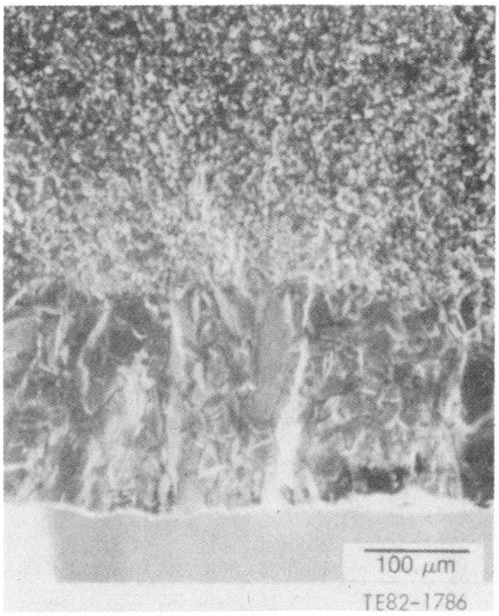

Figure 172. Microstructure of the outside diameter of CBO RBSiC outer vane support ring, showing SiC crystal layer.

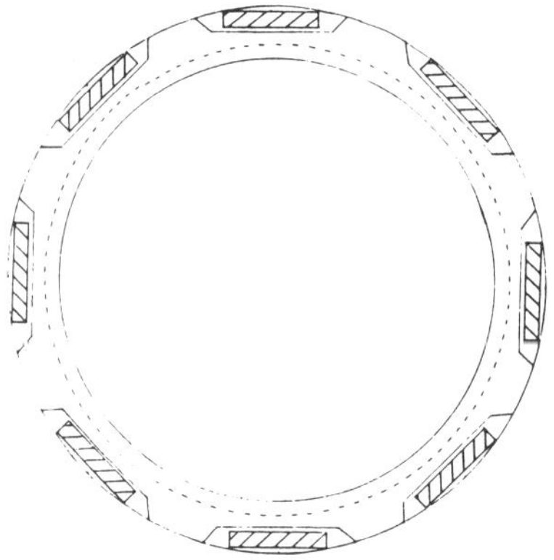

Figure 173. CBO RBSiC shroud ring with extended flange showing location of test bars.

This coarse-grained silicon carbide material was formulated specifically for the plenum to facilitate the slip-casting process and to minimize dimensional shrinkage. Figure 174 shows the microstructure of this material, which has an average grain size of 10.6 μm with numerous large pores present at grain boundary triple points. Visual, FPI, and X-ray examinations of the plenums indicated a generally porous surface, with evidence of discoloration on the inside diameters of two of the plenums.

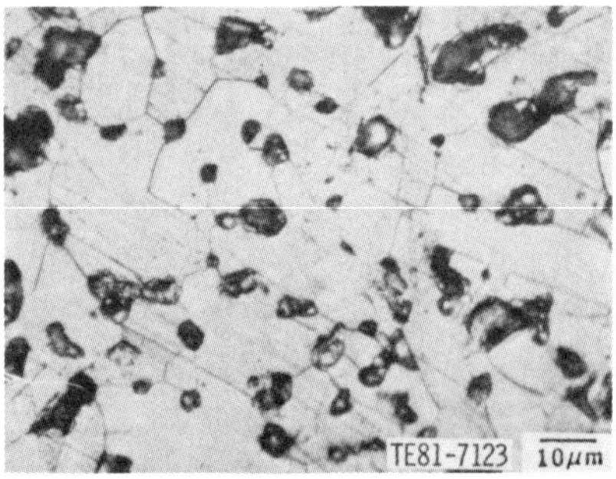

Figure 174. Micrograph of CBO sintered alpha SiC plenum; average grain size is 10.6 μm.

In addition, several large cracks were observed in these components. The average density of these components measured 2.84 g/cm^3.

The accompanying qualification test material possessed an average density of 2.93 g/cm^3. A relatively low strength was observed in the plenum test bars, with an average room temperature MOR for the as-fired surface condition of 220.43 MPa (31.97 ksi) with a standard deviation of 41.92 MPa (6.08 ksi). Failure analysis of these test bars indicated that all fracture origins were located at surface sites, generally at large pores or surface blisters, as in Figure 175. The average MOR for test bars with a machined surface condition measured 333.78 MPa (48.41 ksi) with a standard deviation of 60.12 MPa (8.72 ksi). This represents a significant increase in strength for the machined bars versus the bars with an as-fired surface. Most fracture origins for the machined bars were at surface flaws or cracks, as in Figure 176, as opposed to surface pores and blisters for the as-fired specimens.

Carborundum RBSiC—2070°F-Configuration Vanes

The RBSiC 2070°F-configuration vanes were the subject of extensive process development activities at Carborundum in 1979 and 1980. A number of variations to the base RBSiC material used for the transfer molding of the vanes were investigated. Test bars processed according to these variations were evaluated for both microstructural uniformity and strength characteristics. Table XXIII summarizes the seven variations examined, with the standard material used as a control to give a total of eight materials. These seven changes were a combination of material composition variations (bimodal grain size, silicon doping, high purity silicon, and high silicon content) and process variations (high free silicon outer layer, hammer milled, and hammer milled plus warm molded) which would lead to microstructural changes.

With the exception of Variation 2 (high free silicon outer layer), all the variations were made into test bars without difficulty. Variations 3, 4, 5, and 6 all used the standard mix and process routing established for the baseline vanes, with the variations as noted. Variation 1 (bimodal grain size) used a new mix composition with the silicon carbide component consisting of 65% 400-mesh SiC and 35% 1000-mesh SiC. This gave better molding than the others with good mold filling and a lack of knit lines. Variation 7 (high free silicon content) used a new mix which had no particulate graphite.

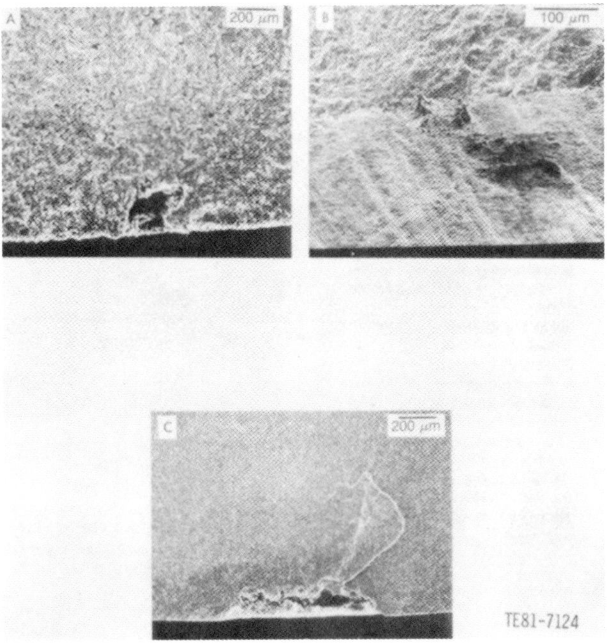

Figure 175. Typical strength-controlling flaws for as-fired CBO alpha SiC plenum test bars: (a) small surface pore, (b) surface blister, and (c) large surface pore.

Variation 2 (high free silicon outer layer) test bars were made by the following procedure:

1. mold standard composition
2. bake at 800°C (1472°F) in inert atmosphere
3. bake at 400°C to 500°C (752°F to 932°F), 4 to 10 minutes, to oxidize carbon and graphite from surface layer
4. siliconize with standard procedure

Seven attempts to make this variation with a total of 72 test bars and 27 vanes resulted only in cracked and warped pieces after siliconization. The difference in thermal expansion between the high free silicon layer and the low free silicon inner core is believed to have caused stresses which were too high. Because of the stress factor, this material variation was abandoned.

Table XXIV summarizes the results of four-point bend tests conducted on the test bars at room temperature. The strongest materials were the standard material, which was hammer milled (Variations 5 and 6), and the bimodal material (Variation 1). The doped silicon material (Variation 3) and the high free silicon material (Variation 7) were clearly weaker than the others. Additional testing has shown that very high free silicon amounts also result in low strengths for compression-molded RBSiC (ring structures); in general, lowering the free silicon increases strength. The high purity silicon (Variation 4) was also higher in strength than the standard material. Photomicrographs of typical microstructures of each of the variations are shown in Figures 177 through 183.

Previous experience with warm compression-molded reaction-sintered materials has indicated that strength is dependent on the uniformity of the dispersion of the free silicon phase. From this study, processing changes appear to affect the occurrence of silicon-rich areas. Hammer milling appears to

Table XXIII.
RBSiC material variation summary.

Variation	Purpose
1. Bimodal grain size, 400-1000 mesh	Improve transfer molding flow, achieve high packing density to increase strength
2. High free silicon outer layer	Obtain residual compressive stress on surface due to high free silicon layer
3. Silicon doping in cover mix	Improve wetting characteristics of silicon on silicon carbide and strengthening of silicon carbide
4. High purity silicon cover mix	Reduction of impurities to improve oxidation resistance
5. Microstructure improvement hammer milled	Eliminate large voids in molding to achieve uniformly small distribution of free silicon and improve strength
6. Microstructure improvement hammer milled and warm molded	Same as above, with warm molded transfer molding pucks to eliminate entrapped air
7. High free silicon content	Improve thermal shock resistance
8. Standard material	Control group

result in an improvement in the dispersion of the free silicon. The additional process step of warm molding with hammer milling (Variation 6) does not appear to gain any microstructural or strength improvement.

The occurrence of silicon-rich areas is also mix composition dependent. The bimodal variation (Variation 1) did not have large silicon-rich areas and had a strength improvement over the standard material. This composition molded better than the others with fewer fill problems such as knit lines and unfilled regions. The uniform distribution of the free silicon phase is a necessary but not sufficient condition to improve strength. The doped silicon material (Variation 3) had a uniform silicon distribution, but the strength was still low.

From the initial test results, several material variations showed promise. The strength results indicated that a hammer-milling step added to the standard mix preparation improved strength by eliminating microstructural nonuniformities. Another variation using a bimodal silicon carbide filler resulted in a higher Weibull modulus and a slightly

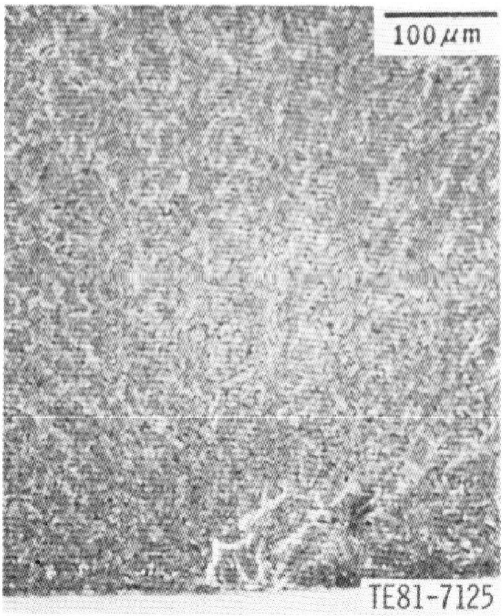

Figure 176. A surface crack—the typical strength-controlling flaw in machined CBO alpha SiC plenum test bars.

Figure 177. Variation 1: bimodal grain size—400-1000 mesh SiC; mean strength—287.86 MPa (41.75 ksi).

Figure 178. Variation 3: Silicon doping in cover mix; mean strength—204.98 MPa (29.73 ksi).

Figure 180. Variation 5: hammer milled; mean strength—321.43 MPa (46.62 ksi).

lower strength. These findings were used as the basis for a second material variation matrix, summarized in Table XXV, of five new materials and a control.

This matrix consisted of a repeat of Variation 5, hammer milling. Next, the hammer-milling step was combined with the bimodal silicon carbide powder mix (Variation 9). In addition to hammer milling, an air impact milling step was added to further reduce the mix agglomerate particle size (Variations 10 and 11). All of these variations reduced the incidence of free silicon-rich areas, which were often the failure initiation sites. One matrix composition with a low free-silicon content (Variation 12) was included to see if the inverse of Variation 7 (high free-silicon—low strength) produces a high strength material.

After appropriate molding conditions were established, all six of the new material variations were successfully molded. This latest matrix used the 2070°F vane mold in which test bars and vanes were molded together. The previous material investigation matrix used a separate test bar mold. Less molding difficulty was experienced with the new tool; the gating configuration differences were believed to be responsible for this improvement. Since different test bar molds were used for the two matrices, comparisons between the material strengths might not have been valid. Variation 5 of the first matrix was repeated as a control to ascertain the effect of mold changes with the new tool.

Variations 9 and 11, both bimodal mixes, did not siliconize well. X-ray inspection indicated density

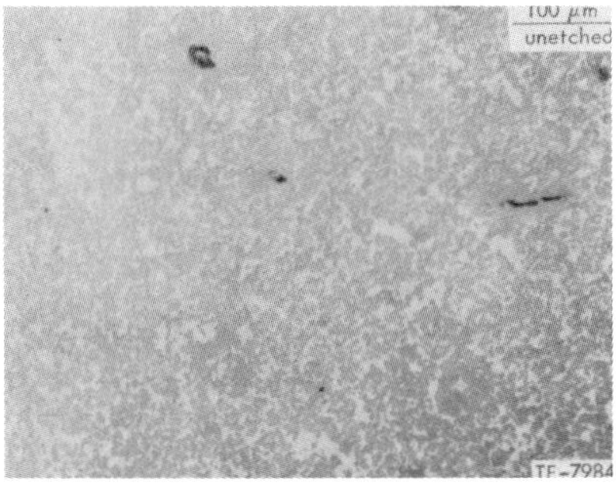

Figure 179. Variation 4: high purity silicon; mean strength—282.55 MPa (40.98 ksi).

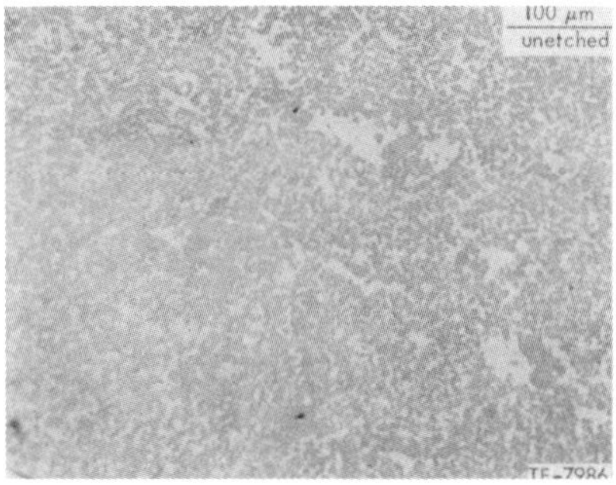

Figure 181. Variation 6: hammer milled and warm molded; mean strength—312.33 MPa (45.30 ksi).

Table XXIV.
Material variation investigation, showing results of four-point bend test (1/3 point loading).*

Material variation	Mean strength		Std dev		Weibull modulus
	MPa	(ksi)	MPa	(ksi)	m
1. Bimodal grain size	287.86	(41.75)	18.34	(2.66)	16.5
2. High free silicon outer layer	—	—	—	—	
3. Silicon doping	204.98	(29.73)	14.89	(2.16)	15.4
4. High purity silicon	282.55	(40.98)	39.58	(5.74)	7.9
5. Hammer milled	321.43	(46.62)	31.99	(4.64)	10.1
6. Hammer milled and warm molding	312.33	(45.30)	31.37	(4.55)	10.8
7. High silicon content	166.72	(24.18)	25.37	(3.68)	7.5
8. Standard material	255.52	(37.06)	32.34	(4.69)	8.6

*Bar size: 50.8mm x 5.08 mm x 2.54 mm (2.0 in. x 0.2 in. x 0.1 in.)

variations resulting from incomplete silicon infiltration. The other four variations, however, were siliconized satisfactorily.

Variation 13, acid treated with a similar technique used for abradable shrouds, exhibited some difficulty. The acid treatment did not remove the silicon from small areas in the center of the bars, apparently because the bars were not left in treatment long enough. (Both X-ray inspection and the fracture surfaces clearly showed this silicon-rich core.)

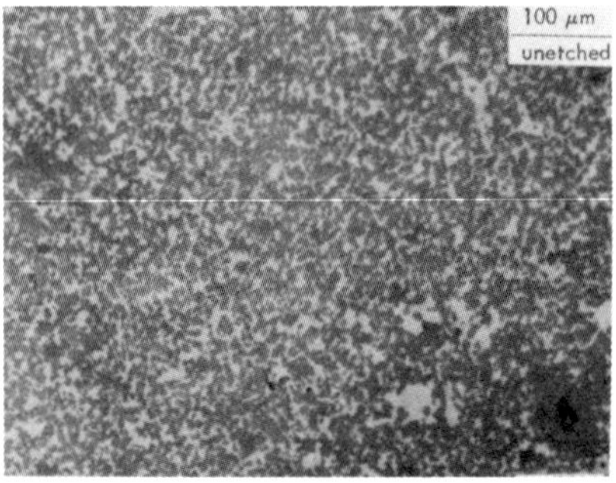

Figure 182. Variation 7: high free silicon content; mean strength—166.72 MPa (24.18 ksi).

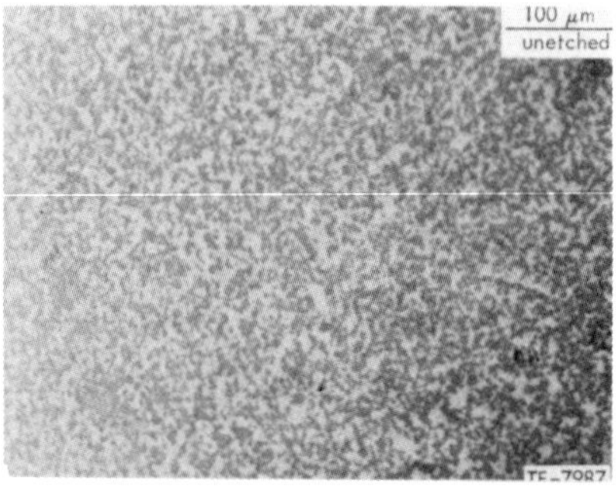

Figure 183. Variation 8: standard material; mean strength—255.52 MPa (37.06 ksi).

Table XXV.
Redefined material variation matrix.

5. Control group—microstructural improvement, hammer milled	Repeat of Variation 5 for a control group
9. Bimodal grain size 400 - 1000 mesh, hammer milled	Improve microstructure uniformity of bimodal material
10. Microstructural improvement, hammer milled, air impact milled	Improve microstructure uniformity more than hammer milling
11. Bimodal grain size 400 - 1000 mesh, hammer milled, air impact milled	Improve microstructural uniformity of bimodal material more than hammer milling
12. Bimodal grain size 400 - 1000, low free silicon, hammer milled, air impact milled	Lower free silicon than Variation 11
13. Variation 12—acid treated to remove all free silicon	Reduce manual labor of cleaning excess silicon from part surfaces

Twenty room-temperature four-point bend tests in air were done on transfer-molded test bar samples for each of the six material variations. Table XXVI presents the results. Two bimodal mix compositions, Variations 9 and 11, did not siliconize completely, resulting in mean material strengths that were quite low—168.14 MPa (24.39 ksi) and 222.95 MPa (32.34 ksi). This correlation between poor siliconization, low-density areas, and low strength has been seen with both warm compression-molded and transfer-molded materials.

The bimodal material with a low free-silicon content (Variation 12) had the highest Weibull modulus (12.0) of the six variations tested. The bimodal composition in the previous matrix (Variation 1) also had the highest Weibull modulus, 16.5, among those

Table XXVI.
Redefined material investigation matrix.

Results of four-point bend test (1/3 point loading)*

Material variation	Mean strength MPa	(ksi)	Std dev MPa	(ksi)	Weibull modulus, m
5. Control, hammer milled	400.75	(58.13)	51.77	(7.51)	8.1
9. Bimodal, hammer milled	168.14	(24.39)	54.88	(7.96)	3.3
10. Hammer milled, air impact milled	391.17	(56.74)	56.88	(8.25)	6.2
11. Bimodal, hammer milled, air impact milled	222.95	(32.34)	76.32	(11.07)	3.2
12. Bimodal, low free silicon, hammer milled, air impact milled	378.14	(54.85)	33.16	(4.81)	12.0
13. Same as Variation 12, acid treated to remove free silicon	56.19	(8.15)	5.24	(0.76)	10.9

* Bar size: 50.8 mm x 5.08 mm x 2.54 mm (2.0 in. x 0.2 in. x 0.1 in.)

eight variations. This seemed to indicate that additional development of bimodal RBSiC materials might be worthwhile.

The addition of an air impact milling step (Variations 5 and 10) did not result in any strength improvement. The acid-treated material (Variation 13) was very weak—56.19 MPa (8.15 ksi). However, the elimination of the difficult step of removing the excess silicon from siliconized vanes might aid in the commercialization of RBSiC engine components.

As a result of these processing studies, Variation 5 (hammer milled) was selected as the optimum process for RBSiC vane fabrication in terms of strength and microstructural uniformity and a run of 56 vanes was produced for characterization and engine qualification. All 56 of these vanes met structural and dimensional requirements, with no objectionable discontinuities. The average density of these vanes was 2.92 g/cm^3.

The average room temperature MOR of vane test bars with a longitudinally ground surface condition measured 413.02 MPa (59.91 ksi) with a standard deviation of 52.88 MPa (7.67 ksi). The associated Weibull modulus was 8.75. The primary strength-controlling flaws observed in the bars were surface flaws and porosity.

Development of Nondestructive Evaluation Techniques

SUMMARY

Nondestructive evaluation (NDE) techniques were developed to detect 25-200 μm failure controlling surface and subsurface flaws in structural ceramic materials. High-frequency ultrasound has been shown to detect 50-100 μm size subsurface voids and inclusions. Scanning photoacoustic microscopy (SPAM) has been shown to detect 25 μm size surface voids and cracks in ceramic MOR bars and turbine blades. Scanning laser acoustic microscopy (SLAM) detected tight surface Knoop indented cracks and surface voids of 70 μm diameter x 20 μm depth. Good correlation was also found between actual fracture controlling cracks and SLAM detected/predicted defects.

Nondestructive evaluation (NDE) is used both as a step in qualification of parts for engine use and a source of data to promote process development at the ceramic supplier. During component development, each material is characterized, NDE is applied to finish parts, and test bars are evaluated. Qualification tests (mechanical and thermal) are run on each part before it is ready for introduction to the engine environment. To make ceramic materials viable in the mass production environment of the automotive product, nondestructive evaluation, failure surface topography, chemical analysis, and microstructure examination serve as data to be related back to ceramic source to revise processing and/or implement process controls to overcome the strength-limiting defects revealed.

Two classes of flaws are important in establishing the performance capability of silicon-base structural ceramics. Class I flaws were those large in degree but small in extent (e.g., pores, inclusions, foreign matter, and microcracks). Conversely, Class II flaws were those small in degree and large in extent. Examples of these types of flaws include long-range variations in microstructure (e.g., phase content, porosity, composition, and grain size). The nature and extent of both classes of flaws are highly dependent on material processing. Both classes of flaws can have a significant effect on material performance. Class I flaws directly control material strength characteristics, particularly fast fracture strength. Class II flaws can also affect strength through control of thermal and elastic properties as well as fracture toughness and oxidation behavior. The primary objective of this task was to develop NDE techniques to detect both Class I and Class II types of flaws in structural ceramic materials and components.

APPROACH

New or improved techniques for detecting and classifying both classes of defects in ceramic material are needed. In addressing this problem, under the CATE program, an assessment of available and emerging techniques was made. McGillem and coworkers, under subcontract to Allison, conducted a state-of-the-art survey of NDE techniques that might be applicable for Class I flaws in silicon base ceramics. They concluded that reflective ultrasound for bulk defects and photoacoustic microscopy for surface defects show maximum promise for near-term exploitation (five years). In addition, Allison's assessment indicated that acoustic microscopy was also a promising technique to detect surface and subsurface defects. For Class II defects, ultrasonic velocity was assessed to be most attractive for near-term application. Based on this assessment, an experimental program was initiated to examine and evaluate the most promising techniques that might impact inspection of CATE hardware. These techniques are listed in Table XXVII.

Table XXVII.
Selected NDE flaw detection techniques.

Flaws	Techniques
Class I (voids, inclusions, cracks)	Reflective ultrasound Photoacoustic microscopy (SPAM) Acoustic microscopy (SLAM)
Class II (microstructural variations)	Ultrasonic velocity

DISCUSSION: HIGH FREQUENCY ULTRASONICS

Ultrasonics is a sensitive technique to detect fracture controlling flaws of 25 to 200 μm in the structural ceramic materials. The sensitivity for flaw detection is, of course, determined by the wavelength (λ), and thus the frequency, of the incident ultrasonic beam and microstructure (grain size, porosity, second phase, etc) of the materials. Because of the necessity for the detection of small flaws, frequencies much higher than those in conventional use have to be employed. The practical application of high frequency ultrasonics, however, is limited by transducer technology, specimen geometry, ultrasonic instrumentation, and attenuation, respectively.

Commercially available state-of-the-art ultrasonic instrumentation was utilized to develop and assemble an in-house 50.0 MHz ultrasonic flaw detection system. The system consists of an x-y scanner, a 75.0 MHz broadband pulser/receiver, and a 50.0 MHz nominal frequency focused transducer. The reflected radio frequency (rf) signals (output of the pulser/receiver) from the specimen were imaged using the following two methods:

1. A Panametrics Model 5052 GPD-1 rf peak detector was used in the interface gate mode. The received signal was peak detected, and the dc peak level out was threshold detected to write the C-scan recording on the electrosensitive paper. The system is shown in Figure 184.
2. The Metrotek Model MC805 Isoscan generator is capable of detecting 50.0 MHz rf signals. The front panel of the system is shown in Figure 185. This system was utilized in conjunction with two display monitors to concurrently generate B- and C-scan images, as shown in Figure 186.

Figure 187 shows the frequency spectrum of the wave pulse reflected from the transducer delay line, water interface, the front surface reflection from a flat alpha SiC specimen, and the first back

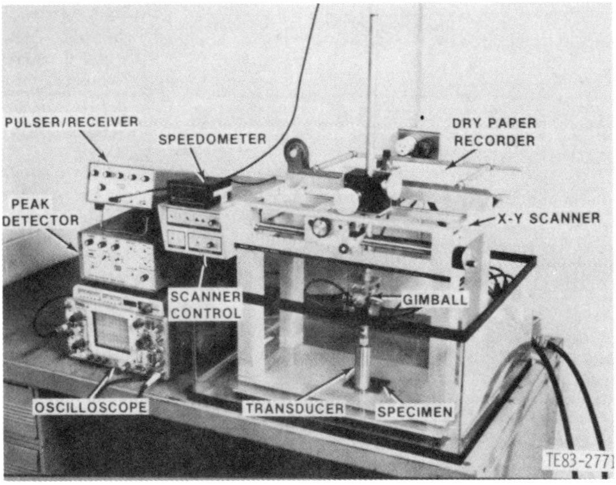

Figure 184. Panametrics Model 5052 GPD-1 system.

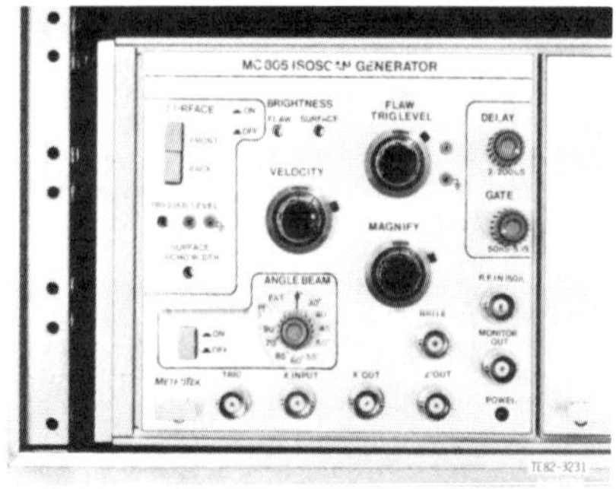

Figure 185. Front surface panel of Metrotek MC805 Isoscan generator.

surface echo of the specimen. The output of the transducer shows high-frequency contents from 30 to 65 MHz with significant power level. Attenuation of the high-frequency portion of this spectrum does occur during transmission through water. However, the center frequency of the pulse as it enters the specimen is still about 38 MHz with a mechanical Q of 2.27. The center frequency of the back surface echo was also about 38 MHz with substantially less amplitude than the front surface echo because of very large impedance mismatch between water and alpha silicon carbide.

Initial studies on seeded (void) alpha SiC disks showed that 50 to 150 μm flaws were detected in the longitudinal pulse-echo mode, as described in EDR 9951 (Ref 3). Similar size inclusions of boron nitride, carbon, silicon, and iron were detected in hot-pressed silicon nitride, as shown in Figure 188. These specimens were not cut to confirm the actual size of the detected flaws.

Three additional sintered alpha SiC circular disks of 3.125 mm (0.125 in.) thickness were fabricated by CBO. These specimens were seeded with three types of inclusions: voids, boron, and carbon. Each disk was prepared with one type of seed ranging from 50 to 125 μm. The disks were evaluated with their as-fired surfaces in the longitudinal mode (L-wave). These specimens were evaluated using both the imaging methods and subsequently sectioned to determine the actual size of the detected flaws.

Figure 189 shows a typical C-scan of the carbon seeded disk using the peak detector. The disk was examined first with the front surface up (see Figure 189a) and then with the back surface up (see Figure 189b). This ensured maximum coverage of the material thickness for flaw detection. However,

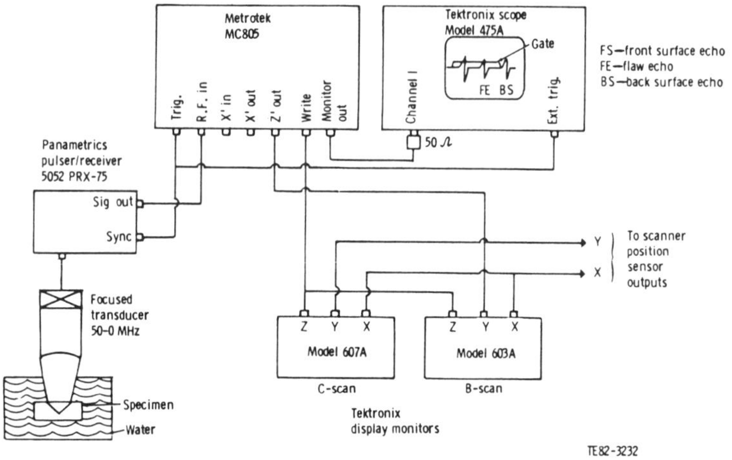

Figure 186. Schematic of high-frequency ultrasonic system using MC805 Isoscan generator for flaw detection in structural ceramics in the longitudinal mode.

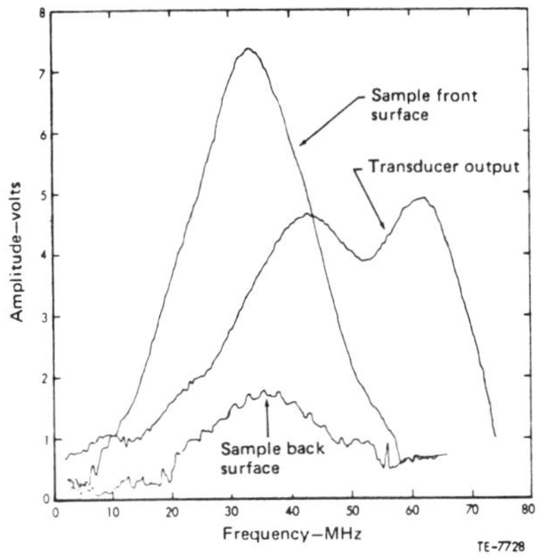

Figure 187. Frequency spectrums of the 50 MHz ultrasonic transducer.

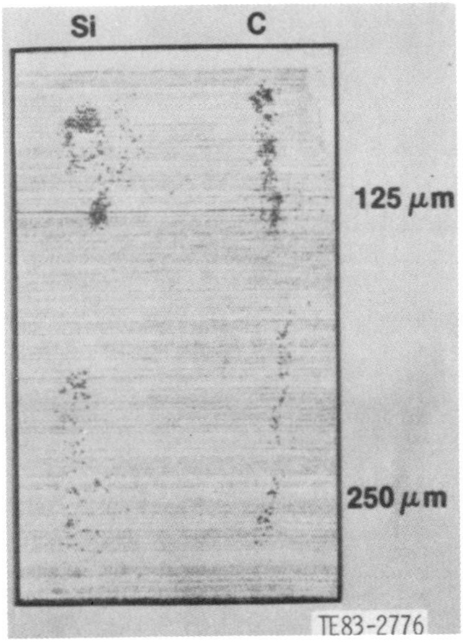

Figure 188. Defect detection high-frequency ultrasonics.

some flaws were detected from both sides of the disk while others were not. The variability in the as-fired surface of the disk causes the front surface echo amplitude and its width to vary continuously. This limits how close the gate can be placed to the front surface echo in the time domain and creates a dead zone. In addition, the unevenness of the as-fired surface causes variable scattering of the front surface echo and changes both its amplitude and frequency component composition. The ultrasonic system detected substantially more flaws than detected by X-radiography. The disk was bisected after the ultrasonic evaluation was completed. The cross section of the disk was polished to a 3-μm finish and examined under an optical microscope. Figure 189 shows a typical correlation between the ultrasonically detected and the actual flaws. The detected flaws were laminar defects of 80 to 100 μm in length. The dimension in the direction of the acoustic wave propagation was approximately 10 μm. The other two disks were similarly evaluated with the peak detector setup.

Figures 190 and 191 show the B- and C-scan recordings, using the MC805, of the disks seeded with voids and boron, respectively. Comparison of Figure 190a and 188b or Figure 191a and 191b shows that variable numbers of flaws were detected from the two surfaces of the disks. Not only were the disks observed to have flaws, but they were also observed to be warped. This varied the water path excessively and further decreased the gate width to avoid the wandering of the back surface reflection into it. A rectangular plate with four adjusting screws to vary the height was fabricated to place the specimens on. This minimized the change in the water path from one point to another and optimized flaw detectability as much as possible. Figure 190a shows that a 125 μm long laminar void was detected from both sides of the disk. The maximum dimension in the direction of acoustic insonification was about 20 μm. Figure 191a shows a 100 μm x 30 μm boron seed that was detected. Figure 191b shows a group of flaws that was found at the site of B- and C-scan recording. The disk with carbon seeds was evaluated similarly.

Both the peak detector and the MC805 Isoscan generator systems were found capable of detecting volume flaws of approximately 100 μm. The advantage of the peak detector over the MC805 is that it is used in the interface gate mode while the MC805 is not. The advantage of the MC805 over the peak detector is that it provides both the C-scan (top view) and the B-scan (side view). The two systems are, therefore, complementary.

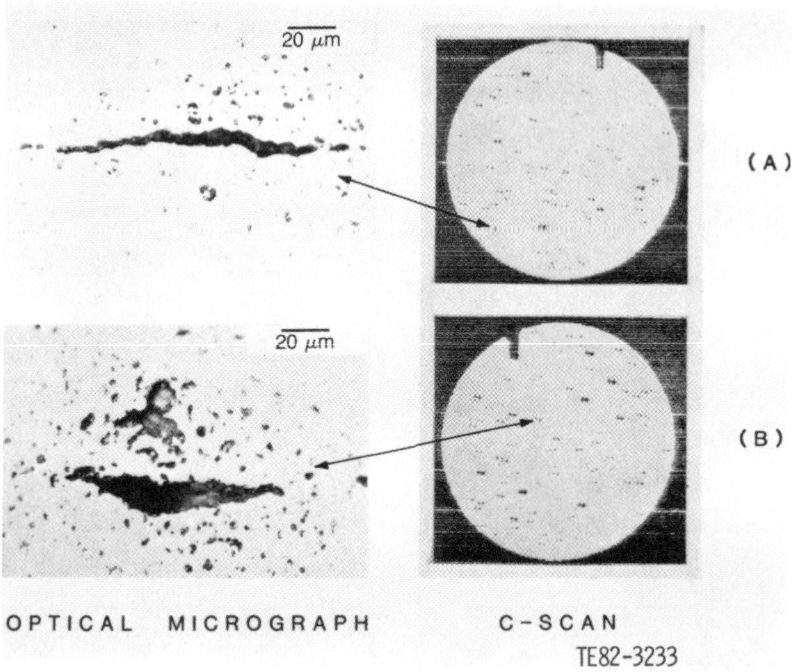

Figure 189. Ultrasonic evaluation of carbon-seeded alpha SiC disk using the peak detector: (a) front surface of the disk up and (b) back surface of the disk up.

DISCUSSION: SCANNING LASER ACOUSTIC MICROSCOPY

Acoustic microscopy is a technique for imaging localized changes in the elastic properties of materials. The physical properties that govern sound propagation are modulus and density. The variability in these two properties directly affects the acoustic properties of attenuation, velocity, and impedance of the material. The variation in the acoustic properties changes the microstructural insonification behavior of the specimen. Microstructural variations and presence of flaws (voids, inclusions, and cracks) both in the bulk and on the surface of the specimen are observed as perturbations in the transmitted acoustic amplitude. The sensitivity and resolution capability of the technique depend on the relative acoustic properties of the parent material and the anomaly.

The scanning laser acoustic microscope, shown in Figure 192, consists of a 100 MHz transducer mounted inside a water stage. The specimen is placed on the stage and insonified by an incident bulk wave. This causes dynamic ripples on the specimen back surface. These surface perturbations are continuously scanned by a focused laser beam. The reflected light is received by an optoacoustic detector and electronically processed to provide a signal whose amplitude and phase are replicas of the surface perturbation. The use of a scanning laser beam to detect surface displacements (i.e., the sound field) allows images of the transmitted and scattered mode converted sound fields to be visualized independently. This results from the fact that the scanning speed is quite rapid, thus creating a Doppler shift between the detected signal and the incident sound field. Therefore, the scattered or mode

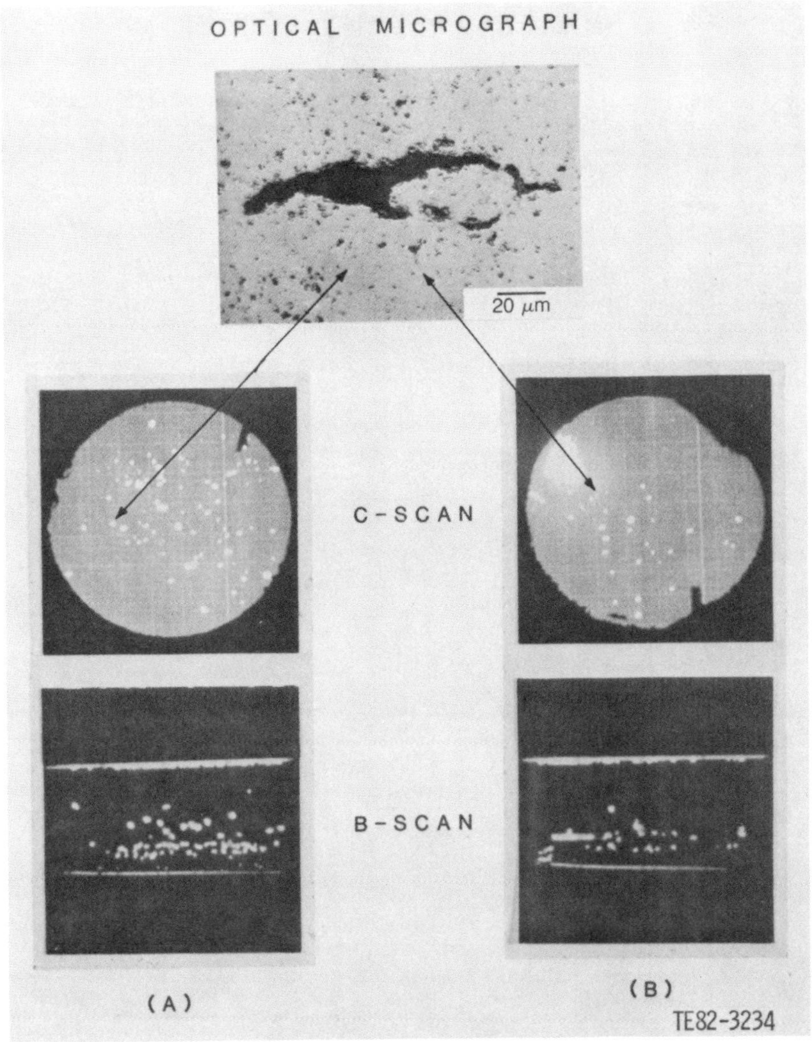

Figure 190. Ultrasonic evaluation of the SiC disk seeded with voids using the MC805 Isoscan generator: (a) front surface of the disk up and (b) back surface of the disk up.

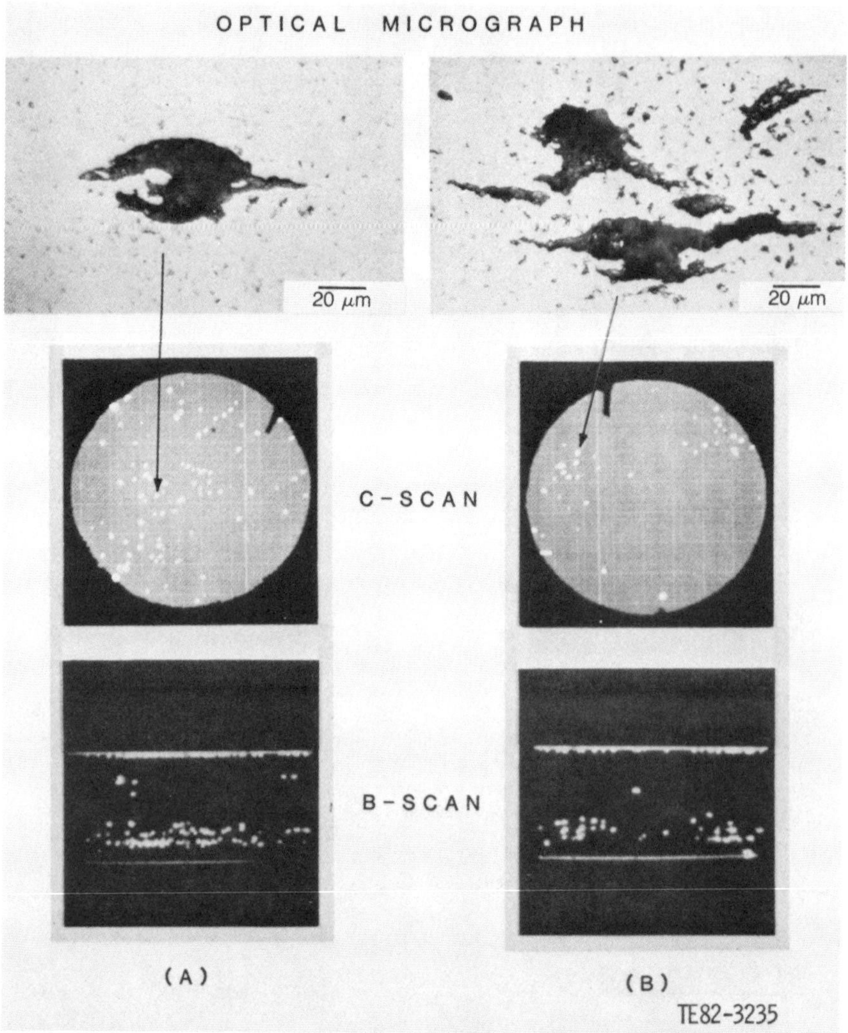

Figure 191. Ultrasonic evaluation of the SiC disk seeded with boron particles using the MC805 Isoscan generator: (a) front surface of the disk up and (b) back surface of the disk up.

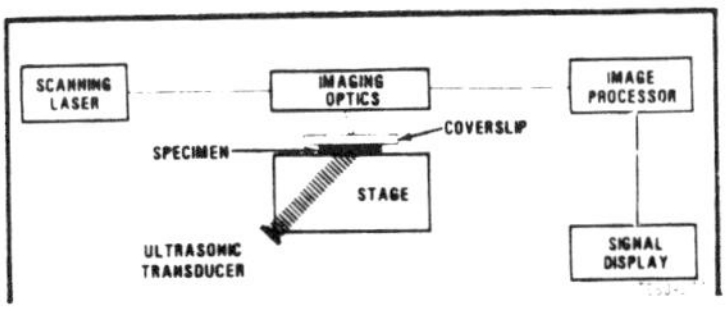

Figure 192. Schematic of the acoustic microscope.

converted waves are perceived at quite different frequencies from those characteristic of the primary insonifying field. Each wave type can then be imaged separately by the application of the proper electronic filtering.

Several different imaging modes are available in the commercial acoustic microscope produced by Sonoscan, Inc. The following two modes are most appropriate for flaw detection in ceramic materials:

- Bright-Field Mode—In this mode, the variation in transmitted energy due to attenuation is imaged. Bright regions show good transmission and dark regions show poor transmission caused by one or more elastic discontinuities within the material under examination.
- Dark Field Interferogram Mode—In the dark field interferogram mode, converted or scattered waves associated with a defect or discontinuity are imaged. The contribution of the transmitted waves to the image is suppressed thus producing an image analogous to the dark field image in optical microscopy. Combining this image with an electronically produced reference signal, a fringe pattern can be developed thus yielding an interferogram. This mode is very useful in detecting surface defects. For further details, see EDR 10383 (Ref 5).

An initial study was conducted at the Indianapolis Center for Advanced Research (ICFAR) to determine the ultrasonic attenuation of sintered alpha SiC, RBSiC, densified SiC (NC-430), sintered silicon nitride, and reaction-bonded silicon nitride. Alpha SiC and RBSiC had the least attenuation (14 dB/cm) of all of the five materials. These materials, therefore, were selected for further flaw detection studies. Attenuation in other materials ranged from 50 to 70 dB/cm. For further details, see EDR 9951 (Ref 3).

Two seeded alpha SiC disks (38 to 44 and 105 to 150 μm voids) were evaluated at ICFAR in the bright field through transmission mode. The examination of the disks revealed that acoustic microscopy can detect small bulk flaws. The limitation, however, is generally imposed by the scattering of surrounding microstructural variations. For more information, refer to Allison EDR 9951 (Ref 3).

Fracture of high-strength structural ceramic materials is often controlled by surface pores or cracks. Therefore, the majority of the flaw detection studies were conducted on such simulated or naturally occurring flaws.

The acoustic microscope was utilized in the dark-field interferogram mode. The following two types of specimens with defects were prepared:

- radiographic step penetrameters of alpha SiC and RBSiC with surface drilled holes
- MOR specimens of alpha SiC with tight surface cracks

Evaluation of Step Penetrameters

Two step penetrameters, one each of alpha SiC and RBSiC, were designed according to the general guidelines of ASTM procedure E142-77. The planned level of inspection was 2-1T, 2-2T, and 2-4T. The penetrameters were fabricated (cold pressed, sintered, and machined) by CBO. The density of the alpha SiC penetrameter was determined to be 3.11 g/cm^3 and that of the RBSiC penetrameter 2.99 g/cm^3. The latter corresponds to a free silicon content of 25% by volume in RBSiC. Holes were ultrasonically drilled by Bullen Ultrasonics. Figure 193 shows the penetrameter geometry and the location of each hole. The thickness and dimension of each step and hole were measured by a micrometer and an optical microscope at 200X, respectively. The holes were numbered 1 to 9 and Steps 1 to 8.

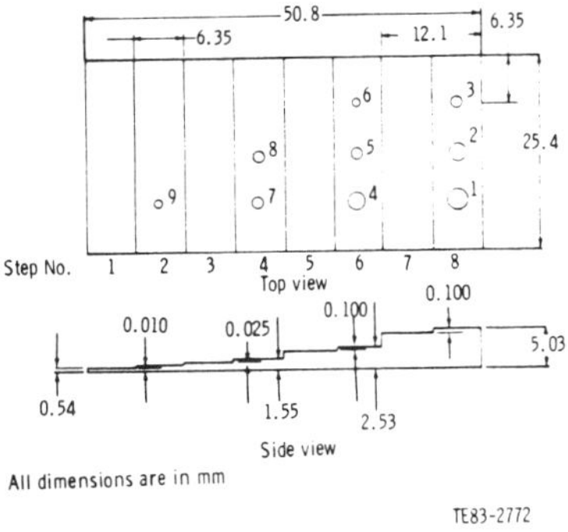

Figure 193. Layout of the radiographic penetrameter.

The results are summarized in Table XXVIII. All holes, regardless of size, were detected in both materials. Specifically, flaws in the 100-125 μm size range can be observed readily in SiC in thicknesses at least as great as 5.0 mm and smaller flaws (75 by 25 μm) can be detected in thicknesses up to 0.5 mm. Discernibility was affected primarily by surface condition. Microscopic examination of each penetrameter revealed varying degrees of surface pitting. The size of some of the pits was of the same order as the smaller drilled holes, and hence the acoustic scattering characteristics of the pits and of the drilled holes were quite similar.

Figure 194 shows an example of the general character of the acoustic images, both bright- and dark-field, in the amplitude mode, developed from the larger drilled holes. The images in both cases show regions of high acoustic transmission (white zone) near the center. For the bright-field mode, the high transmission region resulted because the sound was focused by the bottom of the fluid-filled hole. (The hole fills with water used on the stage of the microscope as a coupling medium.) For the dark-field image, the mode-converted sound is focused by the circular rim of the hole.

Figure 195 shows an example of the acoustic images generated from smaller holes (50-100 μm). In the bright-field mode, the hole appears again as a shadow with a central bright spot. In addition, a distinct ring pattern emanates from it. These rings result from interference between the incident wave and the wave scattered from the flaw. Such interference rings become apparent when the scattered wave is comparable in amplitude to that of the incident wave in the vicinity of the flaw. In the dark-field interference mode, the defect itself cannot be seen. However, a bright cometlike tail emanating from the flaw clearly reveals its presence.

From this study, scanning laser acoustic microscopy can be concluded to be capable of detecting certain fracture-controlling flaws, such as surface voids (holes), in a variety of dense silicon carbide materials. The size (diameter by depth) of the smallest holes present and detectable was 75 by 17 μm in RBSiC and 68 by 25 μm in alpha SiC. Surface conditions, such as pitting, limited the discernibility of the drilled holes. This limitation became important when the pit size and drilled-hole size were comparable. Although the smallest holes were detected in the thinnest step of both penetrameters, near-surface-flaw detectability by SLAM should be independent of specimen thickness. Similar flaws in thicker silicon carbide components should be detected because the bulk acoustic attenuation of both the materials is small. Supporting this point is the observation that small surface pits, the sizes of which were comparable to the smallest hole, were detected on all thicknesses of the penetrameter. Further, their image characteristics were independent of thickness.

Evaluation of Knoop Indented Specimens

An initial study was performed at ICFAR to establish the detectability of tight cracklike defects, such as Knoop indentations, on the polished surface of the MOR specimens of alpha SiC. This study

Table XXVIII.
Measured dimensions and acoustic microscopy (SLAM) of ceramic penetrameters.

	Hole No.								
	1	2	3	4	5	6	7	8	9
αSiC									
Diameter—μm	388	205	108	190	100	52	100	80	68
Depth—μm	95	115	115	55	55	42	35	38	25
Detection*	D,DB	D,DB	D,NDB	D,DB	D,DB	D,NDB	D,NDB	D,NDB	D,NDB
Reaction-bonded SiC									
Diameter—μm	500	210	117	202	105	70	104	102	75
Depth—μm	117	105	123	53	60	60	46	38	17
Detection*	D,DB	D,DB	D,DB	D,DB	D,NDB	D,NDB	D,DB	D,DB	D,DB
Step thickness—mm	5.03	5.03	5.03	2.53	2.53	2.53	1.53	1.53	0.54

*D = detected, DB = discernible with background signal, and NDB = not discernible from background signal and/or surface pits
Note: Both penetrameters were examined with the acoustic microscope at Sonoscan, Inc.

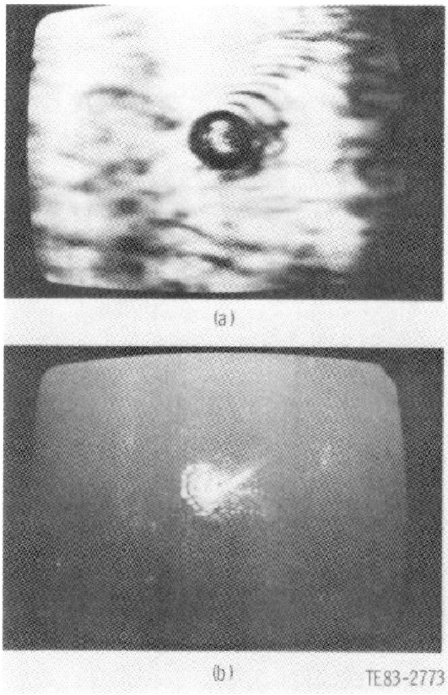

Figure 194. Acoustic micrographs of hole No. 1 in RBSiC penetrameter in (a) bright-field amplitude mode and (b) dark-field amplitude mode.

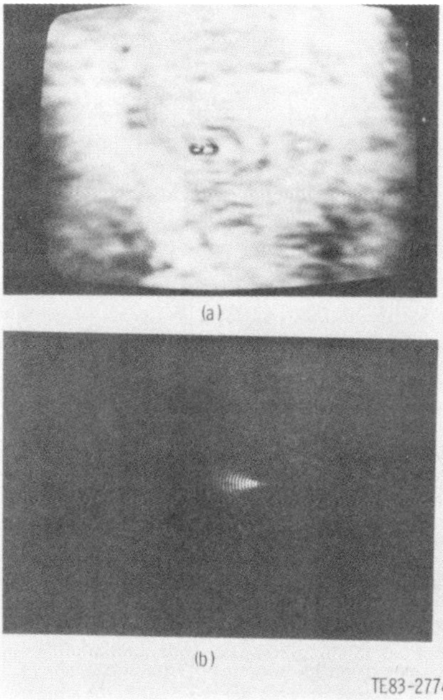

Figure 195. Comparison of acoustic photomicrograph of hole No. 7 in RBSiC penetrameter in (a) bright-field amplitude mode and (b) dark-field interferogram mode.

concluded that such cracks are readily detectable by SLAM in the dark-field mode. For more information, refer to EDR 10383 (Ref 5).

Because of the previously mentioned successful results, a systematic study was initiated at IC-FAR to further establish the ability of dark-field acoustic microscopy to detect surface cracks, to predict failure sites by observation, and/or imaging of the acoustic scattering behavior of the detected flaw and to correlate fracture location with predicted site. The following two types of MOR specimens of alpha SiC were prepared:

- Type A. A specimen surface was polished to a 6 μm surface finish. Twelve Knoop indentations, three each of 23.54-, 13.73-, 4.91-, and 0.98-N (2.4-, 1.4-, 0.5-, and 0.1-kg) loads, were placed within the inner span of 19 mm (0.75 in.) of the four-point MOR bar. These indentations, shown in Figure 196, were equally spaced to minimize any acoustic interference between scattered waves from adjacent flaws.
- Type B. Three specimens of this type per indentation load (23.54-, 13.73-, and 4.91-N [2.4-, 1.4-, and 0.5-kg]) were evaluated. All the specimens were polished to a 6 μm surface finish. Each specimen had three indents of the same load made within the constant stress inner span or gage length, as shown in Figure 197. Therefore, all three flaws (indents) in a specimen were subjected to the same stress and acted as competing flaws during fracture testing in a quarter-point loading condition.

Evaluation of the Type A Specimen

All 23.54- and 13.73-N (2.4- and 1.4-kg) indentations were easily detected and imaged, as shown in Figure 198, with excellent discernibility in the dark-

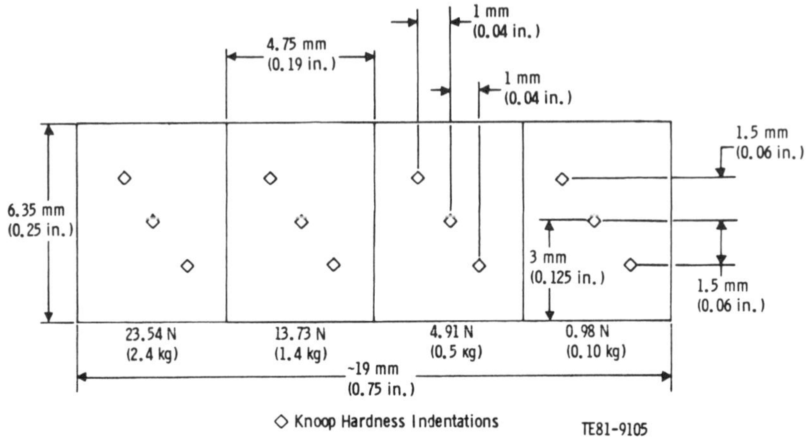

Figure 196. Layout of indentation in Type A specimen.

field interference mode. The 4.91-N (0.5-kg) indentations could be detected and visualized on the TV monitor but could not be imaged on a Polaroid film with sufficient contrast. The 0.98-N (0.1-kg) indentations could not be detected. The acoustic micrographs clearly show the variability in the images of the 23.54- and 13.73-N (2.4- and 1.4-kg) indentations. Scanning electron microscopy (SEM) analysis of the indented surface of 23.54-N (2.4-kg) specimens showed variable damage around each indentation. Similar observations were made for 13.73-N (1.4-kg) load indentations. These variabilities are related to the local microstructural properties that govern both the surface and subsurface indentation-induced damage. Figure 198 also shows that the intensity of the back-scattered energy is greatest in the center and least at the edge. The crack (indentation) appears to have similar acoustic properties to a plane wave oscillator.

Another observable phenomenon is the regularity of the wavefronts. A perfectly planar scattered crack produces regular coaxial wavefronts. When the crack is not planar, it has been observed that mixed-mode scattering occurs, giving rise to different interference patterns with a characteristic fringe spacing and wavelength. This disturbance in the regularity of the wavefronts may be useful for predicting the shape of the scattering crack.

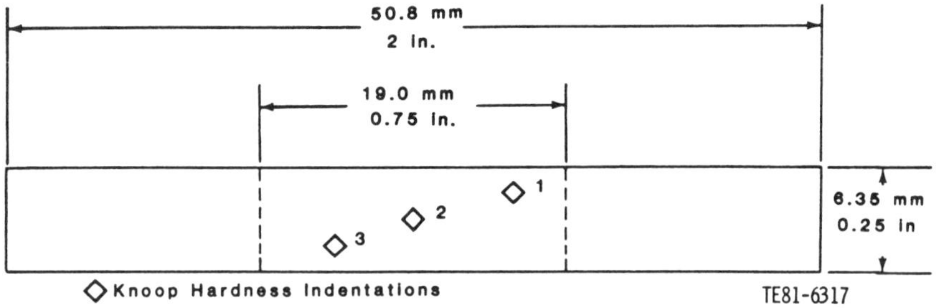

Figure 197. Layout of indentation in Type B specimen.

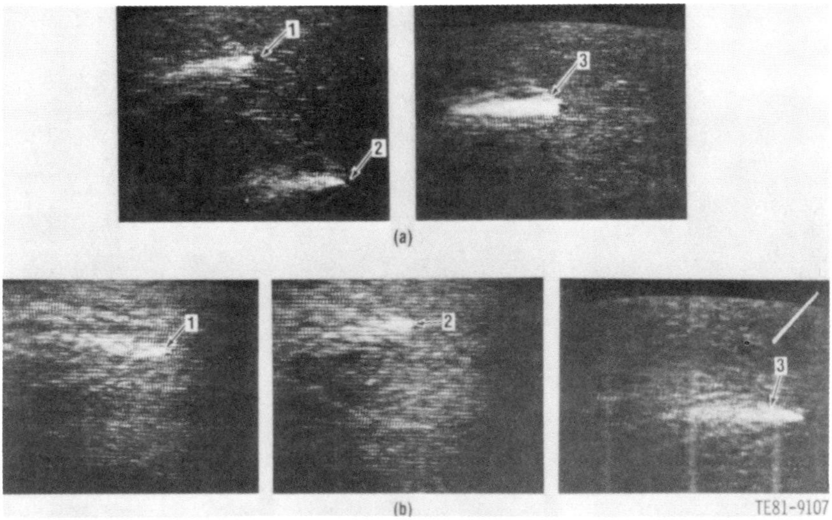

Figure 198. Dark-field acoustic interferograms of three Knoop indentations each for (a) 2.4-kg and (b) 1.4-kg loads in Type A specimen.

Evaluation of Type B Specimens

Figure 199 shows typical dark-field (back scattered) acoustic interferograms of 23.54-N (2.4-kg) indented specimens. The variation in the magnitude of the back-scattered energy from indentation to indentation both within a specimen and between specimens was evident for all indentation loads. Surprisingly, however, the fringe spacing or wavelength (λ) is almost constant, suggesting that the acoustic mode conversion due to the interaction between the incident acoustic beam and flaw remains similar. It has also been found that small rotations of the specimen have significant effects on the defect image. It appears, therefore, that the scattering characteristic of the two edges of the indentation may be different. Note that the acoustic beam is insonifying a different volume of the specimen on rotation, which may also affect the intensity of the acoustic beam interacting with the flaw due to variation in attenuation.

All three indentations in specimen 22.0383 were detected even after the indentor-induced surface damage was removed by diamond polishing. Both the surface and subsurface damage produced by the Knoop indentor are believed to contribute to the acoustic signal and image. Similar kinds of variability in detection of the 13.73-N (1.4-kg) indentation in the dark-field interferogram mode were examined. The 4.91-N (0.5-kg) indented specimens were also evaluated, and indentations along with other naturally occurring flaws were detected. However, no micrographs were possible because of poor photographic quality.

Failure locations and critical flaws for each specimen with three indentations within the inner span were observed by SLAM. The prediction was based on the intensity of the back-scattered energy, its angle of divergence, and other observable features on the TV video monitor from each flow. Specimens were subsequently tested in four-point bending. Table XXIX shows both the predicted and actual failure origins for all nine specimens. Seven specimens failed at predicted locations. Predictions for 23.54- and 13.73-N (2.4- and 1.4-kg) indented specimens were 100% accurate. No prediction was possible in specimen 22.0349 because the acoustic scattering characteristic of all three indentations was similar to that from the large number of natu-

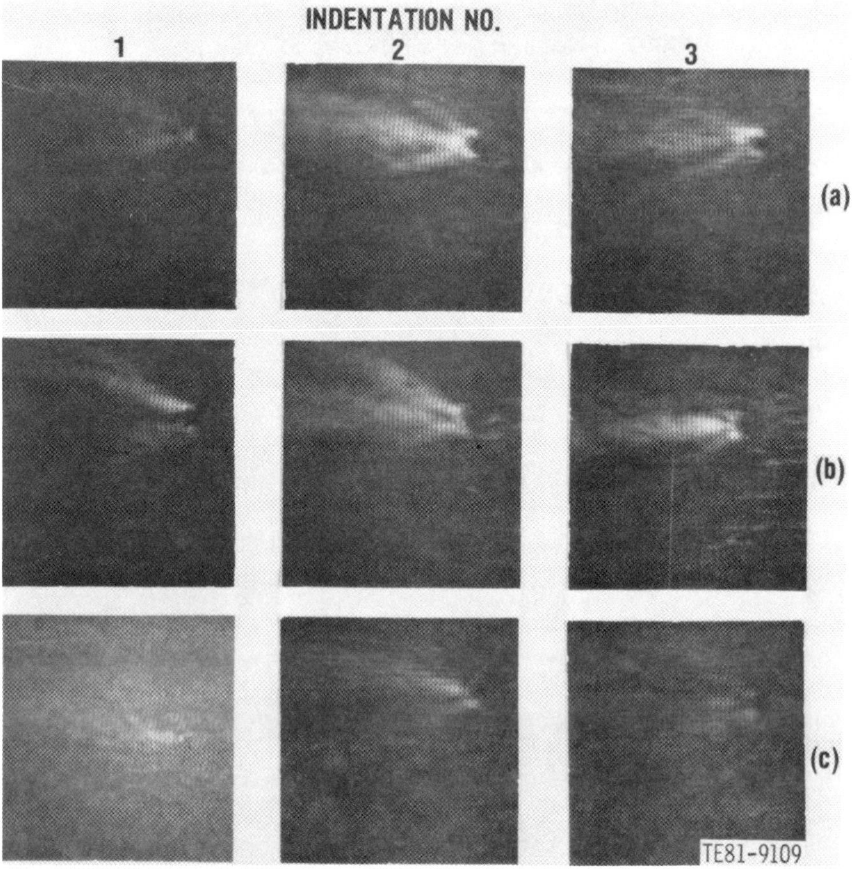

Figure 199. Dark-field interferograms of three 23.54-N (2.4-kg) load Knoop indentations in specimen 22.0383: (a) without removing surface-induced indentation damage; (b) same surface with specimen rotated 180 deg; and (c) specimen oriented as in (a) after removing surface damage.

rally occurring surface flaws, resulting in no discernibility against the background structure. Specimen 22.0353 was predicted to fail at a natural flaw (SEM and optical microscopy revealed multiple flaws in the vicinity from which the acoustic microscopy signal was received) but instead failed at indentation 3. Since the size of the indentation, 4.91-N (0.5-kg), in this specimen is similar to a naturally occurring flaw, the MOR test is realistic and has practical significance. The accurate prediction that specimen 22.0391 would fail at indentation 2 of the same load is remarkable. The 90% accuracy (seven out of eight) for predicting failure origin by acoustic microscopy is excellent.

Table XXIX.
Correlation of acoustic microscopy and MOR test.

Specimen	Knoop indentation load N	(kg)	Indentation No. at fracture Predicted by SLAM	Actual in four-point MOR test
22.0369	23.54	(2.4)	2	2
22.0379	23.54	(2.4)	2	2
22.0383*	23.54	(2.4)	2	2
22.0343*	13.73	(1.4)	2	2
22.0385	13.73	(1.4)	1	1
22.0399	13.73	(1.4)	2	2
22.0349	4.91	(0.5)	NDB**	3
22.0353	4.91	(0.5)	NF**	3
22.0391	4.91	(0.5)	2	2

* These specimens were tested in four-point bending after removal of indentation damage.

** NDB = no prediction possible, no discernibility against background structure; NF = failure predicted at a natural flaw

DISCUSSION: SCANNING PHOTOACOUSTIC MICROSCOPY (SPAM)

Laser-scanned photoacoustic microscopy shows excellent potential for detecting surface and near-surface flaws in opaque ceramics. In this technique the material to be examined is placed in a closed cell and scanned with a modulated laser light. The periodic optical excitation of the sample results in a periodic heating of the specimen and subsequent heat flow into the gas layer immediately adjacent to the specimen surface. This creates pressure variation in the gas at the modulation frequency, which can be detected by a microphone as an acoustic signal. The photoacoustic signal depends on the optical and thermal properties of the material and, in addition, on the details of local microstructure and topography.

To establish the feasibility of SPAM to detect surface and near-surface flaws, an experimental study was initiated at Gilford Instruments. This study was conducted on simulated surface flaws (Knoop indentations). Results indicated that this kind of tight surface defect could be readily detected in various silicon-based ceramic materials (see EDR 9951 [Ref 3]). Because of this initial success and in view of the extensive data base needed to reliably establish the detection sensitivity of this technique to a wider range of flaw types, it was decided to establish an in-house capability to further study and implement the SPAM technique for component evaluation.

Experimental Arrangement

The SPAM arrangement at Allison utilizes a 2.0 W argon laser, as shown in Figure 200. The laser light is modulated by an electromechanical chopper and guided by two mirrors and a beam steerer toward an objective lens, which focuses the light onto the specimen surface. The diameter of the focal spot is less than 30 μm. The specimen is positioned in a SPAM cell, which is mounted on a two-dimensional (2-D) servo-controlled translational stage. The specimen/cell assembly, described in the following subsection, is translated in the x direction and incremented in the y direction. The entire system is situated on a honeycomb suspension table, which is mounted on four vibration isolation legs. A programmable microprocessor-based controller synchronizes all the motions (direction, speed, increment, etc) of the 2-D stage and the x-y recorder. Location of the stage is continuously displayed as front panel digital readout as well as analog voltage for the x-y recorder. The photoacoustic signal is detected by a miniature microphone (Model BT 1759) and lock-in-amplified. The lock-in-amplifier is capable of detecting both the phase and the amplitude of the SPAM signal. Scanning speed can be varied from 2.0 μm/s to about 30 mm/s. Finally, minimum increment is 2.0 μm for each axis.

SPAM Cell Design

A modular photoacoustic cell was designed and fabricated at Allison for the characterization of

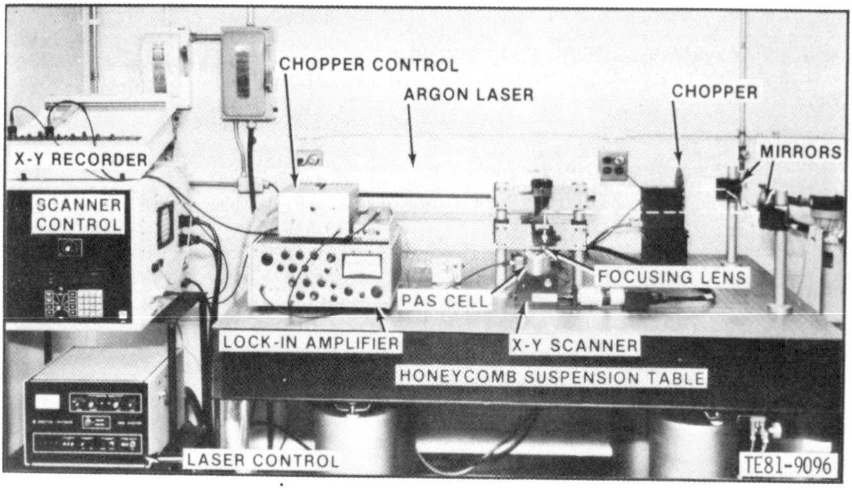

Figure 200. Experimental setup at Allison for flaw detection in ceramics by SPAM.

standard MOR specimens, as shown in Figure 201. The cell has three parts: (1) an aluminum outer shell, (2) a lucite insert with a cavity within which to place the MOR specimen, and (3) an aluminum lid with a 25.4 mm (1 in.) dia window of optically polished fused silica. The lid is attached to the outer shell with three allen screws. An O-ring is used in the top lid to make an airtight seal. This seal prevents the loss of the photoacoustic signal. The air gap between the surface of the specimen and the windown is 0.5 mm. The total volume of the air within the cell is approximately 0.2 cm^3. The microphone is located in the insert.

SPAM Parameter Optimization

To evaluate the sensitivity of the SPAM system, an MOR test specimen of sintered alpha SiC, measuring 50.8 mm x 6.35 mm x 3.18 mm (2 in. x 0.25 in. x 0.125 in.), was polished to a surface finish of 0.12 μm. Two 2.5-kg Knoop microhardness indentations were then placed on the polished surface of the test bar. The indenter-induced damage was not removed.

A parametric study was conducted to obtain optimum scanning conditions. Figure 202 shows a scan of the area containing two hardness indentations. Test parameters used for this mapping were as follows:

- laser power level = 600 mW
- modulation frequency = 200 Hz
- scanning speed = 250 μm/s
- focal spot size = less than 30 μm

The specimen was incremented 25 μm after each scan. The accompanying photographs show the microstructural features corresponding to the observed SPAM signal peaks. Note that features in addition to the indentations are readily detected. For example, at location 2 there is a single peak present only in a single pass. This peak arises from a small pore where the major surface dimension is 30 μm and where depth is estimated to be about 6 μm. At location 1 a pair of peaks originates from a pair of pores 20 μm in diameter and 6 μm in depth. These two pores are separated by 25 μm. Finally, the pair of relatively large Knoop indentations with a major axis of 148 μm was detected in five successive scans.

The reproducibility of detecting all of the flaws both in terms of their SPAM signal and location was excellent. The observations made here confirm that the focal spot size of the laser beam is indeed less than 30 μm. This initial study also shows that both the detection sensitivity and spatial resolution of SPAM for polished alpha SiC is 20 μm for naturally occurring surface pores at 250 μm/s scanning speed. It should also be noted that the Knoop indentations were detected over a length of 125 μm, which

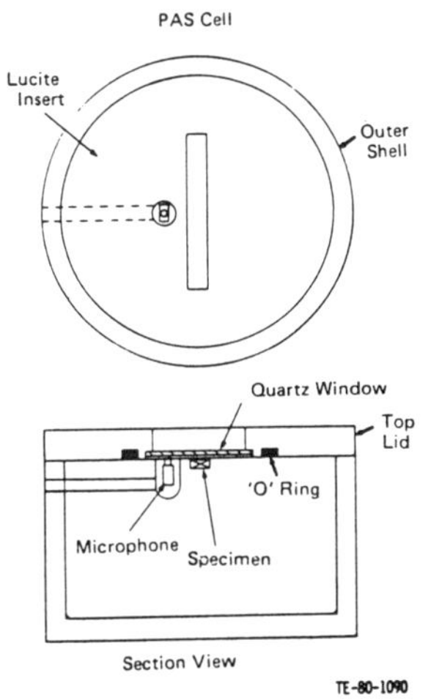

Figure 201. Schematic of photoacoustical cell for MOR specimen.

is within one focal spot diameter of the laser beam when compared with the measured length of 148 μm. There is therefore an excellent quantitative correlation between the major surface dimension of the flaw and distance up to which it is detected by SPAM.

Detection of Surface Cracks

Allison has conducted extensive studies on surface cracks in silicon base structural ceramics under the CATE program. Figure 203 shows an example of a surface crack induced by a naturally occurring process in an injection-molded alpha SiC 1132°C (2070°F) blade qualification MOR bar. The as-fired specimen was examined by SPAM. The length of the crack was measured as 1.0 mm (0.039 in.) by optical microscopy and 1.3 mm (0.051 in.), by SPAM. This shows that the subsurface nature of the flaw was also detected by SPAM. Figure 203 also shows excellent agreement between the actual orientation of the crack and that detected by SPAM.

Figure 204 shows that three different cracks were detected in a reaction sintered silicon nitride specimen. The tightest crack, C_3, was detected by SPAM with good signal-to-noise ratio. Photoacoustic map also revealed the concave shape of the crack. Figure 204 shows the effect of varying the modulation frequency on the SPAM signal on scanning across the cracks C_2 and C_3. Line broadening was observed for crack C_2, indicating that the crack was tilted below the subsurface—an observation that was confirmed by optical microscopy (see Figure 204a). In the case of crack C_3, it appears that the tilt angle with the vertical is small and/or that the variable subsurface crack closure gives two discrete SPAM signals at 40 Hz.

Both the examples given above show that the SPAM technique developed at Allison is very sensitive and can detect and reveal depth profiles for tight surface and subsurface cracks.

Detection of Subsurface Flaws

Subsurface flaw detection sensitivity of SPAM was studied in a wide variety of metallic and ceramic materials. Modulus-of-rupture specimens with simulated holes and slots at various depths were prepared. The following conclusions were reached from this study:

1. SPAM is capable of detecting subsurface flaws in aluminum alloys, carbon steel, sintered alpha SiC, RBSiC, and sintered silicon nitride.
2. Both flaw characteristics (size, shape, and depth) and experimental parameters (e.g., input thermal flux, etc) have a significant effect on the photoacoustic detection of subsurface flaws.
3. Both thermal and microstructural properties (grain size and distribution; phase content and distribution) affect the depth of flaw detection.

Table XXX summarizes the depth of flaw detection in three structural ceramic materials. The defects were detected in the phase and magnitude mode in silicon carbide materials to a depth of about 1 mm (0.039 in.). In the SNW-1000 material, similar flaws were detected in phase mode to about 0.93 mm (0.037 in.) depth and 0.48 mm (0.019 in.)

Table XXX.
Subsurface flaw* detection in structural ceramics.

	Depth of flaw detection, d—μm	
Material	**Phase mode**	**Magnitude mode**
Alpha-SiC	$1080 < d < 1550$	$1080 < d < 1550$
RBSiC	$965 < d < 1550$	$965 < d < 1550$
SNW-1000	$930 < d < 985$	$480 < d < 765$

*Side-drilled holes of 0.5 mm (0.020 in.) diameter

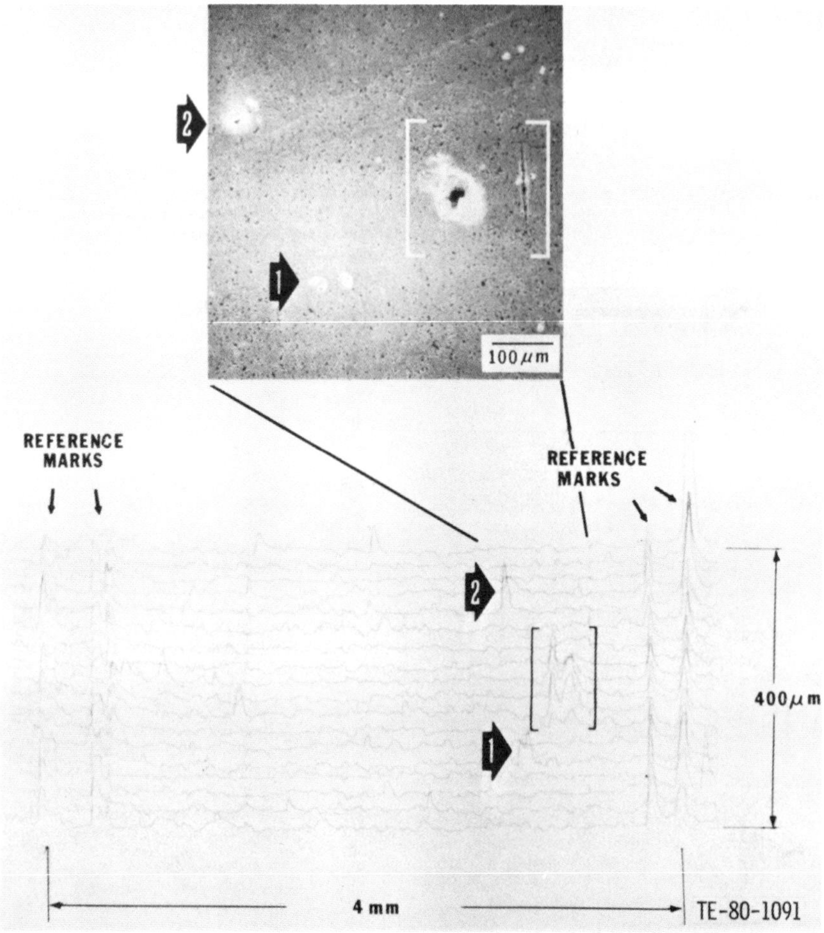

Figure 202. Photoacoustic signals from microhardness Knoop indentations and natural surface pores in alpha SiC.

depth when the magnitude of the SPAM signal is monitored. For details, see EDR 10672 and 10841 (Ref 6 and 7).

Studies to determine the dependence of differential SPAM signal amplitude on modulation frequency were conducted. Specimens with subsurface side-drilled holes in alpha SiC, sintered Si_3N_4, and aluminum were studied. Allison's experimental results (see Figure 205) are in excellent agreement with the recent numerical calculation of

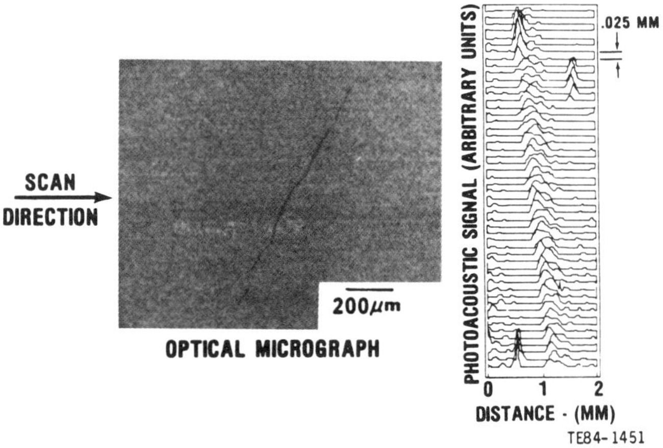

Figure 203. Photoacoustic detection of surface crack in sintered alpha SiC.

Thomas and coworkers (at Wayne State University) for a cylinder beneath the surface (simulating side-drilled holes) both in the thermally thin (μ_s greater than d) and thermally thick (d greater than μ_s) regions. Here d is the depth of flaw from the top surface and μ_s is the thermal diffusion length. It was also concluded from this study that Allison's gas-cell design/fabrication and SPAM procedures were consistent with the SPAM theory.

Characterization of Ceramic Turbine Blades

After successfully detecting surface and subsurface flaws in flat specimens, the Allison SPAM system was further developed and applied to the inspection of actual ceramic engine components (CATE 2070°F-configuration SiC blades). Past spin test experience has shown that blade failure most often initiates either from the as-fired surface of the stalk or the machined surface of the attachment. These two areas correspond to the most highly stressed regions within the blade. A photoacoustic cell (see Figure 206) was designed and fabricated to examine these two critical regions of the blade.

The sensitivity of the cell was evaluated using a net shape silicon carbide blade with one Knoop indentation (2.0 kg) located on the critical region of the attachment. The experimental parameters were as follows: laser power 1.0 watt, scan speed 250 μm/s, time constant 100 ms, and increment 25 μm/scan line. The blades were scanned radially and incremented in the axial direction. Figure 207 shows that the 2.0-kg indentation was detected by SPAM, indicating that the cell design was sufficiently sensitive to inspect ceramic blades of the present design.

The stalk and attachment areas of 12 ceramic blades were evaluated in great detail by SPAM. The experimental parameters were similar to those employed during examination of the Knoop indented trial blade. The amplitude of the SPAM signal was threshold-detected just above the background microstructure signal to facilitate photoacoustic mapping of any flaws.

After the SPAM evaluation, each blade was spun to failure in a cold spin test. The failure speed was recorded and posttest fractographic analysis was conducted on each blade to establish the source and location of fracture initiation.

Table XXXI summarizes the results of the photoacoustic evaluation and spin testing. Fractographic analysis revealed that the failure origins of five blades could be correlated with flaws detected by SPAM. The failure origins of the remaining seven blades were lost, due to the destructive nature of the test. No SPAM correlation was, therefore, possible.

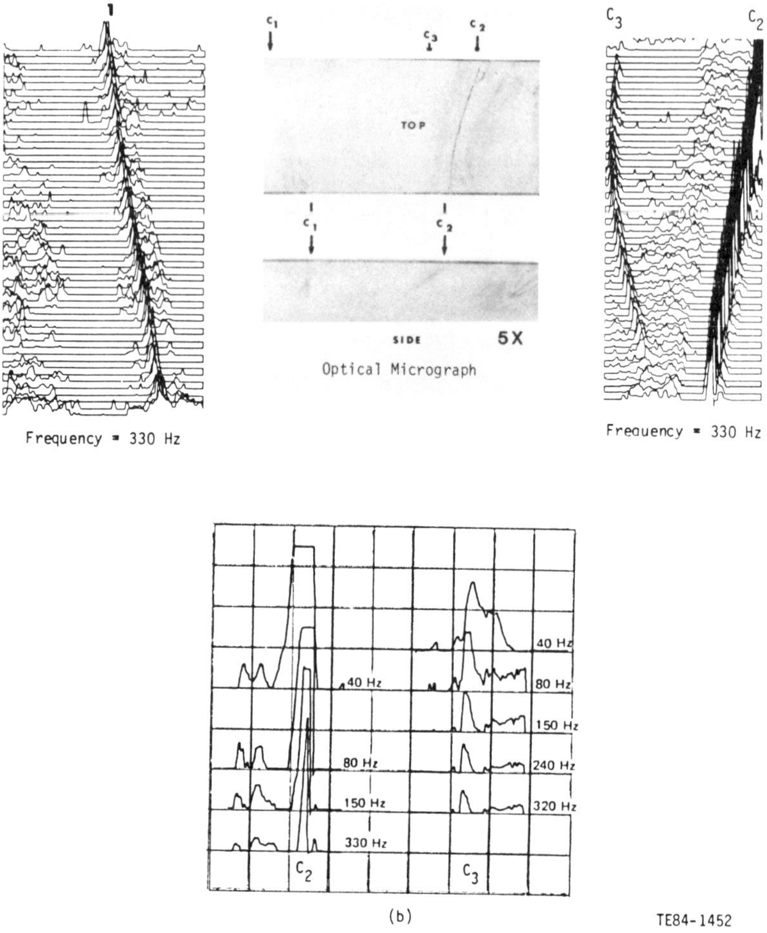

Figure 204. Photoacoustic detection of surface cracks in reaction sintered silicon nitride.

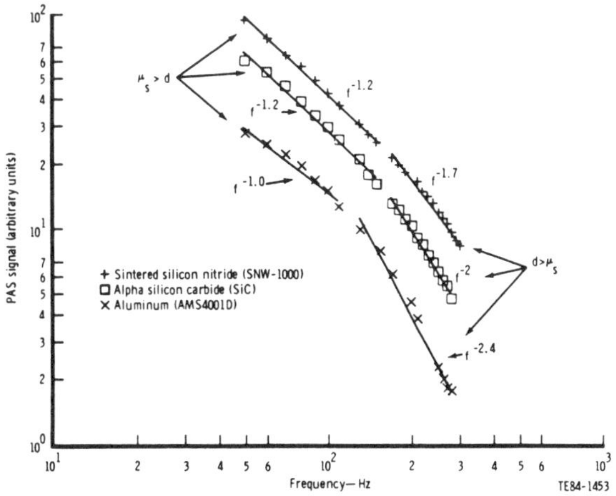

Figure 205. Photoacoustic signal as a function of modulation frequency.

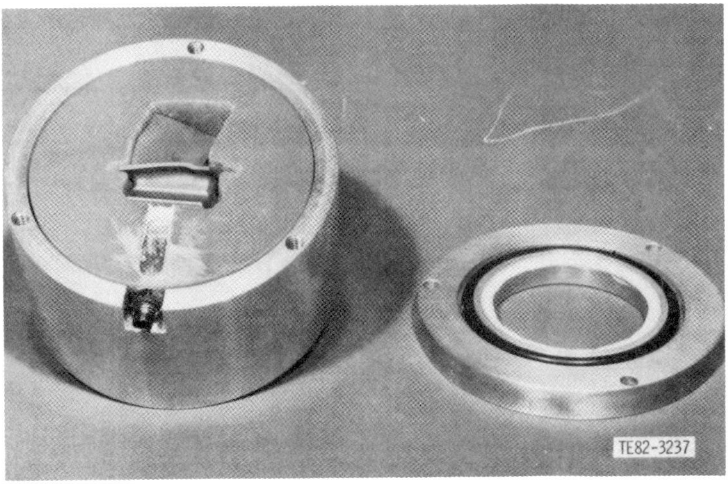

Figure 206. Photoacoustic cell for the evaluation of CATE 2070°F-configuration SiC turbine blades.

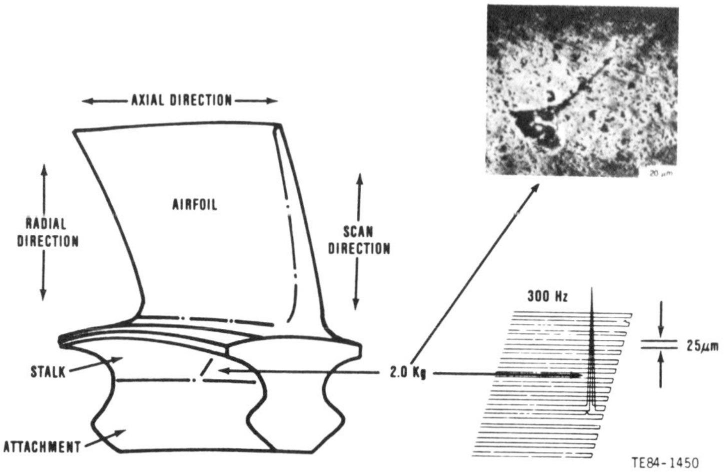

Figure 207. SPAM detection of 2.0-kg Knoop indentation in SiC blade.

Figure 208 shows the SPAM map of one of the failed blades, No. 456, for which a definite fracture origin within the scanning field could be identified. Thirty-three major indications were charted. In addition, a large number of other flaws with lower photoacoustic amplitudes were detected. This blade failed from a surface connected void approximately 35 μm in diameter.

Figure 209 shows the SPAM map of blade No. 1106. This blade was quite clean. Only three indica-

Table XXXI.
Correlation of photoacoustic microscopy and spin testing.

Blade No.	Failure speed—rpm*	Flaw size—μm a′	2C′	Failure origin	PAM results
317	51,000	—	—	Flaw location possible but no sizing	Detected
352	49,100	18	28	Surface pore, attachment	Detected
456	34,750	40	30	Surface pore, attachment	Detected
491	49,150	20	66	Surface pore, stalk	Detected
1106	50,950	50	75	Internal pore, attachment	Detected
319	48,000	—	—	Origin lost, attachment	Not possible**
320	49,400	—	—	Origin lost, attachment	Not possible**
965	54,200	—	—	Origin lost, attachment	Not possible**
318	55,500	—	—	Origin lost, stalk	Not possible**
335	52,800	—	—	Origin lost, attachment	Not possible**
1105	54,200	—	—	Origin lost, attachment	Not possible**
968	53,350	—	—	Origin lost, no location possible	Not possible**

* 100% speed is equal to 36,905 rpm
** No correlation possible because failure origin was lost
Note: a′ is flaw depth; 2C′ is flaw width

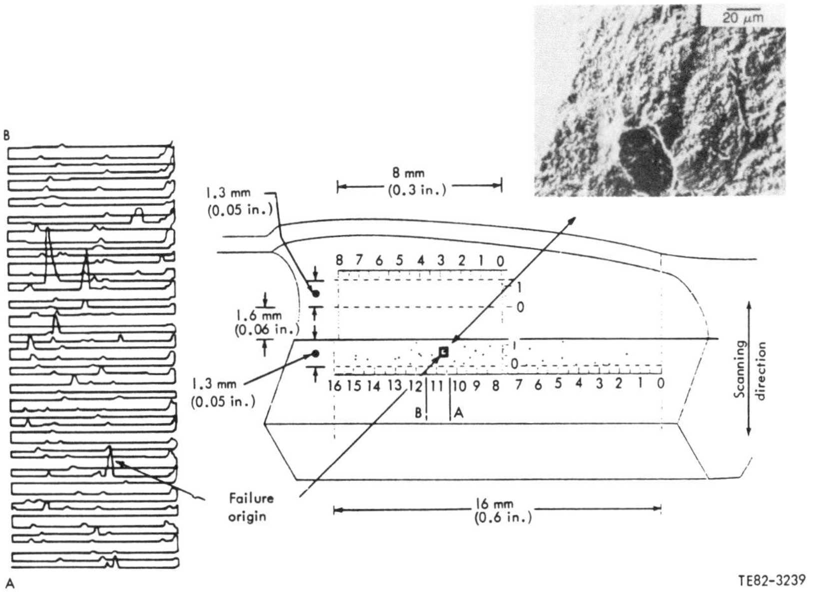

Figure 208. Photoacoustic evaluation of SiC blade S/N 456; details of fracture-controlling flaw shown are (a) location on the attachment, (b) SEM micrograph, and (c) SPAM signal amplitude from flaw.

tions were noted, one in the stalk and two in the attachment. The failure-controlling flaw, located in the attachment, was a subsurface void similar to that found in blade S/N 456. This particular flaw was 75 μm x 50 μm in size and was located 30 μm below the surface. This type of artifact is typical of the strength-controlling flaws in injection-molded SiC. In both of these examples, the critical flaw was detected by SPAM.

The local stress state and flaw strength characteristics will, of course, determine whether failure will initiate from a specific flaw at a given spin speed. Furthermore, both flaw strength and SPAM signal amplitude will be controlled by many of the same parameters (e.g., flaw size, type, orientation, and location). Predicting component load carrying performance from amplitude alone, however, presumes that flaw strength and amplitude are proportional. That this is not the case is confirmed by the limited experimentation described here. Consider, for example, blade No. 456. Even after taking into account the variation in the stress field within the attachment region, failure could not be ascribed to the flaw with the greatest detect amplitude. Thus any notion that components can be characterized and sorted by consideration of amplitude alone is invalid. This result, while not totally unexpected, serves to emphasize that a significant development effort remains before the SPAM technique can be used to predict a priori structural performance of ceramic items.

The following two important conclusions were drawn from this study:

1. SPAM is capable of detecting fracture-controlling flaws in the CATE 2070°F-configuration ceramic blade. This shows that this technique can be successfully applied to evaluate complex shapes.
2. The highest SPAM signal amplitude was not necessarily from a failure-causing flaw. Flaw

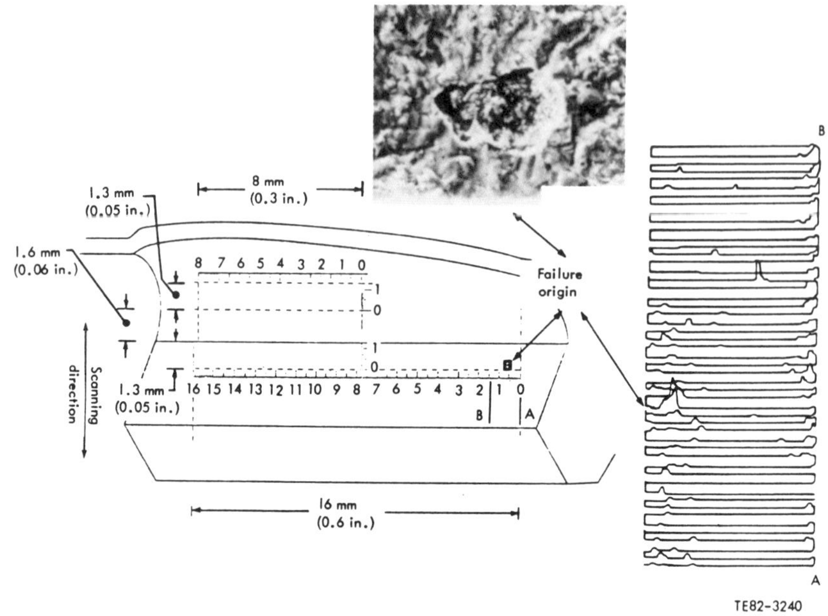

Figure 209. Photoacoustic evaluation of SiC blade S/N 1106; details of fracture-controlling flaw shown are (a) location on the attachment, (b) SEM micrograph, and (c) SPAM signal amplitude from flaw.

characteristics (type, size, shape, and orientation) have a significant effect on the amplitude of the SPAM signal. Substantial development work has to be done to determine the empirical relationship, if any, between the flaw characteristics and the SPAM signal.

DISCUSSION: ULTRASONIC VELOCITY

The ultrasonic properties of solids, including velocity and attenuation, are dependent on microstructure and, as such, are sensitive functions of chemistry, phase content, grain size, and porosity. Characterization of these properties can, therefore, be used as an effective means of monitoring material variability. In particular, ultrasonic velocity can be easily measured with great accuracy. The ultrasonic velocity of waves through a material is related to the density (ρ) and elastic modulus (E) by the following equation:

$$V = \frac{E}{\rho} = \frac{2\eta L}{t}$$

where

- V = wave velocity
- L = thickness of specimen
- t = wave transit time between two selected back surface echoes
- η = 1 when transit time between first and second back surface echoes
- ρ = 2 when transit time between first and third back surface echoes
- E = elastic modulus

By measuring the specimen thickness and transit time of the acoustic wave through the medium, the preceding relationship provides a means of determining the ultrasonic velocity of material.

Figure 210 shows schematically the experimental system assembled at Allison for measuring the ultrasonic velocity of the material. It consists of a Tektronix WP1310 signal processing system (SPS), a Panametric 5052PRX-75 pulser/receiver, and flat delay line (fused silica) transducers—30 MHz for the longitudinal (L) wave and 20.0 MHz for the shear (S) wave for flat specimens. A 20.0 MHz miniature L-wave transducer was used to measure velocity in ceramic blades. The thickness of the specimens was measured using a micrometer.

The transit time (t) between the echoes was measured under software (algorithms were developed at Allison) control from the keyboard. The optimum reflected ultrasonic signals were digitized to the maximum resolution (1024 bits) available with the scope. The oscilloscope sweep rate was adjusted so that only the echoes of interest were visible on the cathode ray tube (CRT) screen. The oscilloscope was programmed to (1) average the signals 20 times; (2) measure the travel time between the echoes of interest by program controlled cursors; (3) store the time in constant memory registers; and (4) repeat steps 1 through 3 ten times to minimize the random variations in the digitization of the signals. The time of flight (t) for a specimen was the arithmetic average of the previous 10 measurements. The ultrasonic velocity (V) for each specimen was calculated using the measured thickness (L) and wave travel time (t) in the previously mentioned equation.

An initial study was conducted to measure ultrasonic velocity of SiC material. Three specimens, measuring approximately 5.0 mm x 3.0 mm (0.197 in. x 0.118 in.), were machined from the attachment of each of two alpha SiC blades, for a total of six test bars. Only the longitudinal and shear wave travel time between minimum amplitude points (negative half cycle) of B1 and B3 were used in the present calculations.

Figures 211 and 212 show both the longitudinal and shear wave velocities plotted as a function of specimen density. The L-wave velocity was observed to increase with increasing density, with the exception of the specimen No. 645-I. The velocity of this specimen, measured on both this apparatus and on NASA-Lewis Research Center's intervalometer, was repeatedly lower than that of other specimens of

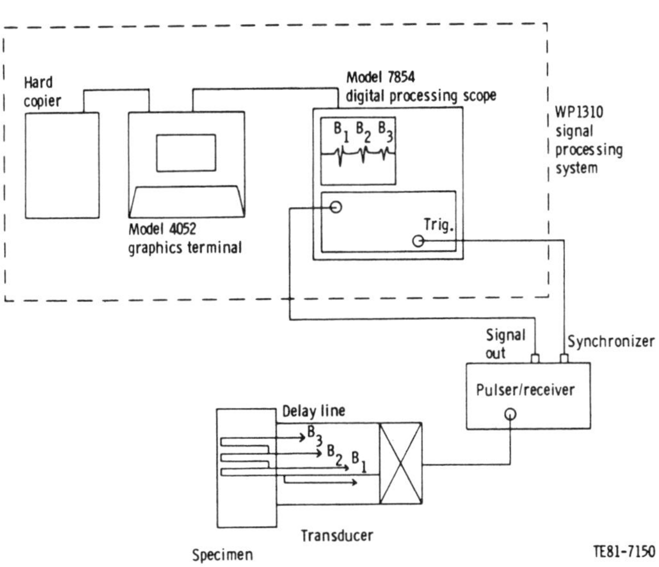

Figure 210. Schematic of ultrasonic measurement apparatus.

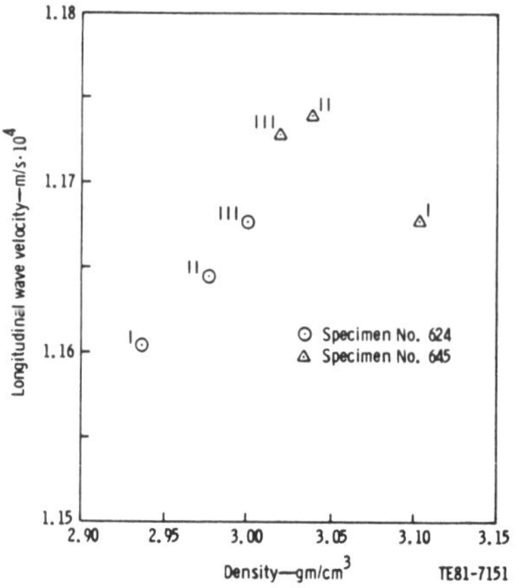

Figure 211. Longitudinal velocity of alpha SiC as a function of density.

lower density. The exact reason for this discrepancy is unknown at this time. The S-wave velocity was also observed to increase with increasing density, except for specimen No. 645-III. Note that the specimens were small compared with the transducer crystal size; therefore, geometrical effects could induce some variations in the transit time measure-

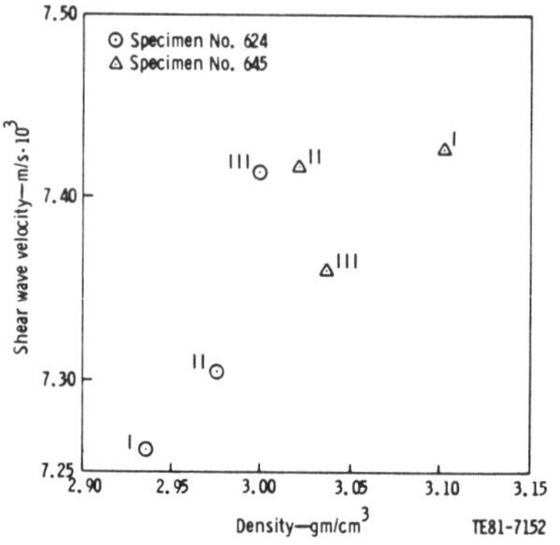

Figure 212. Shear wave velocity of alpha SiC as a function of density.

ments. Also, no diffraction correction was made. It is possible that some unique surface or bulk microstructural feature was also responsible for the previously mentioned deviations in the two specimens.

After successfully developing the methodology to measure velocity in flat specimens, it was applied to 2070°F-configuration ceramic blades. Figure 213 shows the fixture and a 20.0 MHz delay line (polystyrene) transducer capable of measuring the velocity through the stalk and attachment of the blade. The fixture had 4 deg of freedom to optimize the reflected signal from the back surface echoes. This fixture/transducer assembly was used to measure the velocity of blades under the blade matrix program.

Measurement of ultrasonic velocity was included in the matrix program NDE effort both because of its proven ability to detect small density variations and to further characterize each blade recipe. Fifteen blades from each of the 22 recipes underwent velocity measurement through the attachment area in the as-fired state. Additional machining stock had been added to the attachment design to provide a set of parallel surfaces specifically for velocity measurement. Transit time was

Figure 213. Experimental setup for ultrasonic velocity measurement in blade stalk.

measured using the first (B1) and second (B2) back surface reflections. Figure 214 shows the acoustic wave path in the attachment.

A linear correlation was obtained between blade density and longitudinal velocity, as shown in Figure 215. Each point on the graph represents the average density and velocity for each of the 22 recipes. The velocity of a blade reflects the local microstructure whereas its density reflects the entire blade. Therefore, the density used in Figure 215 (average of 15 blades) may not necessarily be the density at the location of velocity measurement and probably is a major contributor to the noticeable scatter.

Velocity of MOR bars supplied with each blade by CBO was also measured. Ten readings per test bar were taken, as before, using the central portion of the bar which is the maximum stress region during the four-point bend test. Excellent correlation was found between the average density and velocity for each recipe, as shown in Figure 216. Each test bar exhibited a markedly narrow range of density. SEM examination also revealed a homogeneous microstructure. The improved correlation exhibited by the bars is undoubtedly a reflection of their material consistency and homogeneity.

The experience gained in measuring ultrasonic velocity on blades and test bars during the matrix program has shown that velocity does indeed provide a valuable tool for detecting density variations and characterizing materials. Velocity is a material property that can be quickly and simply measured, nondestructively, on a wide variety of materials and components. Velocity measurement can act both as a research tool and as a quality control test method.

Turbine Tip Shroud Abradability

SUMMARY

This task is directed toward the development of ceramic abradable seal materials systems compatible with ceramic turbine blading. Interim development of abradable systems compatible with metal turbine blading was also addressed.

Results of the abradable seal/metal blading studies produced a number of seal materials believed to warrant further effort. Notable among these are several silicon carbide and silicon nitride compounds as well as a number of laminated ceramic fabric systems. However, inherent limitations of metallic materials at temperatures approaching 1093°C (2000°F) have shown the inability of unassisted metal blade tips to survive a shroud rub without significant damage. Short-term survivability of metal turbine systems was demonstrated with plasma-sprayed zirconia/eccosphere shroud components; however, these components did not provide extended erosion performance.

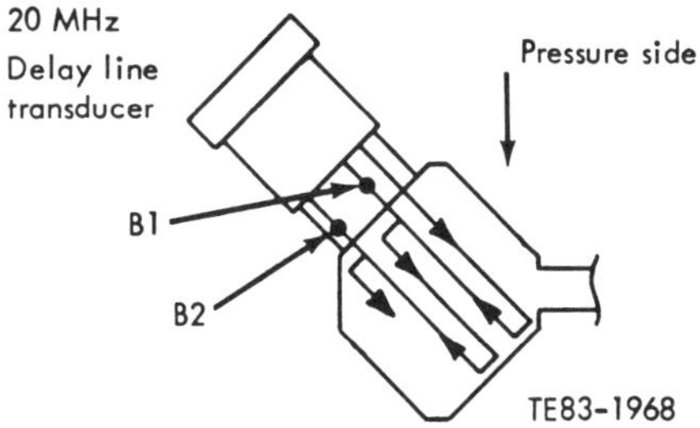

Figure 214. Acoustic wave propagation in matrix blade attachment.

Recent developments in rub energetics studies have shown that a coated blade tip may provide the foundation on which satisfactory blade-tip/abradable surface interactions may be developed. Using this approach, distress at the tip/shroud interface may be avoided for both metal- or ceramic-bladed systems. The feasibility of this concept has been demonstrated during the development of the ceramic bladed/ceramic shroud system. Granular, silicon carbide chemical vapor deposited (CVD) coatings were placed on tips of prototype silicon carbide (SiC) engine blades. The coated blade tips

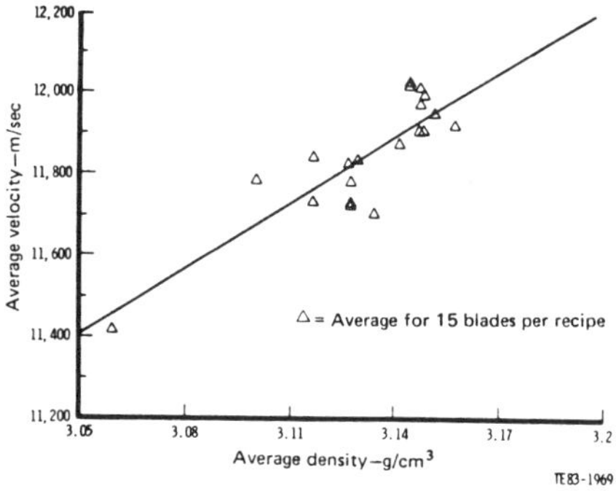

Figure 215. Longitudinal velocity of blade matrix alpha silicon carbide blades as a function of density.

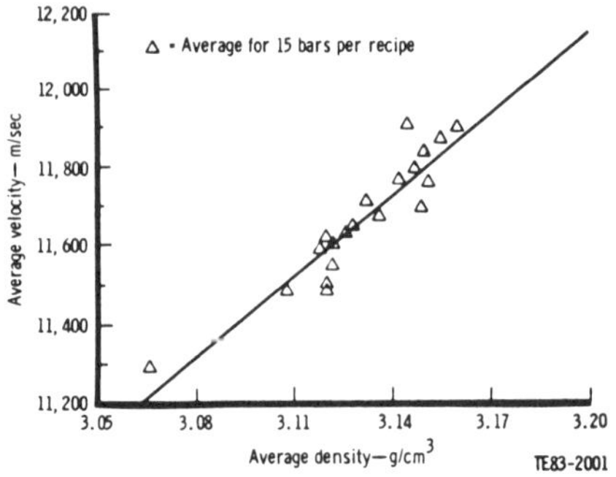

Figure 216. Longitudinal velocity of blade qualification bars (longitudinally ground) as a function of density.

significantly improved the abradability performance of the tip/shroud pair when rubbed against selected plasma-sprayed zirconia/eccosphere systems in high speed rig tests and should definitely be considered as the basis for future development of ceramic-bladed/ceramic-shrouded turbine components.

OBJECTIVE

The purpose of this effort was to develop materials and processing instructions for ceramic-based abradable seals compatible with (1) metallic bladed turbines operating at temperatures of 1038°C (1900°F) and (2) ceramic-bladed turbines operating at temperatures of 1132°C (2070°F) and 1240°C (2265°F). Contractual changes in the workscope subsequently deleted the requirement for 1240°C (2265°F) engine operation. In addition to acceptable abradability performance, these seal materials must also demonstrate a high degree of resistance to hot gas and particulate erosion. Resistance to thermal shock, fatigue, and oxidation is also essential for ultimate acceptance.

APPROACH

A dual-path approach was taken to increase the probability of success in reaching an acceptable solution for using ceramic-structured materials in sealing applications. Parallel but technically different approaches were employed by (1) commercial sources of ceramic engine components and (2) Allison investigators. The commercial suppliers of ceramic engine components applied their materials expertise to chemically similar, but structurally modified, ceramic engine materials that were compatible with abradable materials objectives. Concurrently, materials studies at Allison produced compatible abradable structures for both metallic- and ceramic-blade turbine components.

These materials investigations were extensions of Allison's basic work in the areas of abradable seals, thermal barriers, and composite ceramic materials structures. Techniques such as plasma spraying and the formulation of structures employing ceramic cements and ceramic fabrics were investigated.

All candidate materials were evaluated in a series of laboratory test rigs prior to commitment to engine testing. These tests included low-speed, room-temperature abradability screening, high-speed/high-temperature abradability tests, aerodynamic particulate erosion, and oxidation. Both the low-speed and high-speed abradability test rigs are capable of accepting engine-configured silicon carbide blading as well as Mar-M246 stylized blade elements. Descriptions of the low-speed and high-speed abradability and erosion test rigs are provided in EDR 9951 (Ref 3).

DISCUSSION

Gas turbine engines are highly dependent on close operating clearances between rotating and stationary components to extract the maximum work from the operating fluid. The efficiency improvement sought through the use of higher cycle temperatures afforded by the utilization of ceramic components is measurably increased by the reduction of leakage of the working fluid from its intended gas path. In the specific case of relatively small turbine designs, the blade tip leakage can critically affect engine performance as a result of the leakage, assuming a disproportionate share of the overall losses.

In addition, the proper application of an abradable shroud system can measurably improve the operating lifetime of the rotating components through the reduction in tip material lost as a result of turbine/shroud contacts.

Vendor-supplied materials were adaptations of formulations being considered for engine component items. These included materials such as silicon carbides and nitrides, which were being considered for engine components such as turbine shrouds; foamed lithium-aluminum-silicate (LAS) concepts, which were being employed in the regenerator area; and magnesium-aluminum-silicate

(MAS) compounds, which were being considered for stationary turbine components.

Concurrently, Allison in-house technology was employed to investigate seal structures formulated from ceramic powders such as mullite and zircon, which were subsequently bonded to silicon carbide or silicon nitride substrates. In addition, parallel studies were also carried out on ceramic-bonded systems utilizing ceramic-based adhesives in conjunction with ceramic-fibered compositions. Initial efforts involved felt and papers of the alumina- and alumino-silicate families and zirconia felts and fabrics impregnated with rigidizers. As an alternate approach, plasma-spray techniques were used to cospray dissimilar materials simultaneously on the ceramic matrix. This approach ultimately led to seal matrices that were employed in several engine tests with some success.

Advances in blade tip rub phenomena and progress in understanding the mechanism involved in blade-tip/abradable-shroud interactions led to this realization: smearing of materials at the interface was unavoidable unless the blade tip was assisted in its cutting action. As a result, studies were initiated to include abrasive blade tips as a part of the total system. These studies indicated that significant improvements in blade tip lifetime could be expected for silicon carbide blades.

Materials Development

To maintain a commonality with engine hardware, the primary concern was with those contacts who were supplying or being considered for supplying ceramic component items. Carborundum (CBO), GTE/Sylvania, Corning Glass, and Pure Carbon/BNFL each supplied candidate abradable materials. These materials were investigated for compatibility with both the metal and ceramic-bladed rotor system.

Early in the program, a dual-density silicon nitride material was provided by GTE/Sylvania for evaluation; testing was conducted with Mar-M246 blade tips. Initial results with this material, reported in EDR 9951 (Ref 3), were very promising. Second materials iterations in EDR 10156 (Ref 4) and third generation samples in EDR 10383 (Ref 5) further reinforced the initial conclusions that the GTE concept had merit. Erosion resistance appeared to be lacking, and adequate bonding techniques required further development. However, the GTE silicon nitride material does appear promising as a preliminary evaluation and should be included in future studies.

The Corning Glass Works supplied several variations of a foamed LAS cement for consideration as a turbine abradable material. This material was originally used for bonding the hubs of ceramic regenerators. Test results are reported in EDR 9951 (Ref 3). The results were not very encouraging due to extremely wide variability in the material structure and the poor abradability performance. Therefore, this system did not hold much promise for the intended application.

As an adjunct to determining the suitability of LAS for structures to satisfactorily perform as abradable shroud materials, several shrouds were received from Corning with the abradable section formed of foamed type 1050 cement. Significant variations in the density of the foamed cement were apparent on the shrouds, in that one of the shrouds exhibited an extremely porous structure with large voids. Abradability and erosion performance were judged similar to prior tested samples (see EDR 10156—Ref 4).

Several MAS structures were developed by Coors Porcelain as candidate abradable materials. These were basically solid compact cordierite materials with processing or formulation variations that resulted in hardness and strength changes. Initial test results were reported in EDR 10156 (Ref 4), and second generation samples were discussed in EDR 10383 (Ref 5). Further effort on this material was discontinued due to its extremely poor performance under abradability or erosion test conditions.

Silicon carbide structures were supplied by both Pure Carbon/BNFL and Carborundum. Pure Carbon supplied reaction-bonded material of several variations whose abradability and erosion performance was initially reported in EDR 10156 (Ref 4) and whose structural compositions were shown in EDR 10383 (Ref 5). Although the testing of the Pure Carbon/BNFL materials was limited in terms of the number of samples evaluated, the abradability/erosion performance characteristics of several of the samples were sufficiently promising to warrant being included in future studies.

Early program efforts by Carborundum to provide abradable silicon carbide compacts in the form of component shroud hardware resulted in shrouds that were consistently too dense and hard and consequently produced very poor abradability performance. This resulted in excessive material being removed from the metal turbine blade tips during both rig testing of shroud hardware and actual engine tests.

As a basis for determining the influence of materials variations on abradability and erosion performance, Carborundum prepared materials samples in which the grit size was varied to produce changes in the density. These initial studies were reported in EDR 9951 (Ref 3) and provided guidance for the selection of compositions that were considered to warrant further study. The task of providing abradable engine components based on coupon evaluation was recognized as being extremely difficult since processing technology differed in the manufacture of flat coupons and circular rings. These processing differences were diverse enough to produce shrouds that had no similarity to the original coupons in terms of abradability. An expanded program for providing abradable silicon carbide structures was proposed by Carborundum and presented in EDR 10383 (Ref 5). In this effort, Carborundum supplied four materials systems composed of silicon carbide materials whose densities are varied by changing the method of manufacture. The materials systems can be described as follows:

1. silicon matrix, silicon carbide fiber reinforced
2. close-pore silicon carbide foam on SKT backing
3. fine open-pore silicon carbide foam
4. RBSiC with free silicon in the mix

Samples of these systems were evaluated in an abradability test that was radically different from any testing conducted to date. Up to this point, all abradability and engine tests on candidate silicon carbide (or any other materials candidates) components had been conducted with Mar-M246 blade tips. Some preliminary work had been successfully completed with silicon carbide blades against plasma-sprayed yttria-stabilized zirconia composite structures (discussed later in this section). These were the first tests done with SiC blading versus a silicon-carbide abradable structure. Abradability and erosion testing was conducted on the low-speed and erosion test rigs described in EDR 9951 (Ref 3).

Upon receipt and examination of the Carborundum samples, it was immediately concluded that specimens in Group III were too fragile to be of any value, and these samples were immediately set aside.

Abradability tests of Groups I and II were generally characterized by an inability of the silicon carbide blade tip to effect any penetration of the abradable surface. Virtually all of the samples exhibited severe sparking upon contact by the blade tip accompanied by audible sounds of distress. In many cases, damage to the tip occurred, manifesting itself in chipped or broken tip sections. Out of a total of 11 samples for both groups, only one sample, Sample I-5, exhibited good rub characteristics. The rub path for this specimen is shown in Figure 217. Figure 218 compares this with a typical rub path of one of the other samples tested in Group I. Sample I-5 exhibited a Rockwell R15Y hardness of 40 as compared with a hardness of 89 measured on Sample I-2.

Testing of samples from Group IV revealed that several samples exhibited excellent abradability

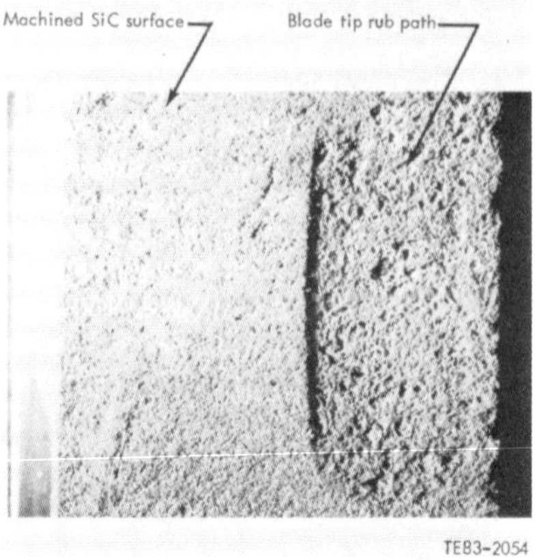

Figure 217. Abradable rub of an untreated RBSiC blade tip into an SiC fiber reinforced matrix (Group I-5 sample); rub depth—0.46 mm (0.018 in.).

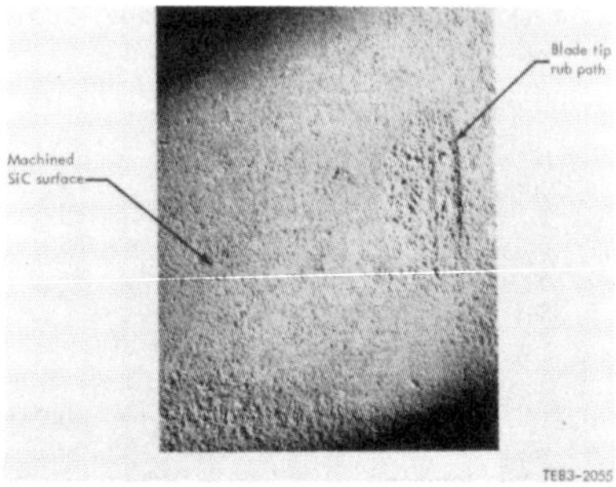

Figure 218. Abradable rub of an untreated RBSiC blade tip into an SiC fiber reinforced matrix (Group I-2 sample); rub depth—0.025 mm (6.001 in.).

characteristics. Of ten samples in this group, eight exhibited excellent rub tracks. One of the better samples, Figure 219, shows the excellent rub track. Figure 220 is the SiC blade tip after several tests. These samples typically had R15Y hardness values ranging from 60 to 90.

Erosion testing proceeded with the results shown in Figure 221. When the results of the erosion tests are combined with the abradability tests, two samples, I-5 and IV-18, shown in Figures 217 and 219, definitely warrant continued investigation.

Materials Studies

Allison's in-house abradable seal materials technology was a natural adjunct to the development of metallic and ceramic compatible abradable seal systems. These in-house investigations encompassed the areas of impregnated and bonded ceramic felts and fibers as well as bonded and plasma-sprayed ceramic powders.

Ceramic cement-bonded systems comprised felts and papers in the alumina- and alumino-silicate families with a parallel effort directed toward using zirconia cements and rigidizers to impregnate zircon and zirconia felts and fabrics. A variety of these experimental coatings were successfully bonded to silicon carbide and silicon nitride substrates (EDR 10156—Ref 4). Composite matrices composed of zirconia fabrics and cements were laminated to form free-standing structures. The problem of attaching laminated ceramic structures to RBSiC shroud elements was thermal mismatch between the SiC shroud material and the zirconia fabric system. The

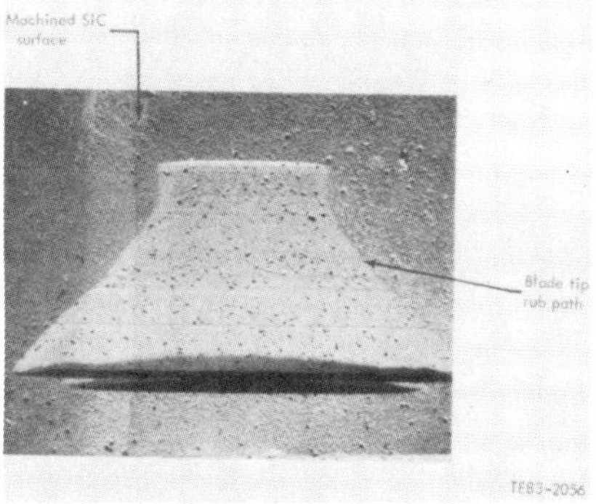

Figure 219. Abradable rub of an untreated RBSiC blade tip into an SiC surface (Group IV-18 sample); rub depth—0.53 mm (0.021 in.).

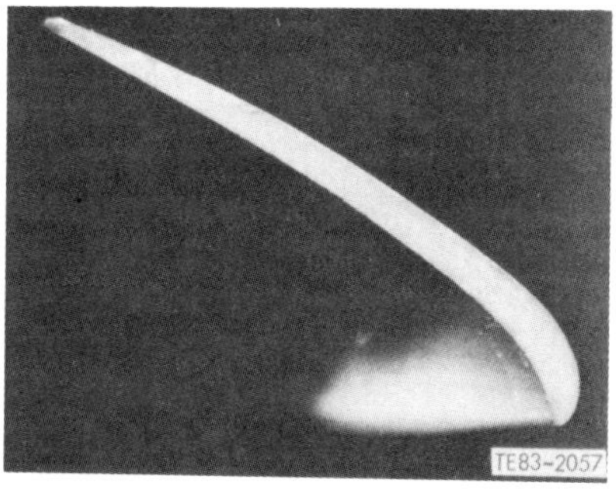

Figure 220. An untreated SiC blade tip after rubs into two Group IV surfaces; rub depth—IV-18, 0.53 mm (0.021 in.), and IV-19, 0.79 mm (0.031 in.).

zirconia-based seal systems have a thermal expansion coefficient approximately twice that of the high-density silicon carbide shroud. This problem becomes evident at high temperatures where differences in expansion cause the seal to dislodge from the cement and substrate. As a result of these studies, it was determined that a zirconia plasma-sprayed layer deposited over the silicon carbide is necessary to provide a semiporous surface to which the cement can bond. This plasma-sprayed layer allows wicking of the Aremco Ultra-temp 516 cement into the pores of the Zircar fabric and at the same time bonds to the SiC substrate. Although bond

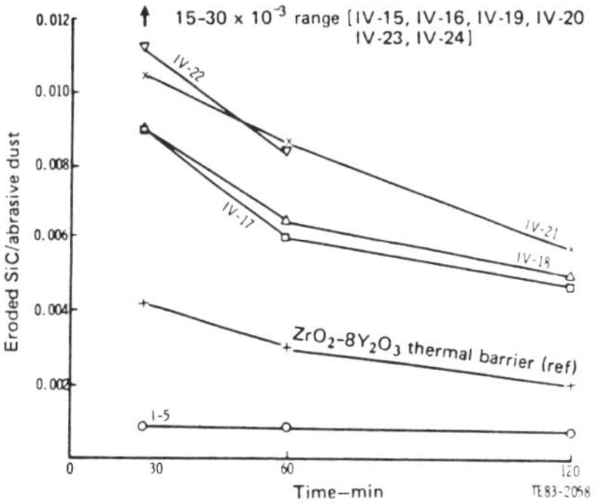

Figure 221. Erosion of SiC abradable materials by coarse Az road dust 15-deg impingement.

strength was sufficient to permit abradability testing, further work is needed before this system can be demonstrated in an engine-operating environment. Results were sufficiently promising to continue this work should laminated zirconia fabrics ultimately become feasible for engine evaluation.

Generally, the cemented felts and paper systems were not as durable as the fabric system in terms of erosion performance. Abradability testing with metal blade tips and erosion testing of the zirconia cemented fabric system—EDR 10383 (Ref 5) and EDR 10672 (Ref 6)—resulted in a number of materials variations exhibiting acceptable abradability characteristics, although erosion performance requires further improvement. Further study of the cemented zirconia fabric systems should be considered.

Sintered mineral compact systems were produced by blending zircon powders with a plasticizing binder. Abradability testing against metal blade tips, erosion, and preliminary thermal shock tests have shown the system to warrant additional study.

The application of plasma-sprayed powders for seal applications has been extended to include ceramic materials which heretofore have normally been used as thermal barriers. The plasma spraying of ceramic materials requires a degree of innovation to reduce the material density to achieve satisfactory abradability. Several approaches have been attempted with varying degrees of success. These approaches employ techniques to reduce the apparent density of the coating through the cospraying of materials that are compatible with the basic ceramic matrix material. Typical of the more successful coatings has been the plasma spraying of an yttria-stabilized zirconia (YSZ) coating system codeposited with either a polyester powder filler or ceramic eccosphere filler (EDR 10383—Ref 5). The YSZ/ceramic eccosphere system has been used in a number of metal bladed engine tests as a turbine shroud seal. Results of these engine tests have emphasized the fact that at 1093°C (2000°F) several factors are causing the blade tip/shroud distress observed. At the indicated temperature, the metal blade tips are relatively soft compared with the ceramic based system, and the tendency toward metal transfer is markedly increased. In addition, the eccosphere portion of the seal starts to deteriorate at these temperatures, thereby increasing the tendency for serious erosion to occur. The optimum structured system in terms of survivability of the blade tip and adequate erosion resistance is a 14% YSZ/80% eccosphere (by volume) cosprayed coating. This combination appears to be the practical limit for use with metallic blades.

Ceramic-bladed systems using prototype silicon carbide blading have resulted in significant improvement of abradability performance over previously tested metal blades. Results of early rubs into various density YSZ/eccosphere plasma-sprayed systems are discussed in EDR 10672 (Ref 6). In these tests, the silicon carbide blade tips produced clean rubs in YSZ systems having more than twice the zirconia concentrations of the 14% system described previously. Erosion resistance necessary for extended service life was still lacking, even with these denser coatings.

Although the ceramic-bladed system has shown a marked improvement in abradability performance over previous metal blade tests, densification of the abradable material at the surface under the moving blade tip still has a tendency to occur, especially as the proportion of zirconia in the mixtures increases. Developments in the field of rub energetics have led to new understandings of the fundamentals of the rub mechanism and possibly explain the densification and glazing that have been prevalent to date (see EDR 10672 [Ref 6] and EDR 10841 [Ref 7]). Simply stated, crushing mechanisms (interaction between the blade tip and the seal surface) must be avoided in nonfriable seal systems because they lead to densification of the material and smeared surface at the rub interface. For material to be removed by "chip" formation and to thereby avoid smearing or densification, the blade tip "cutting edge" must have a radius smaller than the depth of cut of each individual blade tip as it enters the abradable material. This depth of cut is a function of the radial incursion rate, the number of blades, and the rotor speed and is typically about 10^{-6} mm (4×10^{-8} in.). Since the cutting-edge radius of practical systems is many times larger than this, movement of the blade into the seal will always tend to cause densification of the seal material, which leads to smearing of the seal and damage to the blade tip.

One approach toward providing the blade tips with the machining assistance deemed necessary if erosion resistance and abradability are to be in balance is to provide the tip with many sharp, randomly oriented minisurfaces, much like a grinding wheel surface, that will serve as more efficient cutting edges than a single blade tip. This concept of applying a textured coating to the tips of silicon carbide blading has been developed by San Fernando Laboratories and is discussed in EDR 10841 (Ref 7). This is a chemical vapor deposition process that deposits granular silicon carbide upon the blade tips. These textured coatings are slightly harder than the blade substrates and offer a surface that is more analogous to a grinding configuration than the original smoothly machined tips. These blades were evaluated against an 80/20 YSZ/eccosphere system. The results of these tests indicate that the performance of ceramic blade compatible abradable YSZ systems has markedly improved through the use of textured CVD-coated silicon carbide blading. At the same time, erosion resistance of the seal material has been significantly improved through the use of the 80/20 system and is comparable to the 100% YSZ NASA-developed thermal barrier coating system. The favorable results of this preliminary work into coated tip systems certainly merits further study.

Component Development

Early in the program, CBO RBSiC shroud rings were evaluated for abradability performance with metal blades; this was a first attempt at producing engine-oriented abradable components. The initial rings exhibited gross cracking and delamination in the abradable material, as reported in EDR 9519 (Ref 1). However, a slight degree of abradability was present providing the rub depth was not excessive. It appeared that a rotor rub of 0.25 mm (0.010 in.) could be tolerated without excessive tip damage.

Results of engine testing conducted with a silicon carbide shroud similar to the one described above concluded that the shrouds were not abradable. Additional rig tests of the remaining shrouds in stock indicated that none of the shrouds were abradable and further that there was a definite lack of uniformity in abradability performance between each shroud ring. This lack of uniformity was believed to be the result of processing variations.

In addition to the need for correcting errors in processing technology, it was determined that the degree of abradability could be related to the density of the base material. A matrix of various silicon carbide samples comprising particle sizes varying from 1000 mesh to 120 mesh and densities from 2.84 g/cm^3 to 2.9 g/cm^3 were evaluated for abradability and erosion performance. The detailed results are described in EDR 9951 (Ref 3). Based on these results, shroud rings fabricated from 1000 grit 2.84 g/cm^3 and 2.9 g/cm^3 densities appeared most promising for future development.

RBSiC shrouds of the 2070°F configuration were received from CBO with material densities of 2.87 g/cm^3 and 2.88 g/cm^3. These shrouds were to have been fabricated to a density of 2.84 g/cm^3 in accordance with the results noted previously. It is not clear whether these variations are a result of nonuniformities in the density of the green-molded bodies or the possibility of inability to control the density of large diameter ring structures to any degree of accuracy. Rig abradability tests of these two rings indicated that they exhibited rub performance compatible with previously tested 2.88 g/cm^3 samples, as described in EDR 9951 (Ref 3). In all tests of this density (2.88 g/cm^3), both samples and rings exhibited tendencies toward metal transfer to the blade track for rub depths as shallow as 0.178 mm (0.007 in.).

Testing in engine C-4 (BU6) with reduced radial clearances utilizing a shroud similar to those previously described resulted in a severe rub accompanied by heavy metal transfer from the blade tips to the shroud. The problem appears to lie in the processing inconsistencies or unknowns currently prevalent in the fabrication of abradable silicon carbide shrouds.

These variations occur not only between flat materials specimens and supposedly similar shroud rings but also among shroud rings manufactured at the same time and to identical specifications. In the CATE engine, C-4/BU13, a CBO RBSiC ceramic shroud (P/N EX 140019, S/N 23784) and associated ceramic blading experienced a failure. In an attempt to determine the engine shroud abradability, identical shroud rings to those utilized in the engine were rig tested for abradability utilizing ceramic blading. Results of these rig tests indicated one of the comparison rings was definitely not abradable while the other was considered to have some degree of abradability. The net result was that no concrete conclusion could be drawn regarding the abradability characteristics of the actual engine shroud hardware.

Ceramic Component Machining Technology

SUMMARY

The unique properties of fired ceramic materials have made several conventional manufacturing processes ineffective and have made certain unconventional processes effective. Fired ceramic materials properties have also put limitations on the fixturing techniques used during various machining operations. The specific conclusions concerning various manufacturing processes are given in the following paragraphs.

The conventional operations of drilling, milling, and turning are totally ineffective. Conventional form grinding is incapable of machining fired ceramics due to poor surface finishes and excessive chipping. Internal grinding, rotary grinding, and surface grinding with diamond grinding wheels, however, are effective methods for machining fired ceramic components. Rig grinding is an acceptable method of producing good surface finishes, close tolerances, and overall part quality for slots and other radially generated surfaces. Creep feed grinding with crush pressable diamond wheels is an effective method of form grinding complex shapes to close tolerances in fired ceramic components. (See the discussion subsection for further information on the restrictions of grinding ceramic materials.)

Ultrasonic machining has excellent potential for machining fired ceramic components. It is the only method found during this study capable of producing blind or through holes without a cast starter hole in fired ceramic material.

Just as the unique properties of fired ceramics have restricted the methods of machining available, they have also placed severe restrictions on the fixturing techniques used to hold the components, as discussed later.

OBJECTIVE

The objective of this study was to develop tooling and machining parameters for ceramic turbine components.

APPROACH

The study of each machining process was conducted in a similar fashion. First a library study was conducted and any relevant reports were examined. Afterward, the variable parameters of the process were defined and the tests developed.

The first series of tests loosely covered the entire range of machining parameters. Follow-on testing concentrated on those feed, speed, and depth of cut ranges that showed possibilities. Tooling vendors with background in the machining of glass products were helpful in setting parameters for these tests. At this point, testing was discontinued on processes, such as drilling and turning, that did not show sufficient promise. The continued testing was directed toward defining the best machining conditions for each particular part required by the overall CATE program. At this stage, fixturing techniques became the main concern of the machining development. Methods of controlling chipping were also studied at this time. The approach provided the capability to support the CATE program in a timely fashion and defined the general direction for any in-depth machining research that may be undertaken.

DISCUSSION

The properties of the ceramic materials studied during this program influenced the fixturing techniques and suitability of manufacturing techniques. In general, diamond grinding processes worked well. Plated diamond products did not work nearly as well as diamond impregnated wheels and mandrels.

For more information on specific machining processes the reader may reference the thesis "Maching of High Temperature Ceramics." Those who are familiar with diamond grinding, creep feed grinding, or ultrasonic machining could easily adapt to machining fired ceramic components after the proper fixturing techniques are learned.

The most important property of these materials in terms of fixture design is that they will fracture before bending any noticeable distance. This means that the mounting surfaces of the fixture and the part must be in the same plane. For example, a metal casting is normally clamped or bolted down in some manner when the first surface is machined. This works well since the metal will bend some to make up for the natural high and low spots found on any casting. In contrast, a ceramic casting will break when clamped if the clamps force the ceramic to bend to fit. This tendency was noticed when a rough ceramic ring-shaped casting was clamped in four places, as shown in Figure 222. One of the lower supports did not contact the ring. When the machine operator tightened it, the part broke. To overcome this problem, meltable fixturing compounds such as Rigidax®*, Cerro-bend**, or Cerro-cast** are used. These compounds are poured around a part and solidify to form a solid bed that fully supports the part, as shown in Figure 223. This method works well on odd-shaped parts such as turbine blades.

The ceramics materials have compressive strength values of up to 620.5 MPa (90,000 lb/in.^{2}) for Si_3N_4 and 1378.9 MPa (200,000 lb/in.^{2}) for SiC. These

*Rigidax is a registered trademark for a tooling compound of M. Argueso & Co, Inc.

**Cerro-bend and Cerro-cast are low temperature metal tooling compounds.

Figure 222. Rough ceramic ring casting with clamps.

values allow the ceramic components to be tightly clamped and held in place during almost any operation, as long as the ceramic is fully supported, as explained earlier. Figure 222 illustrates an excellent method for fixturing a ceramic part that has a flat, machined surface to clamp against. This illustration, used earlier to show an incorrect method of fixturing an unmachined casting, also emphasizes the need for a machined surface. The six hex bolts are adjusted to support the part against the forces applied by the machined operation. These bolts were set after the part was clamped in place.

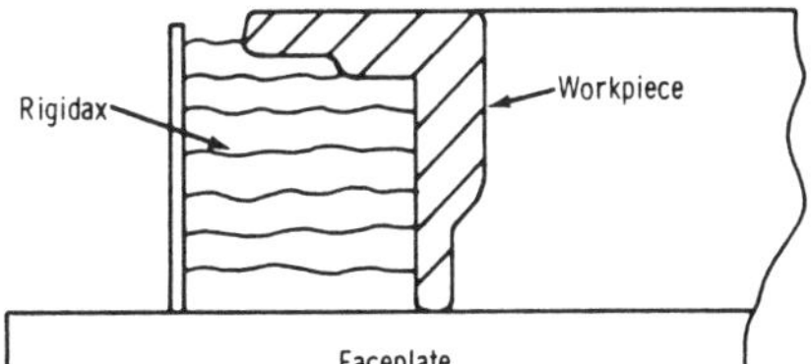

Figure 223. Fixturing compound.

A common fixturing technique called floating works very well on operations that exert mostly compressive forces on the part if and only if the part surface against the machine's faceplate has been machined flat. Surface grinding involves the compressive forces caused by the wheel on the part and the shock caused by the wheel each time it contacts the part as it cycles back and forth. The part should be fully supported against this shock by the metal blocks all around it. Only a small portion of the part should be exposed above these supporting blocks. A setup on a rotary grinder requires less support. A ring is not subject to the shock that a slab is subject to on the surface grinder. For this reason, the ring only needs a few blocks to hold it in location.

The brittle nature of fired ceramic components requires that care be taken in all handling and fixturing operations. The following general cautions will prevent most accidental breakage or chipping:

1. "Pot" all rough castings or odd-shaped pieces in Rigidax or Cerro-bend until a flat surface can be machined to be used for later fixturing.
2. Ensure that the mating surfaces of the part and locating fixture are flat and in contact before clamping.
3. Avoid bracing against sharp corners or thin walls of the part.

4. Utilize an epoxy coating to reduce or eliminate chipping sharp corners during machining.
5. Provide backing on the exit side of a through hole for any ultrasonic machining operations.
6. Provide ample padding or protective containers for the ceramic parts for all shipping or storage.

One final warning concerning the machining of fired ceramics is required. Several investigators have found evidence that the machining of ceramics often causes the formation of subsurface cracking in the part. This cracking does not show up during most common inspection techniques and results in severe weakening of the part. As a result, all machining operations should be studied to determine the amount of damage that they do to the part. An in-depth study of the causes and extent of the machining damage is beyond the scope of this study.

Ceramic Blade Process Development Experiment

One of the major ceramic component process development efforts in the CATE program focused on the ceramic gasifier turbine rotor blade. The initial effort centered around the blade design, followed by early fabrication and rig testing of coupons to verify the attachment concepts incorporated into the design. After satisfactory demonstration of attachment design, injection-molded prototype blade fabrication was initiated. This step's successful completion led to the final design gasifier turbine blade fabrication, development, and final successful engine demonstration.

The final design blade development program can be described as initial blade fabrication, plus follow-on order No. 1 and follow-on order No. 2 development efforts. After evaluation of those three groups of blades produced by the prime silicon-carbide supplier for the CATE program (Carborundum Company), the fabrication process was thought to be well defined; but the blade strength level was somewhat lower than desired, and the yield of acceptable blades was low. To address those problems, an extensive blade development program, referred to as the ceramic blade process development experiment, was conceived. During the planning stages of the experiment, it was decided that each step in the process of fabricating blades should be carefully reviewed and that important variables should be identified. Also determined as necessary was to have a carefully structured experiment defined that modulated the important variables over a reasonable range while holding all other variables in the fabrication process unchanged.

This section of the CATE program final report provides a description of the experiment and its results.

SUMMARY

The ceramic blade process development experiment is a controlled manufacturing of 4800 ceramic blades in 32 groups (150 blades in each group) followed by a thorough evaluation of each group to determine which manufacturing process variables influence blade strength and yield. These results are then useful to provide detailed guidance to future blade processing that would result in higher strength components and higher yields from the manufacturing process.

The experiment was successfully conducted, and all 4800 blades were produced and evaluated by nondestructive inspection techniques, destructive inspection techniques (sectioning to determine microstructure, spin testing to failure, sectioning to obtain MOR test bars from blades, etc), and dimensional checks to define yield. Also a statistical analysis was done on the data to identify significant process variables.

The results from the statistical analysis revealed five variables that significantly influence blade yield, strength, and spin test results. The five variables are the binder system, binder additive, molding temperature, objective furnace type, and postmachining treatment.

OBJECTIVE

The objectives of the ceramic blade process development experiment were primarily to identify the processing variables that control strength and yield. However, several secondary objectives existed, including the fact that the knowledge gained would have application to other injection-molded sintered silicon carbide components. Another objective was to develop the techniques and methods required to conduct this experiment at the component supplier and at Allison.

APPROACH

The approach to the ceramic blade process development experiment focused on the concept of carefully selecting processing variables, identifying a range of modulation for each variable, and fabricating blades with different combinations of these variables while holding all other processing steps invariant. The process also focused on thoroughly evaluating the hardware produced through nondestructive inspection, laboratory evaluations, dimensional inspection, spin testing, finally ending with data correlation by conducting a fractional factorial statistical analysis on all data obtained.

Data analysis identified variables that influenced strength and yield. From that information, follow-on experiments can be designed that will focus only on the significant variables, permitting the exact definition of that variable or those variables that produce the optimum strength and yield from the manufacturing process.

DISCUSSION

The first step in the experiment was the identification of processing variables thought to be important relative to blade strength and yield. The following are the variables selected for the experiment:

- binder system
- binder additives (2)
- compounding
- molding temperature
- molding pressure
- sintering temperature
- heat-up rate
- atmosphere
- fixturing
- furnace
- machining
- postmachining treatment

The next step in the experiment was to define a matrix of the 13 variables that would produce 32 groups of parts, each with a different combination of the variables. The Carborundum Company manufactured these 32 groups of parts in the injection-molding tool developed by the CATE program. Figure 224 illustrates the steps in the experiment, starting with the manufacturing step and concluding with the evaluations at Allison. Also shown is the product of the injection-molding tool as it appears when removed from the tool. Each blade produced had a companion test bar molded simultaneously with the blade. Also shown in Figure 224 is the finished blade ready for spin test evaluation. For each group of blades, 150 parts were injection molded, bringing the total number of parts produced during the experiment to 4800.

The Allison evaluation of the parts can be separated into three activities: laboratory characterization, spin testing of selected blade groups, and data reduction/correlation. Each of these activities is discussed in the following paragraphs.

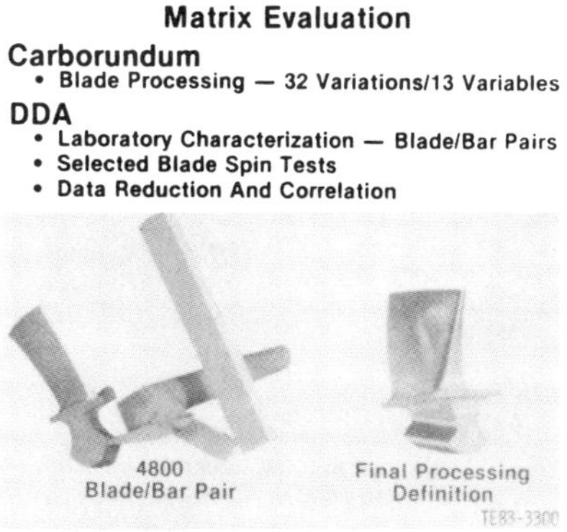

Figure 224. Ceramic blade process development experiment.

Laboratory Characterization

The laboratory work was a major part of the evaluation process. All of the blades and test bars were subjected to nondestructive inspection (NDI), which included X-ray, fluoropenetrant (zyglo) inspection, visual inspection, and dimensional evaluation. In the case of test bars (referred to as full-size test bars) following NDI, 36 bars were selected and divided into three groups. This became the basis for evaluating the machining variable. The three groups are referred to as transverse ground, longitudinally ground, and as-fired. The as-fired bars were not machined; the other two groups were machined, but in different machining directions (transverse and longitudinal). These full-size test bars, with the exception of the longitudinally ground bars, were then exposed to the same post machining treatment as the blades, which is another of the experimental variables. All full-size bars were then MOR tested to provide a measure of bar strength. For blades, 15 parts were selected on the basis of NDI results, and the following laboratory evaluations were conducted on those blades:

- sonic velocity
- C-scan
- PAS stalk
- microstructure
- density

Three of the 15 blades were sectioned to evaluate microstructure. The remaining 12 parts, after the laboratory evaluations, were machined to make three small test bars from each blade (referred to as minibars).

These minibars were machined, postmachining treated, and MOR tested in the same way as the full-size test bars; the MOR results for full-size bars and minibars were input for the data reduction and correlation activity. Following data reduction for the test bars, nine groups of blades were selected for spin testing.

Spin Testing

The nine groups of blades selected for spin testing represented test bar strengths covering the range (low to high) of MOR test data. Blades of acceptable dimensional quality from each of these nine groups were machined (dovetail attachment machining), postmachining treated, inspected with the basic NDI methods, and then tested in the spin pit to failure. The failure speed for each blade within each group was used as input into the data reduction and correlation activity.

Data Reduction and Correlation

All of the NDI, laboratory evaluation results, test bar MOR data, and spin test data for each of the 32 groups of blades were subjected to a statistical analysis to identify the significant variables. During the blade fabrication phase of the experiment, the binder used to fabricate the original blades exhibited sensitivity to changes in the other variables, while the improved binder system exhibited no significant sensitivity to other variables. Consequently, the statistical data analysis results for all groups with the original binder system revealed significant variables that were not confounded with other variables. However, the results for the groups with the improved binder system revealed significant variables, free of confounding. The variables, listed as follows, are found to be significant on an overall basis (i.e., they influence blade yield, strength, and spin test results):

- binder system
- binder additive
- molding temperature
- furnace
- postmachining treatment

IV. Ceramic Component Design and Development

Summary

Structural ceramic gas turbine components have been designed, fabricated, and tested. The methodology necessary to design these ceramic components has also been developed. This section documents the design and development effort on each of the components in addition to summarizing the design methodology.

The design methodology subsection details the probabilistic design techniques used to design and evaluate ceramic components.

The gasifier turbine nozzle subsection presents the design and development effort on ceramic vanes and shrouds for the 1900°F-configuration engine. This subsection also covers the work on a complete ceramic nozzle assembly—vanes, support rings, and shroud—designed for the 2070°F-configuration engine.

The gasifier turbine blade subsection documents the design and development work on the 2070°F-configuration ceramic turbine blade. Spin testing of blades and test coupons is covered as well as the design of an alternate configuration blade with improved reliability.

The gasifier inlet plenum subsection covers the design and limited testing of the ceramic plenum for the 2070°F-configuration engine and the power turbine nozzle subsection presents design concept study results for a cantilevered ceramic stator vane design for the 2265°F-configuration engine.

Conclusions

The conclusions that can be drawn as a result of this work are as follows:

1. The probabilistic design techniques developed are adequate to predict component reliability regarding the fast fracture failure mode. Additional work is required for time-dependent failure modes.
2. Small static structural ceramic components such as vanes, shrouds, and rings can be designed, fabricated, and tested with reasonable confidence of success in a known environment.
3. Large static ceramic components such as plenums require more experience before a statement can be made regarding suitability as a gas turbine component. One conclusion, however, that can be drawn is that much more fabrication process development work is required by the ceramics suppliers if large duct-type structures are to be successful.
4. Rotating ceramic components, such as blades, can be successfully designed and their performance reliably predicted in a spin test rig environment.
5. Proof testing of ceramic components is required prior to engine testing. The proof test should duplicate or exceed the engine environment regarding mechanical and thermal loading.

Design Methodology

SUMMARY

Ceramic components for the CATE program have been designed using methodology developed especially for brittle materials. Linear elastic probabilistic analysis techniques, using two- and three-dimensional finite element models, were used to design and evaluate each component. Extensive heat transfer analyses of critical operating conditions, which normally occur during engine transients, were conducted to select the most severe cases for stress analysis. Component reliability was then assessed by application of Weibull statistics. Methods of interfacing proof test and engine operating conditions have also been developed and incorporated into finite element analysis routines.

OBJECTIVE

The objective of ceramic component design methodology effort was to develop analysis procedures that ensure reliable components in the required operational environment. These procedures have been developed and implemented for the fast-fracture type of failure. Time-dependent failure modes have not been addressed because the ceramic material process development was thought not to have matured to the point where this type of material characterization was warranted.

APPROACH/DISCUSSION

The general approach taken in the design of ceramic components involved the following broad configuration goals:

- minimize structural loads
- utilize interface compliant layers and diffusion barriers
- minimize machining required

Design requirements for each component were established from the known engine duty cycle. Reliability goals (see Table XXXII) for each ceramic component were established on the basis of allowable lifetime warranty cost goals for the CATE IGT 404 engine.

Probabilistic design analysis was conducted on each component. At the center of this process was the finite element model, either two- or three-dimensional, of the part configuration. This model was to calculate temperature and stress distributions as well as overall component reliability. Input data were derived from the engine duty cycle, material characteristics, reliability requirements, and proof-test definition. A schematic diagram of this design process is shown in Figure 225. Thermal, aerodynamic, and mechanical loads for steady-state and transient operation in both the rig and engine environments were considered. Proof-test requirements were also assessed for each component. Proof-test failure rates could be predicted and measured against component operational reliability enhancement.

Table XXXII.
CATE 2070°F ceramic blade design review.

Cost controlling subsystem/component	Design allowable failure probability
Regenerator disk assembly	0.013753
Regenerator seal—inboard	0.012618
Regenerator seal—outboard	0.014577
Regenerator drive	0.026909
Combustor assembly	0.006362
Fuel nozzle	0.093387
Igniter	0.120327
Turbine inlet plenum	0.007911
Gas turbine nozzle assembly	0.001892
Gas turbine rotor assembly	0.003689 (0.000057 for one ceramic blade)
Power turbine nozzle assembly	0.003622
Power turbine rotor assembly	0.003775

Note: Data are based on allowable lifetime warranty cost goals.

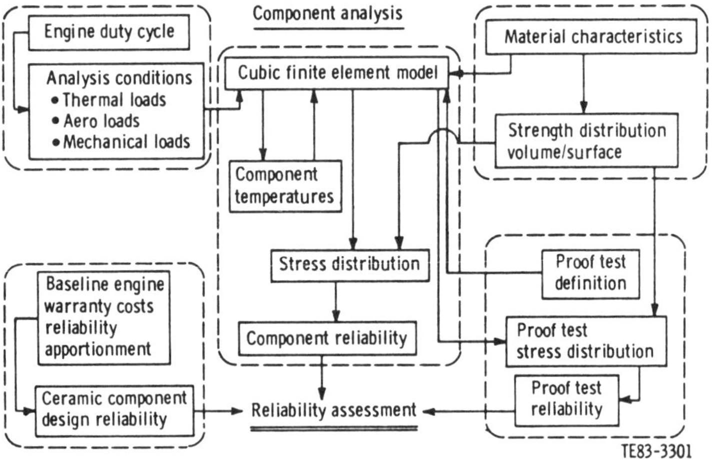

Figure 225. Ceramic component design.

The evaluation of component reliability is based on an integration of failure probabilities using Weibull statistical parameters. A Weibull probability of failure analysis was incorporated into Allison's finite element stress calculation program to permit automatic evaluation of the probability of survival of a component under a given set of loads. The probability of fracture P is given by the following:

$$P = 1 - e^{-R}$$

where R is the risk of rupture. This is divided into two parts—the risk of failure initiated at a surface flaw, R_s, and the risk of failure initiated at an internal flaw, R_v. These are given by integrals over the surface and volume respectively of functions of the stress.

$$R_s = {}_s\int f(\sigma)\, ds$$

$$R_v = {}_v\int f(\sigma)\, dv$$

The form of the stress function has been taken to be the following:

$$f(\sigma) = (\sigma - \sigma_u)^m / \sigma_o$$

where σ_o, σ_u, and m are properties of the material that are related to its likelihood to contain and related to its sensitivity to internal and surface flaws.

The capability to perform these integrations and to compute the probability of survival has been included in the two-dimensional, three-dimensional, and axisymmetric elements. The results are presented by element and are accumulated for ceramic substructures and for the overall structure being analyzed.

To conduct this type of analysis, the Weibull material parameters of characteristic strength, σ_o, and slope, m, are required. The information is obtained from testing fourth-point bending MOR bars. The parameters for both volume flaw and surface flaw strength distributions are determined by applying a suspended item treatment to the least squares linear regression analyses of the test bar strength data.

Weibull Equation

The Weibull equation is given by the following:

$$1 - P_f = e^{-KV(\sigma/\sigma_o)^m}, \text{ or } = e^{-KS(\sigma/\sigma_o)^m}$$

where

$1 - P_f$	= probability of survival
P_f	= probability of failure
V	= volume of test portion of sample failing by a surface flaw
S	= surface area of test portion of sample failing by a surface flaw
K	= loading factor
σ	= strength of sample
m	= slope of Weibull plotted in linear form
σ_o	= characteristic strength of Weibull

Regression Analysis

A least squares regression analysis is used to determine the coefficients of the Weibull equation:

$$\log[-\ln(1 - P_f)] = m \log \sigma - m \log \sigma_o'$$

where σ_o' is the test sample characteristic strength and

$$P_f = \frac{n + \sqrt{2}/2 - 1}{N + \sqrt{2} - 1} \text{ (median rank)}$$

where

n = ranked position
N = total sample number

Note that σ_o' is obtained assuming KS or KV = 1. The Weibull strength characteristic σ_o is obtained from

$$\frac{K(V \text{ or } S)}{\sigma_o^m} = \frac{1}{\sigma_o'^m}$$

where K is obtained from the sample test condition. K for volume flaw failure is $(m + 2)/4(m + 1)^2$ and for surface flaw failures

$$K = \frac{m + 2}{4(\lambda + 1)(m + 1)} \left(\frac{1}{m + 1} + \lambda \right)$$

where λ is the test bar width-to-depth ratio.

Suspended Item Analysis

In any group of samples in which failure occurs either by a surface flaw or a volume flaw, the regression analysis for the volume flaw strength distribution employs the surface flaw failed samples as suspended items, and vice versa. For convenience the flaw types are called X and Y. The suspended item treatment is based on the concept that the best estimate of the strength level at which the suspended item would have failed had it not failed by flaw type X, is given by the strength level at which it most likely would have failed as a type Y flaw.

This equally likely strength level is obtained, as shown in Figure 226.

The failure strength by flaw type X, σ_x, is a suspended item when analyzing the data for the flaw type Y Weibull distribution parameters. If A and B are equal areas, then σ_y is the strength level at which the X failed sample is most likely to have failed with a Y type flaw.

For any X flaw strength value the most likely Y flaw strength at half the remaining area is given by the following:

$$\sigma_Y = \sigma_o \left[-\ln 0.5 + (\sigma_X/\sigma_o)^m\right]^{1/m}$$

In practice the Weibull distribution parameters are first estimated by considering only type Y flaws. All the X flaw failure strengths are adjusted to their most likely Y flaw failure strengths.

The original Y flaw failure strengths and the adjusted X flaws are used to compute a new set of Weibull parameters using the previously described regression analysis. The process of adjusting the X flaws and recomputing the parameters is repeated until the new computed parameters differ from the previous computed parameters by some arbitrarily chosen small values. This process determines the Y type flaw parameters; the process is inverted, and the X type flaw parameters are obtained. These two sets of Weibull distribution parameters are used to evaluate component reliability.

Gasifier Turbine Nozzle

Allison's IGT 404 engine was selected for introduction of structural ceramics into a gas turbine engine. The gasifier turbine nozzle was an ideal component for this introduction. The metal nozzle parts were replaced by ceramics, and the turbine inlet temperature was increased to improve performance. The introduction of structural ceramics was accomplished in three steps, with the rotor inlet temperature determining the configuration.

The first step was the 1900°F configuration

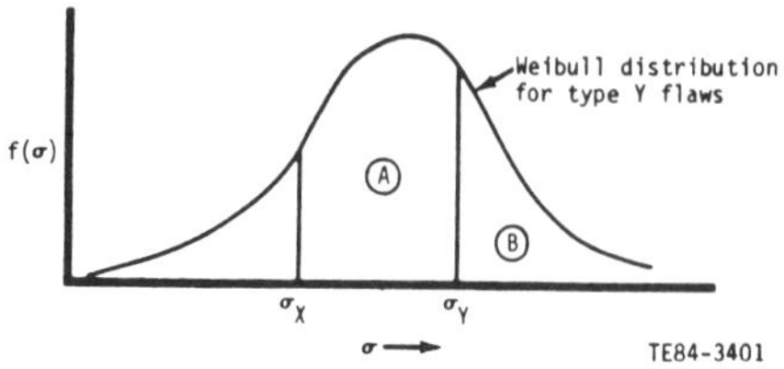

Figure 226. Weibull distribution for type Y flaws.

where every other metal vane was replaced by a ceramic vane. The turbine tip shroud was also replaced by a ceramic ring in an alternate assembly. This configuration addressed the preliminary design, fabrication, and testing problems of structural ceramics with a minimum of change to the basic engine.

The second step for introduction was the design and development of an all ceramic gasifier nozzle—the 2070°F configuration, shown in Figure 5. The lessons learned from design and testing of the 1900°F configuration were carried over to the 2070°F configuration. This second step represented the major effort during the CATE program.

The third step for introduction was the 2265°F configuration. This step used the same configuration as the 2070°F design and was designed to determine the capabilities of ceramics at elevated temperatures. Funding limitations only allowed for preliminary design work to be accomplished.

The following information is divided into discussions of the different configurations. Each subsection will cover the design, development, and testing of the ceramics for a particular configuration.

1900°F CONFIGURATION

Summary

The 1900°F configuration was the initial step for introduction of structural ceramics into a turbine engine. This configuration replaced some metal vanes and the metal shroud with ceramic components to provide a basis for future designs. Finite element heat transfer and stress analyses were performed for each of the ceramic components. Testing was conducted to support the analyses and to qualify parts for engine testing. This subsection will cover a description of the configuration, the analyses performed, and proof testing conducted on the ceramic components.

Objective

The objective of the 1900°F configuration was to introduce ceramic materials quickly into the engine operating environment with a minimum of risk to the engine and a minimum of change to the existing hardware. This introduction would address the basic fabrication, analysis, and testing problems and give valuable experience for the capabilities of structural ceramics.

Approach

Ceramic parts were substituted for metal parts in the 1900°F configuration. The components selected to achieve these objectives were the gasifier nozzle vanes and the gasifier turbine tip shroud. Each ceramic component was used in separate assemblies; therefore, the ceramic vanes and tip shroud could not be tested simultaneously. Testing concentrated on only one component at a time. Heat transfer and stress analyses were performed on each component to determine critical operating modes and high stress areas. Finally, before installing parts in an engine, thermal shock rig testing was conducted to expose each component to simulated engine operational stress levels.

Discussion: Design

Vanes

The initial design for introducing ceramics into the engine utilized the air-cooled metal gasifier turbine nozzle from the IGT 404-4 engine. Every other air-cooled vane was removed creating a recess in the inner and outer bands. Ceramic vanes of identical shape to the metal vane were placed in the recesses and retained with sheet-metal caps, as shown in Figure 227. This geometric configuration provided simply supported ceramic vanes and subjected them only to aerodynamic and thermal loads. The normal vane function of reacting pressure load on the outer band and centering structure for the tip shroud were achieved by the remaining 10 air-cooled metal vanes.

Vane temperatures, temperature gradients, and stresses were calculated by utilizing finite element techniques. Both two-dimensional (2-D) and three-dimensional (3-D) finite element models were used in EDR 9346 (Ref 9), pp 89-91.

The conditions selected for transient heat transfer analysis included an acceleration (dynamic braking to maximum gas temperatures) and a deceleration (from maximum gas temperatures to stabilized dynamic braking). Analyses indicated that the maximum temperature differences in the parts and thereby the worst thermal conditions occurred at 3 sec into acceleration and 8 sec into deceleration.

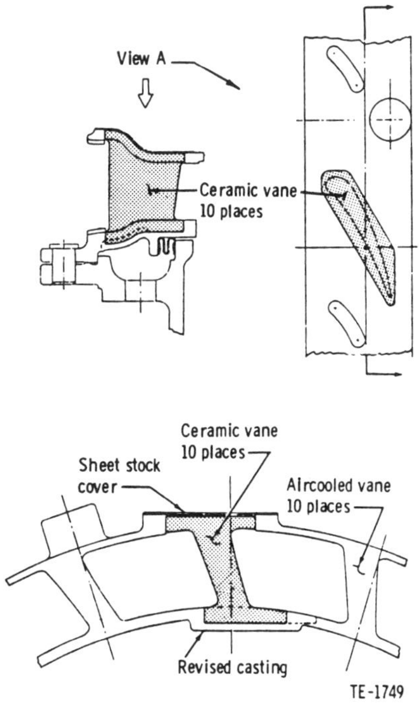

Figure 227. Gasifier nozzle with ceramic test vanes.

Table XXXIII shows the results of the analyses for two candidate materials: reaction-bonded silicon carbide and reaction-bonded silicon nitride. The deceleration was not analyzed for silicon nitride because the SiC vane peak stress occurred during acceleration.

Figures 228 and 229 show typical stress distributions for acceleration and deceleration respectively. Based on the reported strength of the ceramic materials, the analyses indicated that some vane failures would be expected during thermal shock rig

Table XXXIII.
Results of stress analysis for 1900°F configuration vanes.

Material	Peak radial stress—MPa (ksi) Acceleration	Deceleration
Reaction-bonded silicon carbide	382.62 (55.50)	50.84 (7.13)
Reaction-bonded silicon nitride	227.71 (33.03)	—

testing. Additional information on the analyses were also reported in EDR 9346 (Ref 9), pp 88-97.

Shroud

The incorporation of a ceramic shroud in the gasifier turbine section also used a reworked, air-cooled metal nozzle assembly from the IGT 404-4 engine. The assembly utilized the existing vane row casting and two metal support structures in Figure 230. Equally spaced lugs on the ceramic shroud ring outer diameter centered it by meeting with five equally spaced pegs on the support structure. Both sides of the ceramic ring were sealed with fiber-

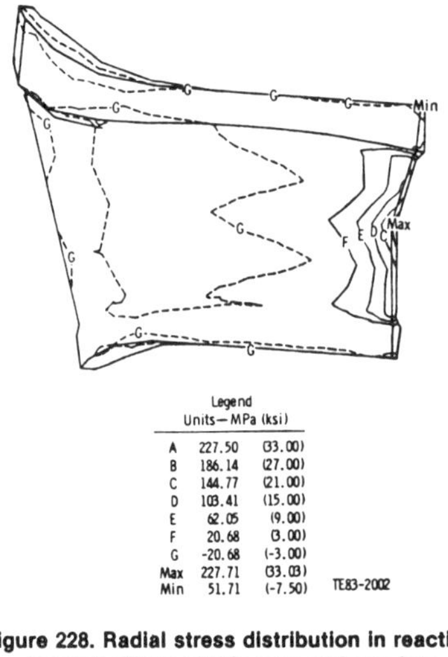

Legend
Units—MPa (ksi)

A	227.50	(33.00)
B	186.14	(27.00)
C	144.77	(21.00)
D	103.41	(15.00)
E	62.05	(9.00)
F	20.68	(3.00)
G	-20.68	(-3.00)
Max	227.71	(33.03)
Min	51.71	(-7.50)

TE83-2002

Figure 228. Radial stress distribution in reaction-bonded silicon nitride vane at 3 sec into acceleration from dynamic braking to maximum gas temperature.

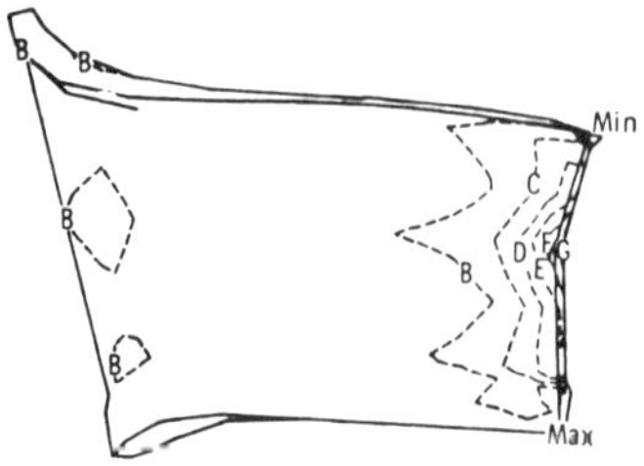

Legend
Units—MPa (ksi)

A	48.26	(7.00)
B	-6.89	(-1.00)
C	-62.05	(-9.00)
D	-117.20	(-17.00)
E	-172.35	(-25.00)
F	-227.50	(-33.00)
G	-282.65	(-41.00)
Max	49.15	(7.13)
Min	-331.53	(-48.09)

TE83-2003

Figure 229. Radial stress distribution in RBSiC vane at 8 sec into deceleration from maximum gas temperature to dynamic braking.

glass rope packing, which positioned it axially. Thus, all the primary load was carried by the metal support structure.

A shroud inner diameter (i.d.) abradable coating was incorporated in several parts. This coating was designed to reduce blade tip clearance and improve performance, which was an overall program goal.

Shroud Analysis

Finite element analysis was used to predict thermal gradients and stresses during engine operation. Both 2-D and 3-D finite element models were used in EDR 9346 (Ref 9), Figures 53, 60, and 61. The conditions selected for transient heat transfer analysis included an acceleration (rapid fire-up from cold start to maximum gas temperature) and a deceleration (from maximum gas temperature to stabilized dynamic braking). Analyses indicated that the maximum gradients occurred at 7 sec into acceleration and 8 sec into deceleration. Stress analysis was performed for five candidate materials with an abradable coating. The results are presented in Table XXXIV.

Typical stress distribution plots for RBSiC abradable shrouds are shown in Figures 231 and 232. Additional analysis was performed on RBSiC and

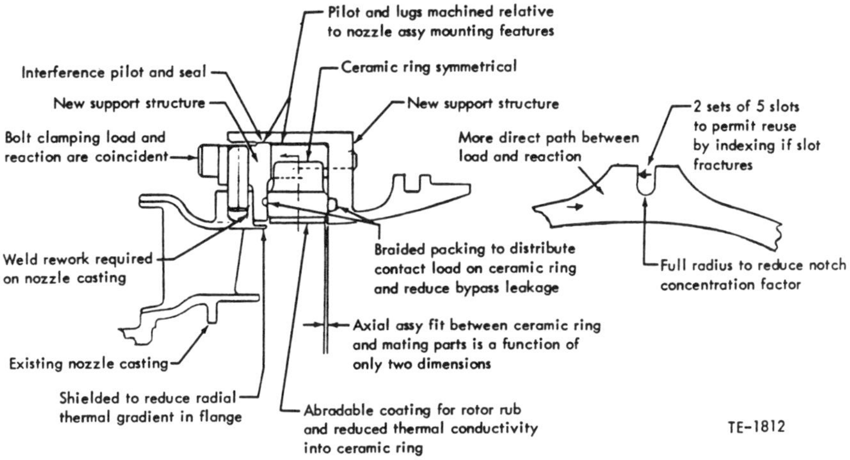

Figure 230. Gasifier turbine tip shroud design.

Table XXXIV.
Gasifier turbine rotor ceramic tip shroud 2-D stress results.

	Stress: max/min—MPa (ksi)							
	Fire-up				Braking			
	Base material		Coating		Base material		Coating	
Shroud material*	Tangential	Radial	Tangential	Radial	Tangential	Radial	Tangential	Radial
Reaction-bonded silicon carbide	29.6/-13.8 (4.3/-2.0)	0.0/-38.6 (0.0/-5.6)	0.0/-104.8 (0.0/-15.2)	-5.5/-33.1 (-0.8/-4.8)	27.6/-38.6 (4.0/-5.6)	22.8/2.8 (3.3/0.4)	75.8/0.0 (11.0/0.0)	19.3/2.9 (2.8/0.4)
Sintered silicon carbide	31.7/-13.8 (4.6/-2.0)	0.0/-27.6 (0.0/-4.0)	0.0/-88.9 (0.0/-12.9)	3.5/-29.0 (-0.5/-4.2)	34.5/-81.4 (5.0/-11.8)	122.0/13.8 (17.7/2.0)	181.3/0.0 (26.3/0.0)	96.5/-1.4 (14.0/-0.2)
Sintered silicon nitride	46.2/0.0 (6.7/0.0)	0.0/-68.9 (0.0/-10.0)	0.0/-165.5 (0.0/-24.0)	-6.9/-71.7 (-1.0/-10.4)	13.8/-29.6 (2.0/-4.3)	40.0/5.5 (5.8/0.8)	100.0/-13.8 (14.5/-2.0)	38.6/11.0 (5.6/1.6)
Cordierite (CD-1) (MAS)	17.9/0.0 (2.6/0.0)	0.0/-48.3 (0.0/-7.0)	0.0/-111.7 (0.0/-16.2)	0.0/-48.3 (0.0/-7.0)	6.9/-16.5 (1.0/-2.4)	27.6/3.4 (4.0/0.5)	70.3/-6.9 (10.2/-1.0)	24.1/6.9 (3.5/1.0)
CER VIT C132 (LAS)	0.0/-2.1 (0.0/-0.3)	4.1/0.0 (0.6/0.0)	9.0/0.0 (1.3/0.0)	4.1/0.0 (0.6/0.0)	1.4/0.7 (0.2/-0.1)	0.0/-2.1 (0.0/-0.3)	0.0/5.5 (0.0/-0.8)	0.0/-2.1 (0.0/0.3)

* Bore coated with low-density abradable coating

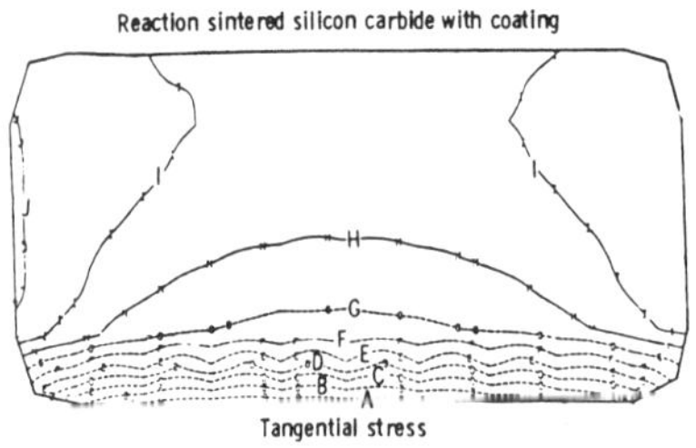

Tangential plot legend

	MPa	(ksi)
A	- 96.5	(-14.0)
B	- 82.7	(-12.0)
C	- 68.9	(-10.0)
D	- 55.2	(- 8.0)
E	- 41.4	(- 6.0)
F	- 13.8	(- 2.0)
H	0.0	(0.0)
I	13.8	(2.0)
J	27.6	(4.0)
Max	29.6	(4.3)
Min	-104.8	(-15.2)

TE-1817

Figure 231. Ceramic shroud ring isostress plot at 7 sec into fire-up.

sintered Si_3N_4 nonabradable shrouds, and the results are shown in Table XXXV. The acceleration is not shown because the nonabradable shroud peak stress occurred during deceleration.

Figure 233 shows the stress distribution for an RBSiC nonabradable shroud.

This analysis shows that the nonabradable shrouds develop higher stresses than the abradable shrouds. However, based on the reported material strengths of the ceramics, a shroud would probably not be broken during thermal shock rig testing.

Discussion: Process Development

Vanes

The 1900°F configuration vanes were initially ordered from two suppliers: RBSiC from Carborundum Company and reaction-bonded silicon nitride from AiResearch Casting Company.

Initial RBSiC vanes contained internal voids, pits, cracks, and blisters. Revisions of molding and binder burnout parameters reduced the severity of these problems. An additional problem was the existence of a low strength silicon rich layer on the vane surface. Carborundum revised the processing

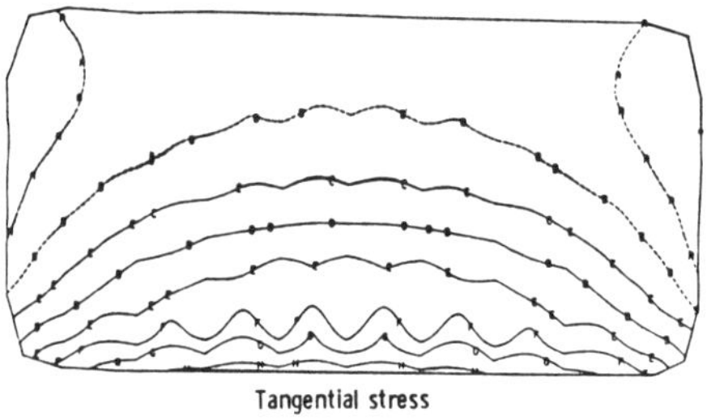

Stress legend

MPa	Symbol	(ksi)
-27.6	A	(- 4.0)
-13.8	B	(- 2.0)
0.0	C	(0.0)
13.8	D	(2.0)
27.6	E	(4.0)
41.4	F	(6.0)
55.2	G	(8.0)
68.9	H	(10.0)
75.7	Max	(11.0)
-38.5	Min	(-5.6)

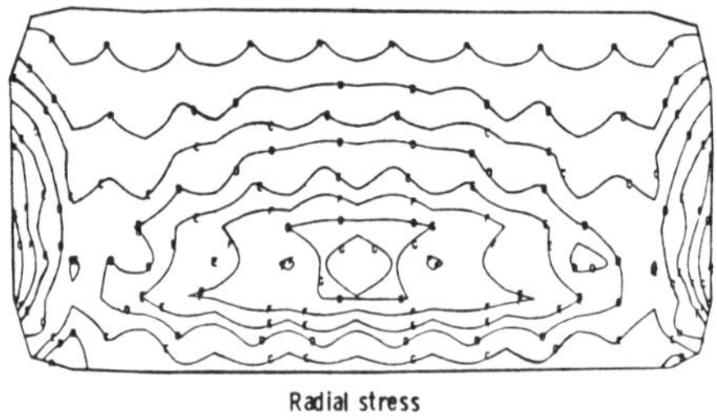

Stress legend

MPa	Symbol	(ksi)
2.8	A	(0.4)
5.5	B	(0.8)
8.3	C	(1.2)
11.0	D	(1.6)
13.8	E	(2.0)
16.5	F	(2.4)
19.3	G	(2.8)
22.1	H	(3.2)
22.5	Max	(3.3)

TE-1825

Figure 232. Ceramic shroud ring isostress plot at 8 sec into braking.

Table XXXV.
Peak stress for nonabradable shrouds at 8 sec into deceleration.

Stress	Sintered Si_3N_4	Reaction-bonded SiC
Tangential max/min—MPa (ksi)	55.2/-24.1 (8.0/-3.5)	162.0/-58.6 (23.5/-8.5)
Radial max/min—MPa (ksi)	12.4/-1.4 (1.8/-0.2)	41.2/5.5 (6.0/0.8)

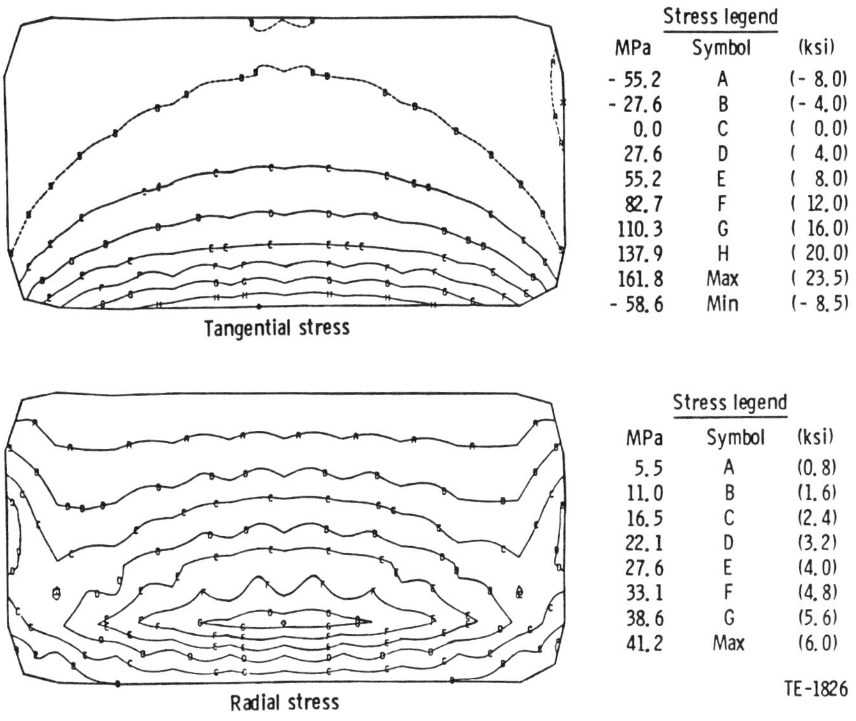

Figure 233. Stress distribution for RBSiC shroud ring: no abradable bore coating (8 sec into braking).

parameters to minimize this layer and to minimize the laminations and voids resulting from release of gas in the binder burnout cycle. These changes increased the mean MOR strength from 262 MPa (38 ksi) to 331 MPa (48 ksi).

Additional information on the processing and NDI results can be found in past reports (EDR 9346—Ref 9, pp 150-154, 195-201; EDR 9519—Ref 1, p 69; and EDR 9951—Ref 3, pp 79-80, 152).

The initial silicon nitride vanes received contained internal voids, surface scratches, porous trailing-edge areas, cracks, and high-density inclusions. Processing changes improved the quality and surface finish of the vanes. Additional information on the processing and NDI results can be found in the following reports: EDR 9346 (Ref 9), pp 152-53, 195-201; EDR 9722 (Ref 2), pp 57-59; and EDR 9951 (Ref 3), pp 73-74, 79.

Additional vanes were ordered from Pure Carbon Company, Rosenthal Technik AG, and Annawerk Company. Only four sample vanes from Pure Carbon were received in time for testing. The other orders were received after testing was completed on the 1900°F configuration.

Shrouds

Initial 1900°F-configuration shrouds were ordered from two suppliers. Carborundum made both abradable and nonabradable RBSiC shrouds while GTE Sylvania made a nonabradable sintered silicon nitride. Abradable coating development is discussed in Section III.

Early shrouds received exhibited numerous areas of porosity, edge cracking, and circumferential lamination. These shrouds were processed by cold die pressing. This procedure was abandoned in favor of warm compression molding, which greatly reduced the extent of porosity and cracks. The only further process development was to vary the density of the part to determine the effect on the abradable surface (see Section III). Additional information on fabrication and NDI results are reported in EDR 9346 (Ref 9), pp 152, 194-201, and EDR 9519 (Ref 1), pp 29-34, 69.

The shroud rings received from GTE Sylvania were generally of good quality. Occasional high-density inclusions formed by impurities in the base powder were found by X-ray. Subsequent screening of the powder removed these inclusions. Additional information on the NDI results can be found in the following reports: EDR 9346 (Ref 9), pp 155, 195-201, and EDR 9519 (Ref 1), pp 31-38.

Three additional abradable shroud rings were ordered from Corning. These rings were made from Corning code 9458 glass ceramic with a foamed ceramic cement (Corning code 9457) abradable surface. No quality problems were observed with these rings, and no process development was conducted.

Discussion: Rig Testing

As current NDI techniques can only detect gross material defects as far as critical size flaws are concerned, it is necessary to eliminate nondetectable flaws through a series of proof tests. These proof tests subject the ceramic component to thermal and mechanical loads and stresses similar to the engine operating environment. Ceramic components that fail in rig proof test because of flaws and reduced material strength would be those that had a high probability of failure in the engine. Thus the probability of failure during engine test is greatly reduced. These proof tests, however, do not take into account the effect of long-term failure modes of the ceramic material. The following tests qualify a ceramic component for engine test:

- cold flow rig—measures flow capacity of vane row for performance evaluation and aerodynamic proof load
- hoop pressurization fixture—mechanical proof load evaluates shroud material strength
- vibration rig—evaluates response of metal support structure and effect on ceramic components
- thermal shock rig—induces maximum thermal and mechanical proof loads experienced in engine testing

The vibration rig was run only for one set of vane and shroud hardware to check for vibrational problems. The other three tests were run on all ceramic components that were to be engine tested.

Cold Flow Rig

The purpose of the cold flow rig is twofold. First, it provides a measure of mass flow versus inlet total-to-exit static pressure ratio and represents the flow characteristics of that gasifier vane row—a primary component in control of engine pressure ratio match. The flow curve becomes an element in engine performance analysis. Second, the cold flow test subjects the vanes to an aerodynamic proof load similar to that present in the engine. Note that the gasifier nozzle assembly was rig tested to a total-to-static pressure ratio of 2.3:1. This subjected the vanes to the same pressure differential that occurred during engine operation at a 1.5:1 nozzle pressure ratio.

The cold flow rig is a large chamber with inlet and exit features as shown in Figure 234. Pressurized air at room temperature is admitted in the inlet. The gasifier nozzle assembly is attached over the exit. All air leakage points are sealed except for the nozzle vane row—similar to the engine installation. By measuring the inlet total-to-exit static pressure ratio at different mass flow rates, the nozzle flow characteristic curve can be generated.

The aerodynamic proof load previously mentioned did not produce the critical stresses in the vanes. A simple analysis of vane aerodynamic bending stresses showed the maximum stress was 3.1 MPa (0.45 ksi) in tension. This stress was well below the average material strength of 310 MPa (45 ksi) for silicon carbide. As was stated in the design section,

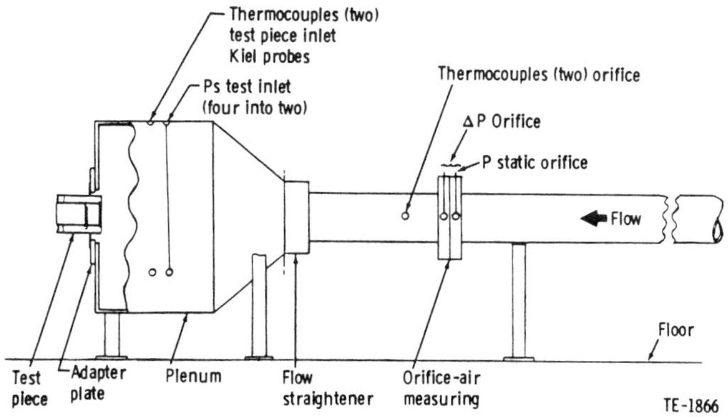

Figure 234. Nozzle cold flow rig.

the transient thermal stresses are critical. The cold flow rig aerodynamic loads on the shroud are similar in that they are not critical.

A typical flow curve appears in Figure 235. Two nozzle assemblies have been tested. The silicon carbide vane assembly was measured at 0.775 kg/s (1.71 lb/sec) at the design pressure ratio of 1.50:1. The ceramic shroud nozzle assembly flow was 0.737 kg/s (1.625 lb/sec) at the same conditions. The difference between the flow values is caused by two effects. The ceramic vane assembly has additional leak paths around the ends of the vane platforms and permits additional airflow. The ceramic shroud assembly has a full set of 20 air-cooled metal vanes in which the vane cooling exit windows have associated with them an aerodynamic loss that reduces the nozzle flow.

Hoop Pressurization Fixture

The purpose of this test was to subject the shroud ring to a stress level that the part will be exposed to during engine testing. Another purpose of the fixture was to determine component strength during a burst test. The fixture loaded the inside diameter of the shroud ring with an internal pressure applied through a thick-walled rubber bladder pressurized with hydraulic fluid. The internal pressure produced hoop stresses around the shroud circumference.

Since the test fixture-generated hoop stresses are mechanically induced and the engine stresses are thermally induced, they have different stress distributions. A 2-D finite element analysis was used to compare the distributions and match the desired stress level with the corresponding internal pressure.

Thirteen shrouds were tested in this fixture. The results are presented in Tables XXXVI and XXXVII. Additional information on these tests can be found in the following reports: EDR 9346 (Ref 9),

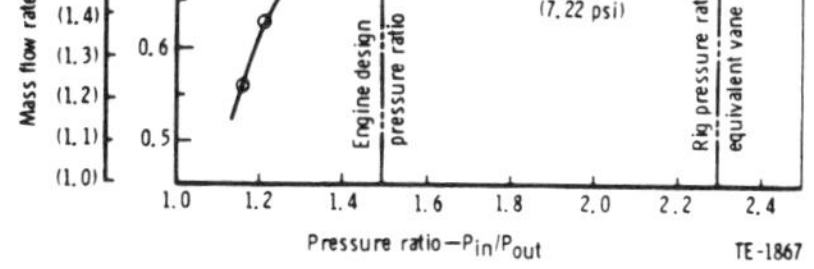

Figure 235. Typical mass flow rate curve—gasifier nozzle assembly.

pp 155-57, 205-207; EDR 9519 (Ref 1), pp 31, 34; and EDR 9722 (Ref 2), pp 52-55.

Vibration Rig

The purpose of the vibration test was to determine if the ceramic components displayed any distress when subjected to vibration levels equal to the engine vibration limits. This was accomplished by mounting the ceramic components in the metal support structure and vibrating the assembly in four different axes through a range of frequencies. The axes tested were one axial and three radial, and the frequencies scanned were 20 Hz to 2000 Hz. The driving force was 1 g from 20 Hz to 131 Hz and 7.6 mm/s (0.3 in./sec) average velocity from 131 Hz to 2000 Hz. Accelerometers were mounted at various locations on the assembly to determine the response of the structure to the driving force. This test is not a material strength proof test but a check of the stability of the metal support structure and its interface with the ceramic component. For this reason, only one vane assembly and one shroud assembly were tested to qualify the design.

When the vane assembly was vibrated, only two ceramic vanes (silicon carbide) were installed because of vane availability at the time of test. The shroud assembly used a silicon nitride shroud. The shaker tables used were standard industrial test equipment.

For both the vane and shroud assemblies the results of the frequency scan indicated there was no excessive response of the ceramic components or the metal support structure to the driving forces imposed. FPI examination of the parts after test disclosed no cracks or distress. Thus the ceramic components and their metal support structure were vibrationally acceptable for experimental use in the 1038°C (1900°F) CATE engine.

Table XXXVI.
Ceramic 1900°F shrouds passing bladder proof test.

Material	Manufacturer
Three nonabradable SiC	Carborundum
Four abradable SiC	Carborundum
Two nonabradable Si_3N_4	GTE

Thermal Shock Rig

The purpose of the thermal shock rig is to combine all factors (mechanical, aerodynamic, and thermal) that will occur in engine testing into a single rig proof test. The airflow, fuel-flow rate, and air temperature were varied to simulate steady-state and transient engine operation. Thus, the ceramic components were subjected to a maximum engine condition proof load prior to actual engine testing. The test cycle is shown in Figure 236.

Table XXXVIII is a summary of 25 rig builds for 1900°F-configuration vane proof testing over a period of 3.5 years. In addition to testing vanes from three different suppliers that met inspection standards, rejected vanes were also tested to determine tolerance to defects when exposed to thermal cycles.

Failure analysis was conducted on the 34 failed vanes to determine correlation with the finite element analysis performed. Fifteen vanes failed from the predicted high stress area at the trailing edge of the vanes. Eleven additional vanes failed from preexisting flaws or mechanical loading; the cause of failure for the eight remaining vanes was not determined. Additional detailed information on this testing can be found in the following reports: EDR 9346 (Ref 9), pp 209-11; EDR 9519 (Ref 1), pp 71-73; EDR 9722 (Ref 2), pp 77-81; EDR 9951 (Ref 3), 153-59; and EDR 10156 (Ref 4), p 71.

A total of 11 shrouds were tested in the thermal shock rig with only two failures. Both failures were attributed to improper installation of the fiberglass rope packing in EDR 9346 (Ref 9), p 158. Both failed rings were nonabradable RBSiC shrouds manufactured by Carborundum. Table XXXIX shows the shrouds that successfully passed five cycles (see Figure 236) in the thermal shock rig.

Table XXXVII.
Results of burst testing of ceramic 1900°F shrouds.

Supplier	Material	Burst pressure kPa	Burst pressure (psig)	Peak stress MPa	Peak stress (ksi)
Carborundum	Nonabradable SiC	6,480	(940)	235.0	(34.1)
Carborundum	Abradable SiC	6,308	(915)	228.91	(33.2)
GTE	Nonabradable Si_3N_4	9,169	(1330)	266.8	(38.7)
GTE	Nonabradable Si_3N_4	11,099	(1610)	323.3	(46.9)

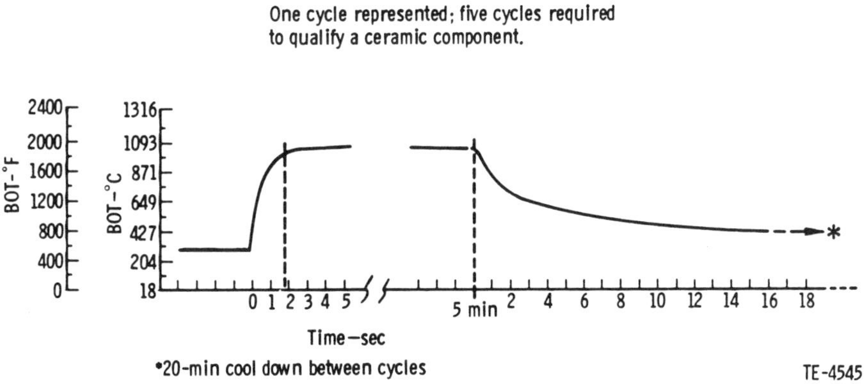

Figure 236. Typical burner outlet temperature trace.

Table XXXVIII.
Summary of 1900°F-configuration vane thermal shock rig proof testing.

Vendor and material	Quantity tested	Failed	
		Quantity	Percentage
Carborundum RBSiC	135	21	16
Carborundum RBSiC (rejects)	26	4	15
Pure Carbon RBSiC (Refel samples)	4	0	0
AiResearch $RBSi_3N_4$	75	7	9
AiResearch $RBSi_3N_4$ (rejects)	5	2	40
Totals	245	34	14

Table XXXIX.
Ceramic 1900°F-configuration shrouds passing thermal shock rig testing.

Ring	Manufacturer
1 Nonabradable SiC	Carborundum
4 Abradable SiC	Carborundum
2 Nonabradable Si_3N_4	GTE
2 Abradable LAS	Corning

2070°F CONFIGURATION

Summary

The 2070°F configuration was the intermediate step in the CATE program for introduction of structural ceramics into a gas turbine engine. This configuration included a full ceramic flow path for the gasifier turbine to accommodate the higher rotor inlet temperature. Finite element heat transfer and stress analyses were performed on each of the ceramic components to determine peak stresses and probability of survival. Thermal shock rig testing supported the analyses and was used to qualify the parts for engine testing. This subsection will cover a description of the configuration finite element analyses performed and proof testing conducted on the ceramic components.

Objective

The objective of the 2070°F configuration was to design and develop an all-ceramic gasifier turbine nozzle assembly. The design included selecting appropriate materials and configuration to attain a desired probability of survival. The development included fabrication and proof testing of the components to provide qualified parts for engine testing.

Approach

The approach used for the design of the 2070°F-configuration components was to perform finite element analyses on each component to determine critical operating modes and high stress areas. These results determined what, if any, modifications were necessary for the configuration. Following completion of the analyses, the parts were installed into a thermal shock rig to expose them to simulated engine operational stress levels to support the analyses and to qualify the parts for engine operation.

Discussion: Design

The 2070°F configuration was designed and developed using the following guidelines:

- The design must incorporate ceramic material for the gasifier stage—inlet duct, vanes, rotor blades, and rotor tip shroud.
- There should be a minimum of change to the non-flow-path structure of the baseline engine.
- The design should be adaptable to the 1241°C (2265°F) engine design with a minimum amount of change.
- Space must be allowed for the 1241°C (2265°F) engine two-stage power turbine.
- The mechanical load in the ceramic parts should be minimized.
- More than one type of material should be considered for each ceramic component.
- The ceramic vane exit annulus must match the inlet annulus of the baseline metal rotor, since initial operation of the 1132°C (2070°F) ceramic design will be with the existing baseline engine metal rotor.
- The ceramic rotor exit annulus is compatible with the existing gasifier/power-turbine transition duct.
- The number of vane airfoils is compatible with the ceramic blade vibration characteristics.

Figure 237 presents the final 2070°F-configuration ceramic gasifier nozzle assembly design derived by using the guidelines. This design consists of an inner metal support structure (-1) attached to and centered on the gasifier bearing support. Four equally spaced hollow metal struts (-2) engage four radial holes in the inner metal support structure to provide centering. The outer end of the struts are bolted to an outer metal support ring (-3). The outer metal support structure is centered on the gasifier shaft center by the four metal struts. The slip fit on the inner end of the struts allows for differential thermal expansion. The back side of the outer

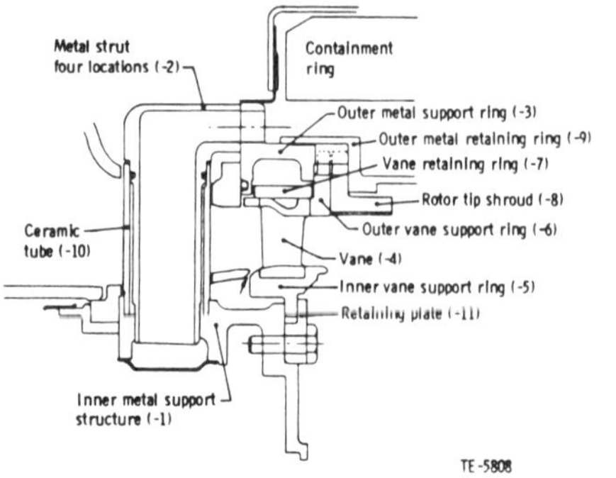

Figure 237. Ceramic gasifier nozzle assembly for 2070°F configuration.

metal support structure has radial keys to provide a centering feature for the ceramic tip shroud (-8). This metal substructure supports and centers the stationary ceramic components.

The ceramic vanes (-4) are airfoils with small load-reacting feet at each end. The ceramic inner vane support ring (-5) is centered on the inner metal support structure and retained by a plate (-11). The ceramic inner vane support ring contains 28 pockets on its outside diameter to accept the vane inner feet. The ceramic outer vane support ring (-6) is cross-keyed to the outer metal support ring (-3) and contains 28 holes to locate the vane outer feet. To retain the vanes radially, a ceramic vane retaining ring (-7) is slipped over the ceramic outer vane support ring. The gasifier rotor ceramic tip shroud (-8) is centered off the outer metal support ring (-3) by a cross-key arrangement.

All vane aerodynamic and component diaphragm pressure loads are transmitted through the ceramic support rings to the metal support rings that are interconnected by the hollow metal struts. Thus, the rotor tip shroud is centered off the same support as the gasifier rotor. Since the metal struts pass through the hot gas path, a thermal barrier (-10) is provided around the struts. The barrier selected was a separate ceramic shell with air gap around the strut. To further control the strut metal temperature, secondary cooling air is ducted through the struts and is later merged with the primary flow.

The basic configuration described here eliminates any significant mechanical loads in the ceramic components, provides unrestrained freedom for thermal growth, and centers the tip shroud to the same center as the rotor.

Component temperatures, temperature gradients, and stresses were calculated utilizing finite element techniques. Both two-dimensional (2-D) and three-dimensional (3-D) finite element models were used. The analyses were modified and updated when additional information became available. No finite element analysis was conducted on the retaining ring or strut shells because of their simple shapes.

The conditions selected for heat transfer analysis for the components were steady-state operation, acceleration (dynamic breaking to maximum gas temperature), and deceleration (maximum gas temperature to stabilized dynamic braking).

The heat transfer analysis indicated the maximum temperature differences in the vane occurred at 2 sec into the acceleration and at 4 sec into the deceleration. Stress analysis predicted a maximum principal stress of 317.1 MPa (46.0 ksi) at 2 sec into acceleration, as shown in Figure 238. This stress is twice the peak stress calculated for deceleration. The probability of survival associated with the acceleration is 0.98281, assuming σ_{os} = 774.1 MPa (50 ksi), σ_{ov} = 626.2 MPa (27 ksi), and Weibull modulus (m) = 8. Additional information on the analyses can be found in EDR 9722 (Ref 2), pp 61-65; EDR 9951 (Ref 3), pp 122-28; and EDR 10383 (Ref 5), pp 73 and 74.

Finite element analyses were conducted on the inner and outer vane support rings and the shroud ring. The initial analyses were modified and updated based on the results of cascade rig testing, thermal shock rig testing, and slight geometry changes. Major modifications to the analyses were based on the results of instrumented nozzle assembly testing.

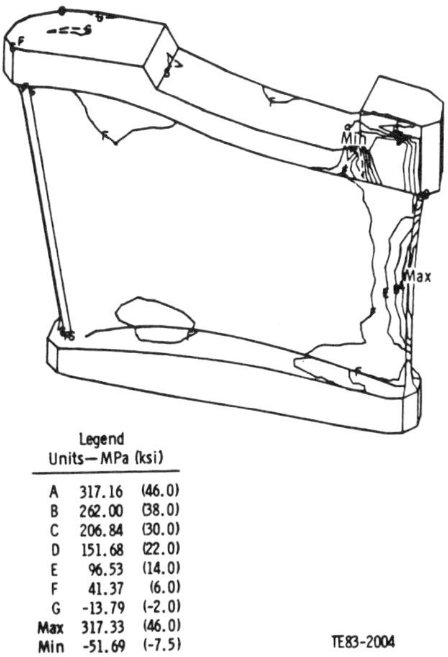

Figure 238. Maximum principal stress distribution in RBSiC at 2 sec into acceleration from dynamic braking to maximum gas temperature.

This testing measured the environment in the thermal shock rig and the engine.

The final stress results are shown in Table XL. The component probabilities of survival were con-

Table XL.
Calculated maximum principal stress and probability of survival for 2070°F conditions.

Component	Max principal stress MPa	(ksi)	Probability of survival*
Inner vane support ring accel at 18 sec	118.6	(17.2)	0.99526
Outer vane support ring accel at 18 sec	131.90	(19.0)	0.98551
Shroud ring decel at 12 sec	51.6	(7.45)	0.9999987

* This probability of survival assumes σ_{os} = 774.1 MPa (50.0 ksi), σ_{ov} = 626.3 MPa (27.0 ksi), and Weibull modulus = 8 for both.

sidered acceptable for testing in this program. Stress distribution plots showing the location of the maximum principal stress are presented in Figures 239, 240, and 241. The following subsection details the results of the instrumented nozzle assembly testing and additional analysis resulting from the failure in C-4 engine.

Discussion: Testing

A metal nozzle assembly, with components geometrically identical to the ceramic parts, was instrumented and tested in the thermal shock rig and the engine to measure the thermal environment and component temperatures for comparison with analytical predictions. The analytical predictions involve heat transfer correlation for the metal rings, heat transfer analysis for the ceramic rings, and the resulting stress analysis during both steady-state and transient operation. This subsection will detail the data reduction from the thermal shock rig testing along with the results of subsequent testing with modifications to the assembly in the engine.

To ensure that the instrumentation would be the most useful, locations were selected by the heat transfer group (see Figure 242). All thermocouple locations within the parts represent skin temperatures while the thermocouples located in the airstream and surrounding cavities represent air temperatures. The instrumentation was located at three positions circumferentially around the rings to provide an adequate data base. The locations were chosen to be 60 deg, 180 deg, and 300 deg clockwise, viewed from the rear. The 300-deg location is typically the hottest section based on the rotating probe data. The 60-deg location is the area where the shroud fractured into many pieces on BU6 of the thermal shock rig. The 180-deg location is a location where the inner and outer vane support rings have fractured during testing.

A test plan was established to subject the instrumented nozzle to the qualification cycle conditions shown in Figure 243. The tests were not run to the full 1132°C (2070°F) conditions because of the metal vanes being used.

The results of the instrumented nozzle assembly testing were interpreted in the following manner. The metal rings were first analyzed at steady-state rig conditions assumed for the previous 2-D analysis. The steady-state measurements recorded were compared with the analysis to determine if there

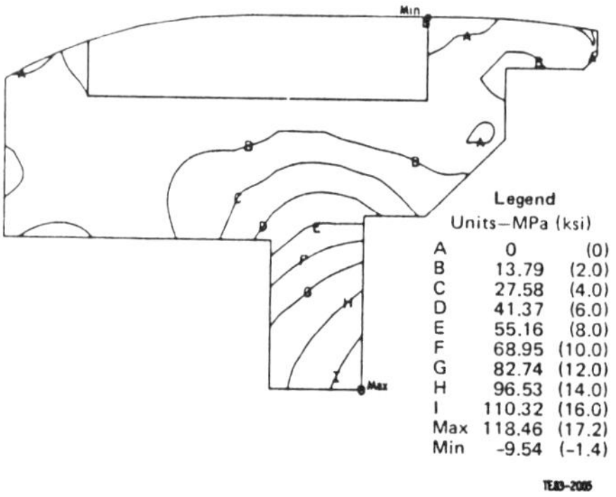

Figure 239. Maximum principal stress distribution in RBSIC inner vane support ring at 18 sec into acceleration from dynamic braking to maximum gas temperature.

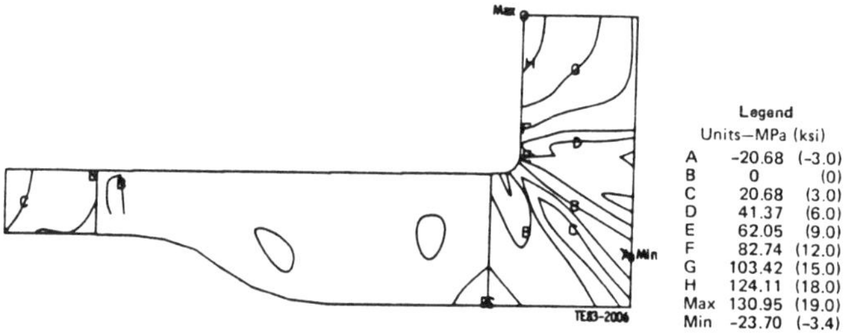

Figure 240. Maximum principal stress distribution in RBSiC outer vane support ring at 18 sec into acceleration from dynamic braking to maximum gas temperature.

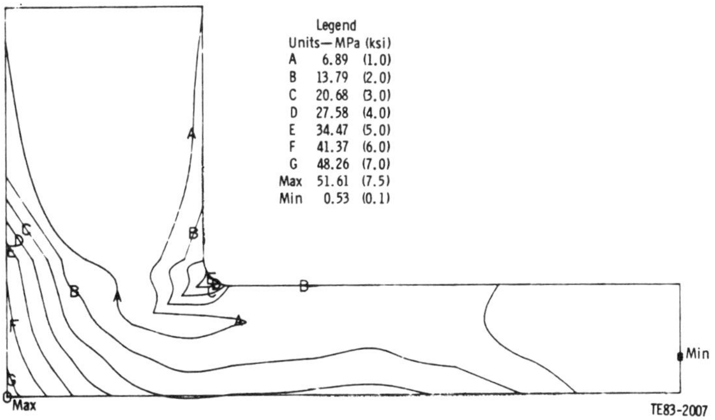

Figure 241. Maximum principal stress distribution in RBSIC shroud ring at 12 sec into deceleration from maximum gas temperature to stabilized dynamic braking.

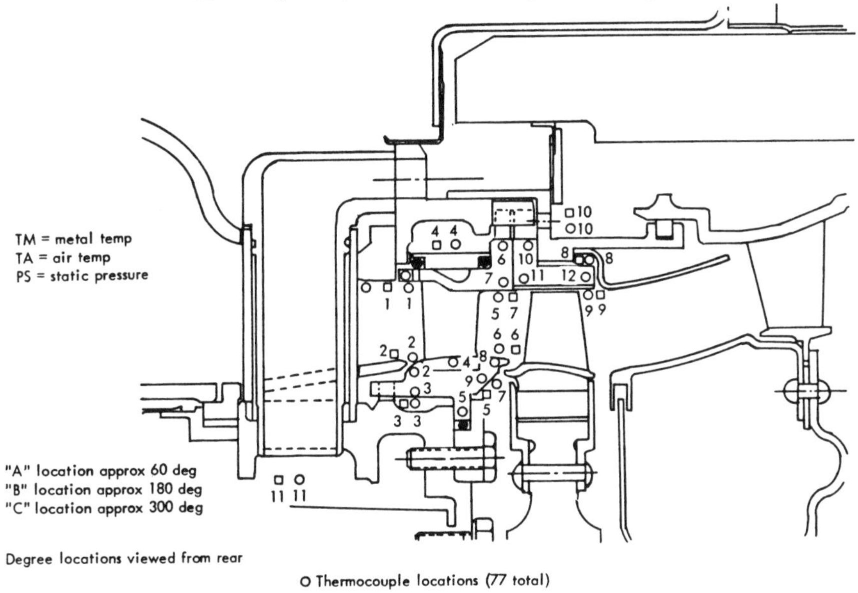

Figure 242. Location of instrumentation for testing.

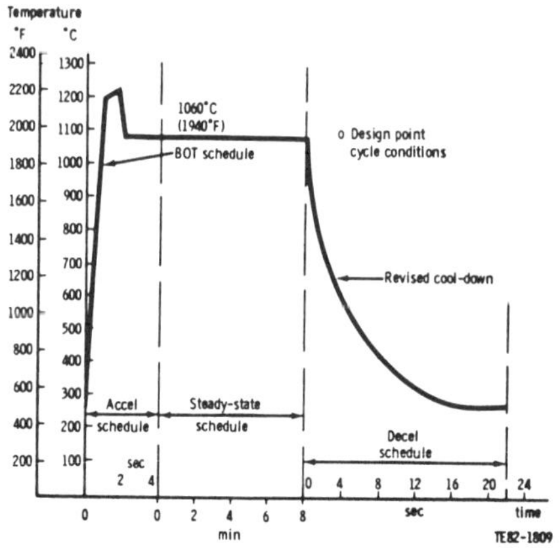

Figure 243. CATE program thermal shock rig qualification test schedule.

were any discrepancies. When a discrepancy was found, heat transfer coefficients around the ring were modified until the revised analysis matched the measured results. The changes involved adjusting heat transfer coefficients, changes in cooling airflow rate due to temperatures different from those assumed, adjustments to conductivity and heat capacity of the metal rings, and application of the measured boundary temperatures. These changes can all be verified by endwall cascade rig data or variations in material properties. A detailed description of this analysis process for each ring will follow.

Some general observations about the thermal shock rig were made after an examination of the data. The temperature gradient in each of the rings was significantly higher than was originally predicted. To calculate these gradients, the heat transfer coefficients in the flow path had to be increased significantly. For the outer vane support ring and the shroud ring, the revised cooling circuit had a more severe effect on the rings than was originally thought. In the inner vane support ring, increased gradients were due to leakage underneath the inner flange of the ring. The problems associated with each of these observations were corrected and the assembly was installed into engine C-4 to verify the results. Each ring will be discussed separately with a general summary following these discussions.

The steady-state temperatures, at specific locations calculated for the metal inner vane support ring at 1038°C (1900°F) inlet temperature, are shown in Figure 244. The boundary conditions shown for the original prediction are those used in the 2-D sensitivity study presented in EDR 10841 (Ref 7), pp 57-63. These were the best estimates of conditions in the thermal shock rig prior to this testing. The analysis assumes some leakage underneath the ring because this was seen during the testing. The calculated axial and radial gradients in the part were 0.6°C (1°F) and 195°C (351°F), respectively.

The measured steady-state temperatures used to compare with the original calculated temperatures are also shown in Figure 243. The three temperatures at each location correspond to the three circumferential locations. In some areas, the thermocouple was damaged or inoperable and no reading was obtained. The measured axial and radial gradients were 5.6°C to 68.3°C (10°F to 123°F) axially and 197.2°C to 275°C (355°F to 495°F) radially. The measured gas-path air temperatures were lower than originally predicted, while the cavity temperatures were significantly higher than predicted. The higher-than-predicted cavity temperatures indicated that gas-path air was leaking under the ring. Upon investigation, it was determined that the rope packing, normally inserted at the inside diameter of the flange, had been inadvertently omitted. Therefore, all analyses were performed with leakage assumed under the ring.

The revised steady-state prediction for the metal inner vane support ring is also shown in Figure 249. The prediction used the measured rig air temperatures at the appropriate positions around the ring. The revised heat transfer coefficients are also shown in this figure. The flow-path heat transfer coefficients increased from 607.6 W/m^2-°C to 760.9 W/m^2-°C (107 Btu/hr-ft^2-°F to 134 Btu/hr-ft^2-°F) at the leading edge of the ring and from 715.5 W/m^2-°C to 1396.8 W/m^2-°C (126 Btu/hr-ft^2-°F to 246 Btu/hr-ft^2-°F) at the trailing edge of the ring. These were obtained by an extrapolation of endwall cascade rig data. The data showed that the heat transfer coefficient increased rapidly downstream of the vane exit. This information was not used in the initial analysis because the data were thought to have been erroneous. The higher-than-predicted measured metal temperatures indicated that the cascade data were correct. The new heat transfer coefficients around the inside diameter flange were calculated from a revised flow calculation based on higher temperatures and a more restrictive area than originally assumed.

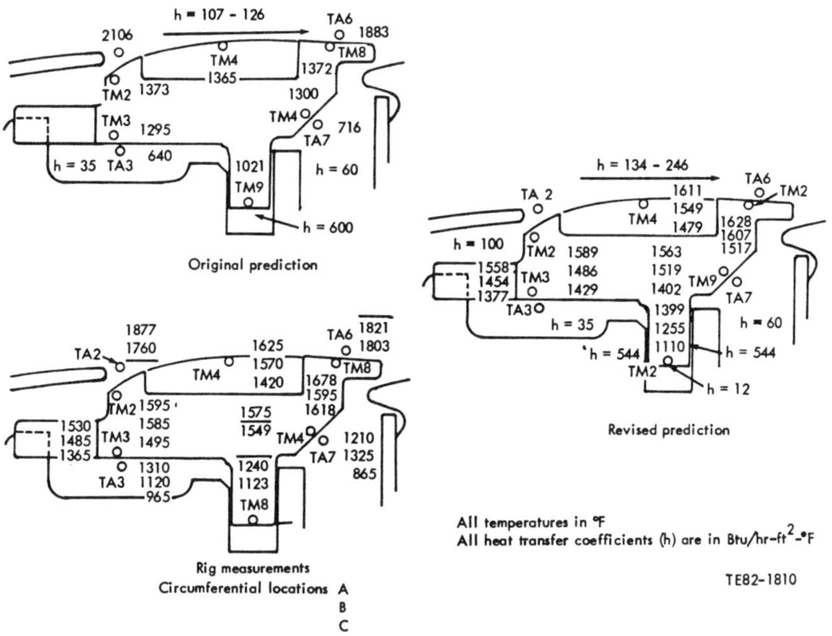

Figure 244. Metal inner vane support ring temperatures: original, measured, and revised analysis at 1038°C (1900°F) steady-state conditions with leakage.

The revised prediction, shown in Figure 244, indicates good correlation with the measured temperatures at steady-state conditions. The revised predicted axial gradient is 21.7°C to 67.2°C (39°F to 121°F) while the measured gradient was 5.6°C to 68.3°C (10°F to 123°F). The revised predicted radial gradient is 127.2°C to 226.1°C (229°F to 407°F) while the measured radial gradient was 197.2°C to 275°C (355°F to 495°F).

After correlation was obtained at steady-state conditions, efforts were applied toward matching calculated with measured transient temperatures. Measured transient temperatures were plotted and compared with the calculated temperatures. Comparisons at rig circumferential location "A" are shown in Figure 245. The measured radial gradient cannot be determined at that location, as the flange inside diameter thermocouple (TM5) failed. Therefore, the comparison for the acceleration transient was also examined at circumferential locations "B" and "C" (see Figure 246). Although correlation was not as close as at steady-state conditions, reasonable correlation had been obtained.

Once the measured temperatures had been matched, the new boundary conditions were applied to an RBSiC model of the inner vane support ring to provide temperature data for stress analysis. The results of this heat transfer analysis at 1132°C (2070°F) steady-state conditions are shown in Figure 247. The steady-state axial gradient in the part increased from 3.3°C (6°F) to a range of 19.4°C to 38.9°C (35°F to 70°F). The radial gradient changed from 135.6°C (244°F) to a range of 106.1°C to 171.7°C (191°F to 309°F). The highest gradients in the part occurred at steady state for the ceramic ring. This was contrary to what was measured and calculated for the metal ring. This was attributed to the higher thermal conductivity of RBSiC, as compared with

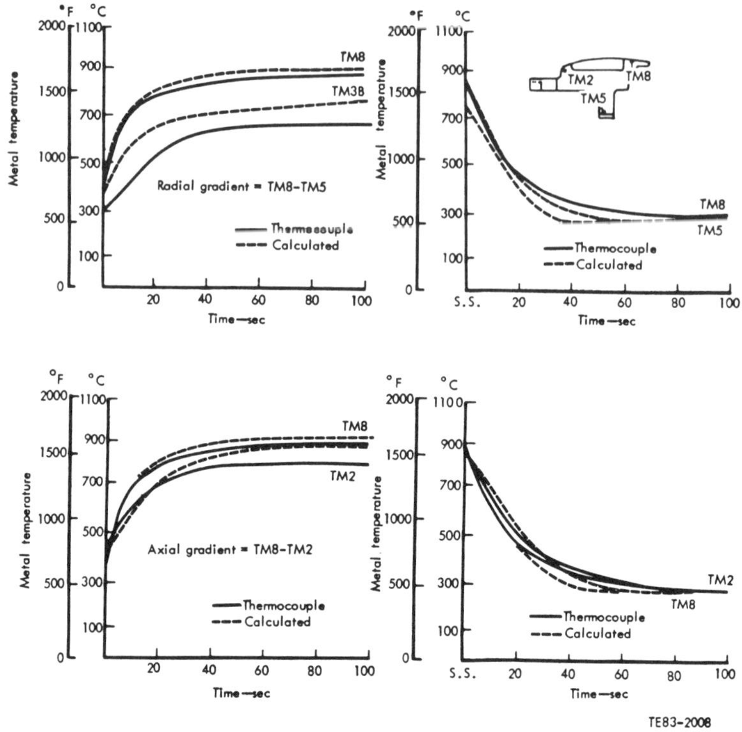

Figure 245. Metal inner vane support ring: comparison of measured and calculated transient temperatures at rig circumferential location "A."

NX 188, which allows heat to be transmitted through the ring at an increased rate, thereby reducing the gradients during transients.

The stress analysis was performed for 1132°C (2070°F) steady-state rig conditions at the three circumferential locations at which measurements were taken. The results showed the lowest probability of survival (P_s = 0.43285) at location "C," the location of highest measured and predicted radial gradient, with the maximum principal stress of 225.4 MPa (32.7 ksi). The maximum principal stress is located at the forward edge of the flange inside diameter, corresponding to failure origins observed in parts.

This probability assumes strength characteristics of σ_{os} = 774.1 MPa (50.0 ksi), σ_{ov} = 626.3 MPa (27.0 ksi), and a Weibull modulus of 8 for each. When the surface characteristic strength is changed to σ_{os} = 474.7 MPa (30.66 ksi) with m = 8, representative of an as-fired and oxidized surface, the P_s becomes 0.01783. These probabilities help explain why only one inner vane support ring had passed thermal shock rig testing before the modifications explained later.

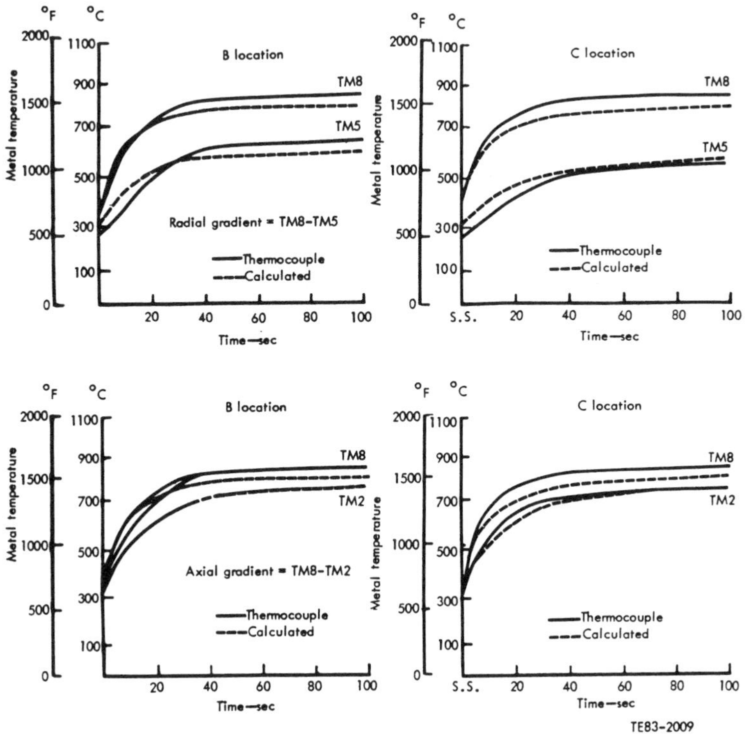

Figure 246. Metal inner vane support ring: comparison of measured and calculated transient temperatures at rig circumferential locations "B" and "C."

Table XLI shows the maximum principal stresses and probability of survival for the acceleration case. This transient case was chosen because higher gradients were observed during acceleration than during deceleration.

The probabilities shown in this table were calculated assuming σ_{os} = 774.1 MPa (50.0 ksi) and m = 8. The results indicate that the lowest P_s occurs at steady state, which is the same time that the highest temperature gradients are now calculated.

To improve the survivability of the ring, leakage underneath the ring must be stopped. To determine the effect of sealing the leakage, a metal plate was clamped between the inner flange of the inner vane support ring and the retaining plate (see Figure 248), and the assembly was installed into engine C-4. This seal is only for the metal substitute for the ceramic ring, and alternate seals have been designed for use with the ceramic ring.

Two designs have been considered to seal leakage under the flange of the inner vane support ring. The preferred method is to insert an "E" seal between the flange and the inner support (Figure 249). The selected "E" seal material is L605 to resist adhesion to the ceramics at temperature.

An alternate method of sealing the flange with ceramic rope packing was designed in the event that the "E" seals were not received in time for the next thermal shock rig test. This method involves the use of ceramic rope packing in two grooves, for-

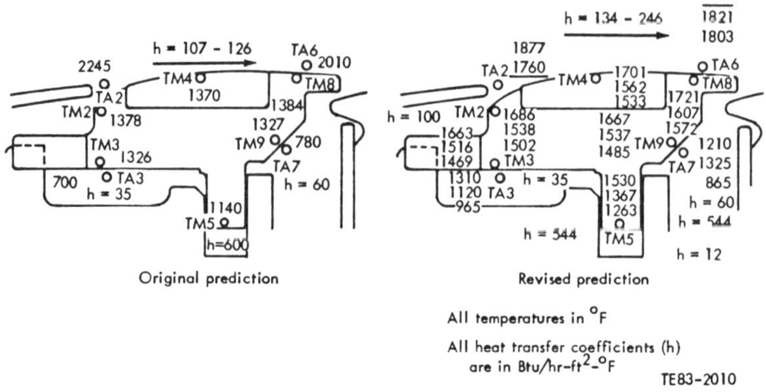

Figure 247. Ceramic inner vane support ring temperatures: original versus revised analysis for 1132°C (2070°F) steady-state conditions with leakage.

Table XLI.
Results of an acceleration transient stress analysis for the inner vane support ring at 1132°C (2070°F) rig conditions.

Time—sec	Maximum stress MPa	Maximum stress ksi	Probability of survival
12	104.8	(15.2)	0.99817
20	117.9	(17.1)	0.99486
28	149.6	(21.7)	0.97161
36	177.2	(25.7)	0.89260
140	225.4	(32.7)	0.43758
S.S.	225.4	(32.7)	0.43285

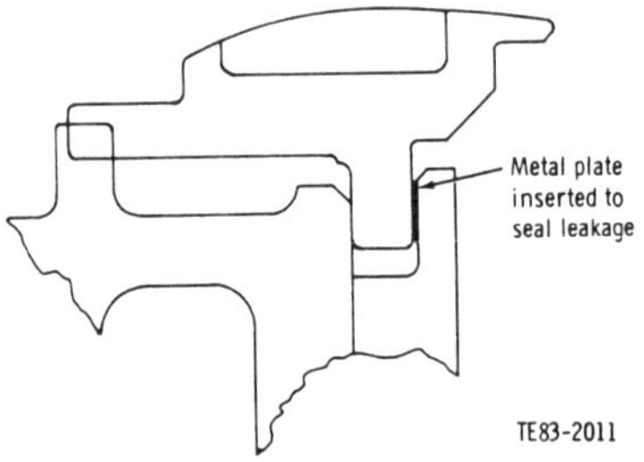

Figure 248. Sealing of leakage around flange of metal inner vane support ring using metal plate.

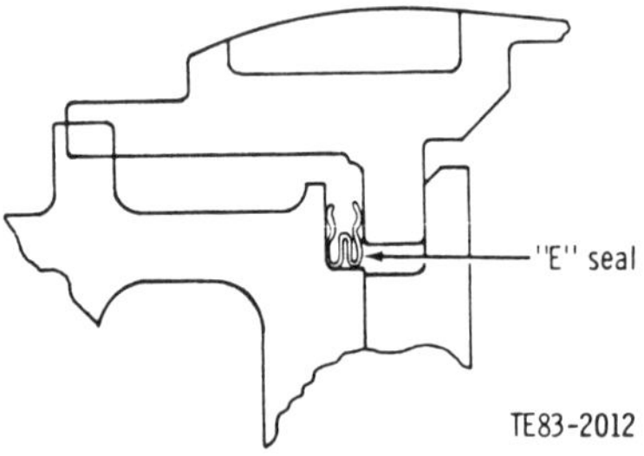

Figure 249. Sealing of leakage around flange of ceramic inner vane support ring using "E" seal.

ward and aft of the flange (see Figure 250), and larger diameter rope packing installed in the inside diameter of the flange. The "E" seal is preferred because of its known sealing qualities. How well a piece of ceramic packing will seal is unknown, although the three pieces together are thought to prevent a large flow.

With the instrumented nozzle assembly installed into engine C-4, the engine was operated through a series of tests to determine the gradients developed in the rings during steady-state and transient operation. The test plan for the engine was to run a qualification cycle consisting of an acceleration from braking to 1038°C (1900°F) conditions, steady-state operation for 8 minutes, and then a deceleration into braking. During the acceleration phase, the thermal response of the engine was good with the nozzle inlet temperature increasing from 316°C (600°F) to 1232°C (2250°F) maximum nozzle inlet in 2 sec. However, the engine gasifier speed did not exceed 90% when 100% speed was expected. The reason for the lack of performance was the excessive clearance between the metal gasifier turbine and the metal tip shroud sized for the larger ceramic rotor.

The initial acceleration phase was conducted using a T6 control system. To try to attain 100% speed with a temperature acceleration, a manual fuel control was used with the speed being controlled by the motoring dynamometer. When this test was conducted, the speed remained at 100%, but the temperature did not reach 1232°C (2250°F) maximum nozzle inlet until 10 sec. Between the snap acceleration (without speed increase) and the constant speed maximum fuel increase, acceleration transient gradients were thought to be predictable and analyzable.

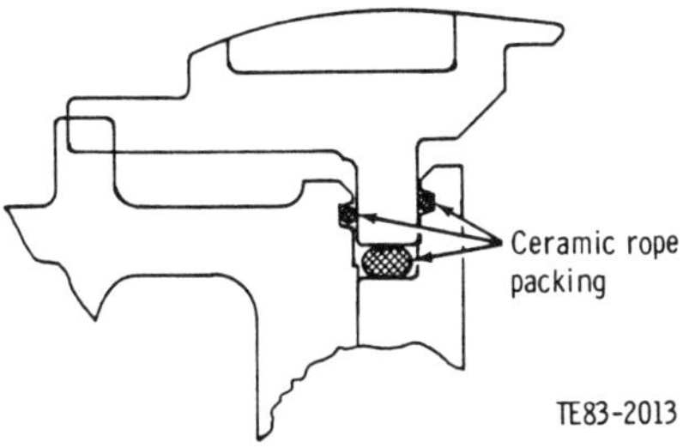

Figure 250. Sealing leakage around flange of ceramic inner vane support ring using ceramic rope packing.

To analyze the steady-state and deceleration conditions, the engine was operated using a manual control at 100% speed and 1060°C (1940°F) average nozzle inlet temperature for 8 minutes and then shutting the fuel off. The fuel shutdown produced the maximum deceleration rate possible. This test was conducted twice with two different gasifier deceleration speeds. The first test was at 100% speed during the temperature deceleration to simulate the conditions in the thermal shock rig. For the second test the gasifier speed was reduced to 85% to simulate actual braking in the engine. The data from these tests looked reasonable, and the engine was removed from the stand.

The reduction of data from the engine was done in the same manner as the reduction of data from the thermal shock rig. The steady-state conditions were reviewed and matched before attempting the transients. The measured results were compared with the predicted results and changes were made to correlate.

The inner vane support ring measured temperatures are shown in Figure 251. Note that all the thermocouples on the inside diameter flange were not operating during this test. To determine if there was leakage under the ring, the response of the air temperature under the ring was examined and compared with the testing on the thermal shock rig. Figure 252 shows the air temperature beneath the ring responded as fast as the gas-path temperature in the thermal shock rig while it did not respond with the gas-path temperature in the engine. This indicates that the air was not leaking under the ring in the engine, and this is assumed for correlation.

Figure 253 also shows the predicted temperatures for the inner vane support ring using measured temperatures for boundary conditions, and the correlation is good. The predicted axial gradient is 13.3°C to 33.3°C (24°F to 60°F) while the measured gradient was 16.7°C to 133.3°C (30°F to 240°F). The 133.3°C (240°F) gradient is thought to be erroneous due to thermocouple problems. The predicted radial gradient is 54.4°C to 74.4°C (98°F to 134°F), but there is no measured gradient with which to compare this.

The heat transfer coefficients used for the prediction are also shown in Figure 253. These coefficients are the same ones used to correlate the thermal shock rig data, with the exception of changes due to lack of leakage. These changes are the coefficient at the leading edge of the ring, which decreased from 567.8 $W/m^2\cdot{}^\circ C$ to 198.7 $W/m^2\cdot{}^\circ C$ (100

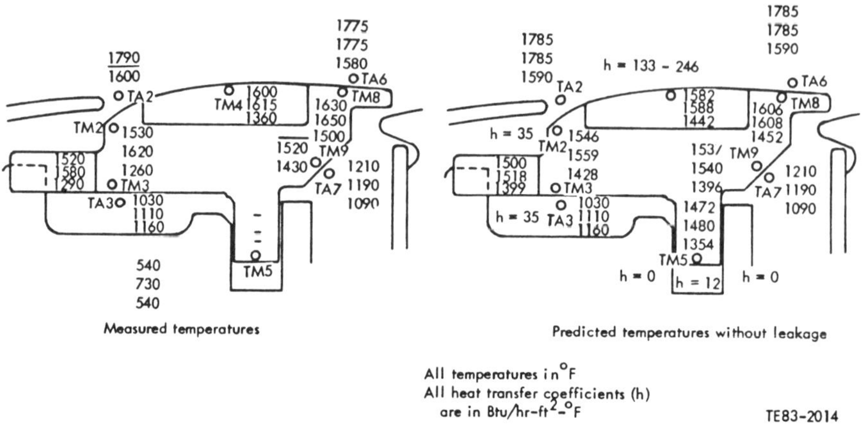

Figure 251. Metal inner vane support ring temperatures: measured versus predicted at 1038°C (1900°F) steady-state conditions without leakage.

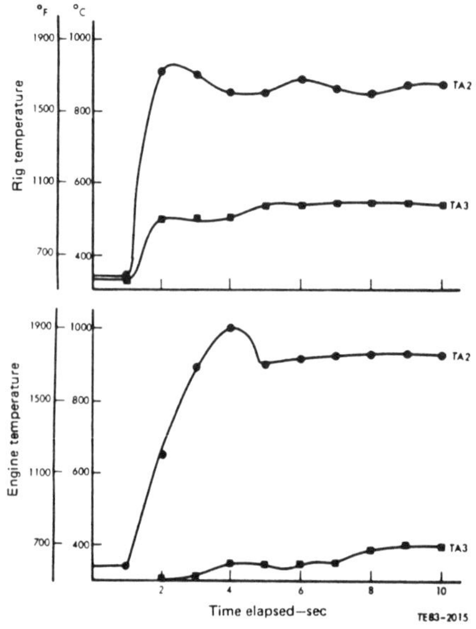

Figure 252. Comparison of responses of gas air path (TA2) and cooling air (TA3) around inner vane support ring for the thermal shock rig and engine C-4.

Btu/hr-ft²-°F to 35 Btu/h-ft²-°F), and the heat transfer coefficient around the flange, which decreased from 3088.9 W/m²-°C to 0 W/m²-°C (544 Btu/hr-ft²-°F to 0 Btu/hr-ft²-°F).

After correlation was obtained at steady-state conditions, efforts were directed toward matching calculated with measured transient temperatures. Measured transient temperatures were plotted and compared with the calculated temperatures. Comparisons at engine circumferential location "A" are shown in Figure 253, for the inner vane support ring. Location "A" was chosen because the maximum transient gradient occurred at this location. Similar to steady-state conditions, it can be seen that correlation is good for the inner vane support ring transients.

Once the measured temperatures had been matched, the new boundary conditions were applied to an RBSiC model of the inner vane support ring to provide temperature data for stress analysis. The results of the heat transfer analysis at 1132°C (2070°F) steady-state conditions are shown in Figure 254. The steady-state axial gradient increased from 1.1°C (2°F) to a range of 10°C to 27.2°C (18°F to 49°F). The steady-state radial gradient changed from 66.7°C (120°F) to a range of 36.1°C to 55.6°C (65°F to 100°F).

The stress analysis was performed for 1132°C (2070°F) steady-state engine conditions at the three circumferential locations where measurements were taken. The results showed the lowest probability of survival (0.9999871) at location "A," with a peak

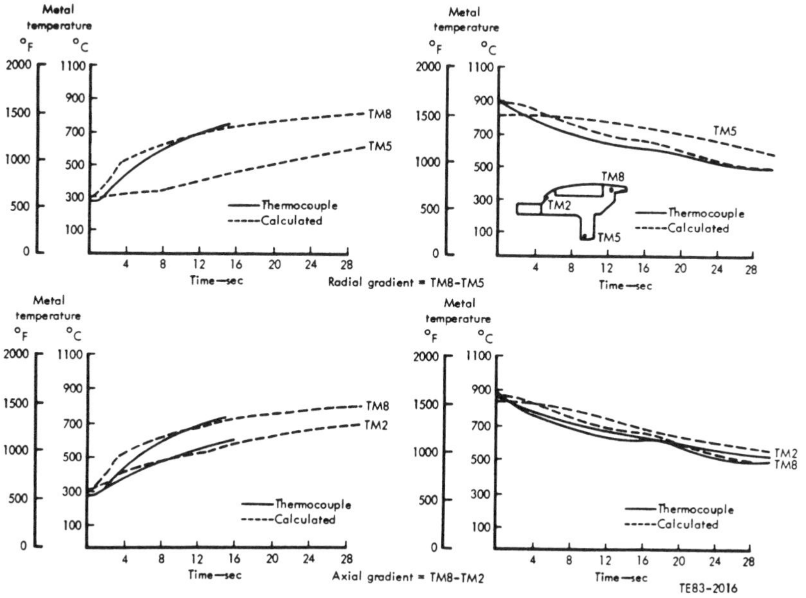

Figure 253. Metal inner vane support: comparison of measured and calculated engine transient temperatures at circumferential location "A."

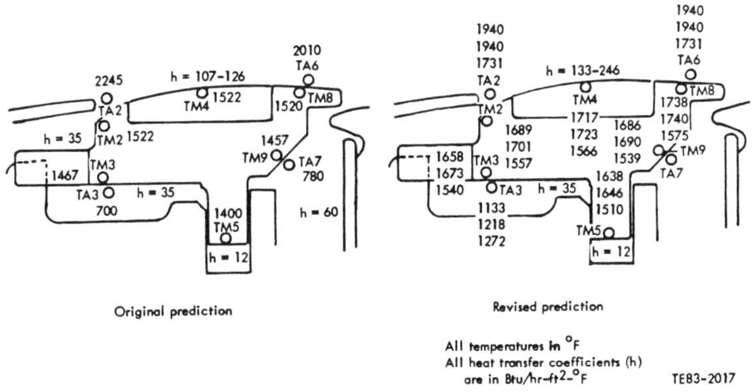

Figure 254. Ceramic inner vane support ring temperatures: original versus revised analysis at 1132°C (2070°F) steady-state conditions without leakage.

maximum principal stress of 57.6 MPa (8.35 ksi). The maximum principal stress is located at the forward edge of the flange inside diameter, corresponding to failure origins observed in parts. The probability of survival assumes strength characteristics of σ_{os} = 774.1 MPa (50.0 ksi), σ_{ov} = 626.3 MPa (27.0 ksi), and the Weibull modulus is 8 for both.

Tables XLII and XLIII show the peak maximum principal stresses and probability of survival for acceleration and deceleration conditions. The probability shown in these tables assumes σ_{os} = 774.1 MPa (50 ksi), σ_{ov} = 626.3 MPa (27 ksi), and m = 8 for both. The results indicate that the lowest P_s (0.99526) occurs 18 sec into the accel with a peak principal stress of 118.6 MPa (17.2 ksi). When the surface characteristic strength is changed to σ_{os} = 474.7 MPa (30.66 ksi) with m = 8 (representative of an as-fired and oxidized surface), the P_s becomes 0.98219 for the worst case.

These results show that sealing the leakage underneath the flange reduces the stress level during steady-state operation. The stress level and P_s remain the same for transient operation with and without leakage. Therefore, the acceleration transient will be the critical operation during all future testing.

Steady-state temperatures, at selected locations, calculated for the metal outer vane support ring at 1038°C (1900°F) are shown in Figure 255. The boundary conditions used in the calculations assumed changes for the revised cooling circuit and some revised flow-path heat transfer coefficients. The heat transfer coefficients were revised from the original analysis based on data from the endwall cascade rig. The calculated axial and radial gradients in the ring were 17.8°C (32°F) and 22.8°C (41°F), respectively.

The measured thermocouple data are also shown in Figure 255. The gas-path temperatures followed the same pattern as the inner ring measurements by being lower than predicted. The cavity temperatures above the outer vane support ring were also lower than predicted. These lower temperatures caused the outer flange of the ring to be significantly lower in temperature than predicted with this effect carried throughout the ring. Since the cooling air is the only source available to cool the ring, the effect of the revised circuit was considerably more severe than was originally thought. The cooling air causes a measured radial gradient of 196.1°C to 236.1°C (353°F to 425°F). The measured axial gradient ranged from 33.3°C to 227.8°C (60°F to 410°F). There was a question about a low reading at the leading edge of the outer vane support ring that showed the axial gradient to be higher than it probably was.

The revisions to the boundary conditions around the outer vane support ring included increasing the heat transfer coefficients along the flow path and across the outer flange. A thin boundary film effect was applied to the leading edge of the

Table XLII.
Peak maximum principal stresses and probability of survival for inner vane support ring engine acceleration.

Time—sec	Maximum principal stress MPa	(ksi)	Probability of survival
12	112.4	(16.3)	0.999644
14	115.8	(16.8)	0.996879
16	118.6	(17.2)	0.995934
18	118.6	(17.2)	0.995260
20	117.2	(17.0)	0.995647

Table XLIII.
Peak maximum principal stresses and probability of survival for inner vane support ring engine deceleration.

Time—sec	Maximum principal stress MPa	(ksi)	Probability of survival
10	79.3	(11.5)	0.999896
12	89.6	(13.0)	0.999676
14	89.6	(13.0)	0.999658
16	88.2	(12.8)	0.999685
18	86.2	(12.5)	0.999730

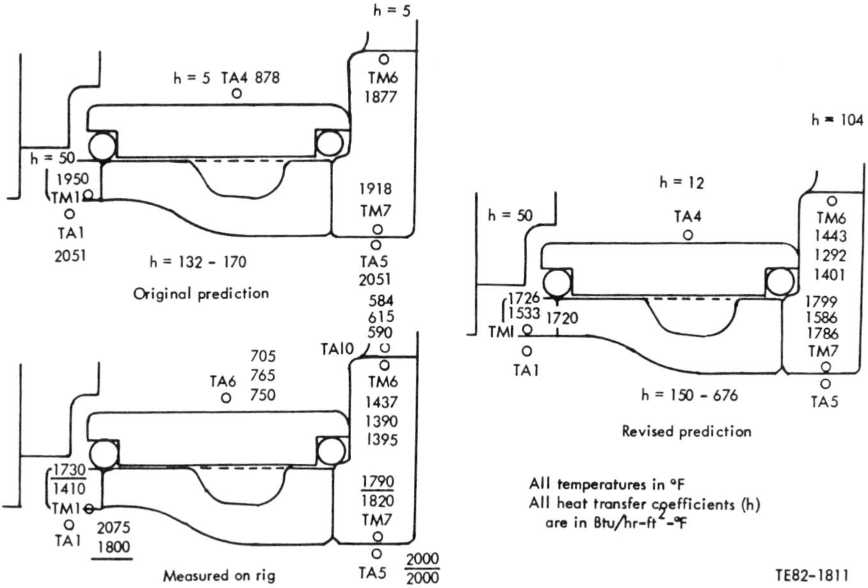

Figure 255. Metal outer vane support ring temperatures at 1038°C (1900°F) steady-state conditions with impingement on outer flange.

ring to account for the cooling air reentering the gas path at that point. The coefficients in the gas path were increased from 749.5 W/m²-°C to 851.7 W/m²-°C (132 Btu/hr-ft²-°F to 150 Btu/hr-ft²-°F) at the leading edge of the ring and from 965.3 W/m²-°C to 3838.5 W/m²-°C (170 Btu/hr-ft²-°F to 676 Btu/hr-ft²-°F) at the trailing edge of the ring. The extremely large increase at the trailing edge is necessary to correlate measured with calculated temperatures and can be verified by data from the endwall cascade rig. The heat transfer coefficient on the outer flange was increased from 28.4 W/m²-°C to 590.5 W/m²-°C (5 Btu/hr-ft²-°F to 104 Btu/hr-ft²-°F). Most of the increase was due to impingement on the flange caused by the thermal shock rig metal shroud assembly, which required a geometric change from the engine part to install the nozzle assembly in the rig. This change made it necessary for the cooling air holes to be drilled at a 45-deg angle toward the flange (see Figure 256). With the air being directed onto the flange, the impingement and scrubbing effect increased the heat transfer coefficient significantly.

Figure 244 shows that the revised calculations predict much higher temperature gradients than the original analysis and correlate well with measured temperatures. The revised predicted axial gradient is 29.4°C to 38.9°C (53°F to 70°F), and the measured gradient is 33.3°C (60°F). The revised predicted ra-

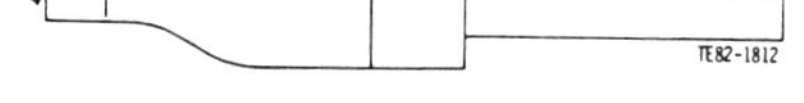

Figure 256. Details of thermal shock rig outer ring cooling circuit.

dial gradient is 163.3°C to 213.9°C (294°F to 385°F), and the measured radial gradient is 196.1°C to 236.1°C (353°F to 425°F).

Following the steady-state correlation, the transient cases for rig circumferential position "A" were analyzed. This position was selected because the largest measured transient gradients occurred at this position. A comparison of measured versus calculated temperatures is shown in Figure 257. The correlation between measured and calculated transient temperatures, although not as good as at steady state, is considered acceptable.

Once the measured temperatures had been matched, the revised boundary conditions were applied to an RBSiC model of the outer vane support ring to provide a temperature distribution for stress analysis. The results of this heat transfer analysis at 1132°C (2070°F) steady-state conditions are shown in Figure 258. The steady-state axial gradient in the part increased from 7.2°C (13°F) to a range of 18.9°C to 35.6°C (34°F to 64°F). The steady-state radial gradient changed from 13.3°C (24°F) to a range of 121.7°C to 177.8°C (219°F to 320°F). As was seen for the inner vane support ring, the highest gradient in the outer vane support ring occurred at steady state for the ceramic ring, even though the measured gradient for the metal ring was approximately 1.5 times as high during the transient. Once again, the higher conductivity of RBSiC reduced the gradient during transients by allowing more rapid heat conduction.

The stress analysis performed indicated the lowest P_s (0.94627) for steady state at rig location "C." This location is the position at which the maximum measured and calculated gradient occurs. The maximum principal stress—167.5 MPa (24.3 ksi)—is located on the forward edge of the flange outside diameter, which correlates to failures seen during thermal shock rig testing. This P_s assumes σ_{os} =

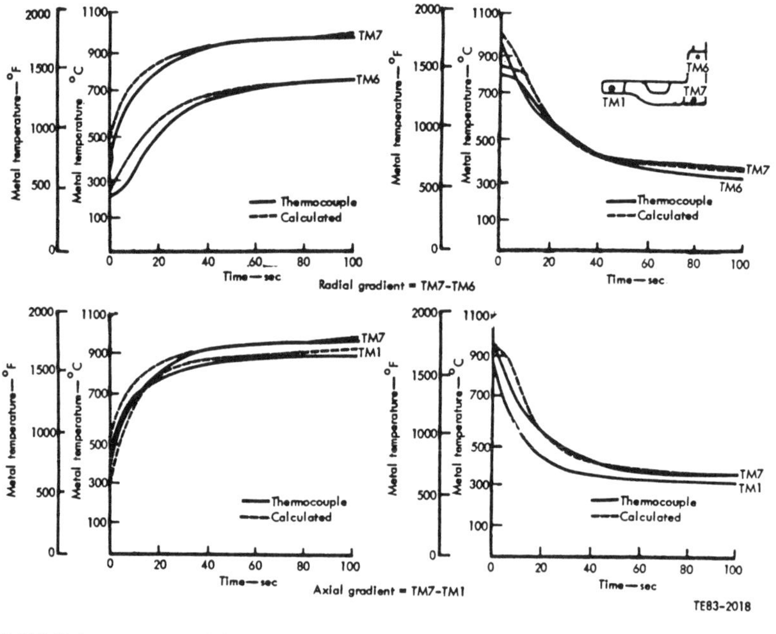

Figure 257. Outer vane support ring: comparison of measured and calculated transient temperatures at rig circumferential location "A."

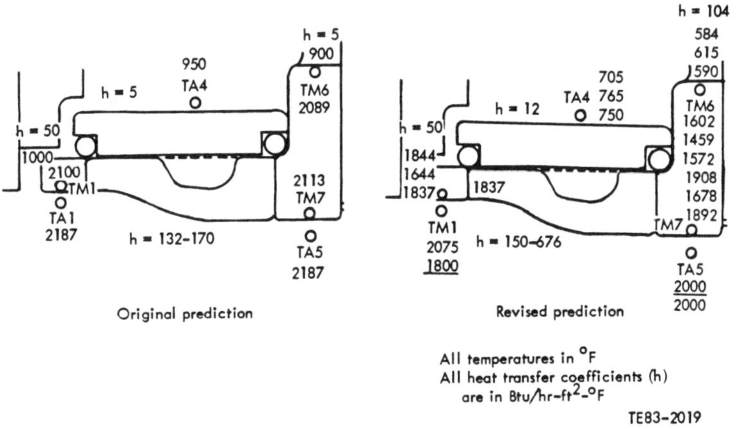

Figure 258. Ceramic outer vane support ring temperatures: original versus revised analysis of 1132°C (2070°F) steady-state conditions with impingement on outer flange.

774.1 MPa (50.0 ksi), σ_{ov} = 626.3 MPa (27.0 ksi), and m = 8.0. When the surface characteristic strength is changed to reflect as-fired and oxidized surfaces, the P_s decreases to 0.64199.

The highest measured transient gradient occurred during an acceleration at rig circumferential position "A." Table XLIV shows the maximum principal stress and probability of survival for the acceleration case.

The probability calculations assume σ_{os} = 774.1 MPa (50.0 ksi), σ_{ov} = 626.3 MPa (27.0 ksi), and m = 8.0. As was the case for the inner vane support ring, the lowest P_s and highest stresses occur at steady-state conditions. The significance of this result is that previously the transient was assumed to be the most stringent case. During testing of BU5 and BU6 on the thermal shock rig, parts failed when subjected to nonsevere conditions (slow accelerations and decelerations). In actuality, the parts were subjected to the most severe case since they reached steady-state conditions.

To improve the survivability of the outer vane support ring, the cooling air flowing across the flange was redirected. The method chosen was to seal the cooling air holes in the shroud assembly and direct the air into the gas path behind the gasifier turbine, as illustrated in Figure 259. The reworked nozzle assembly was installed into engine C-4 and run through the cycles described in the inner vane support ring subsection.

Figure 260 shows the measured and predicted temperatures for the outer vane support ring at 1038°C (1900°F) steady-state conditions. The measured axial gradient is 5.6°C to 27.8°C (10°F to 50°F)

Table XLIV.
Results of an acceleration transient stress analysis for the outer vane support ring at 1132°C (2070°F) rig conditions.

Time—sec	Maximum stress MPa	(ksi)	Probability of survival
12	122.0	(17.7)	0.99494
20	137.2	(19.9)	0.98968
28	143.4	(20.8)	0.98906
36	146.2	(21.2)	0.98865
140	147.5	(21.4)	0.98958
S.S.	162.0	(23.5)	0.95974

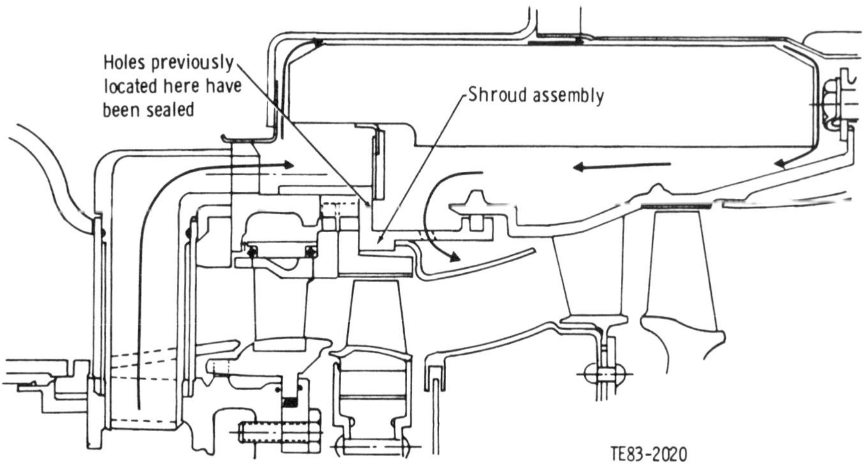

Figure 259. Flow of cooling air after redirection by sealing holes in shroud assembly.

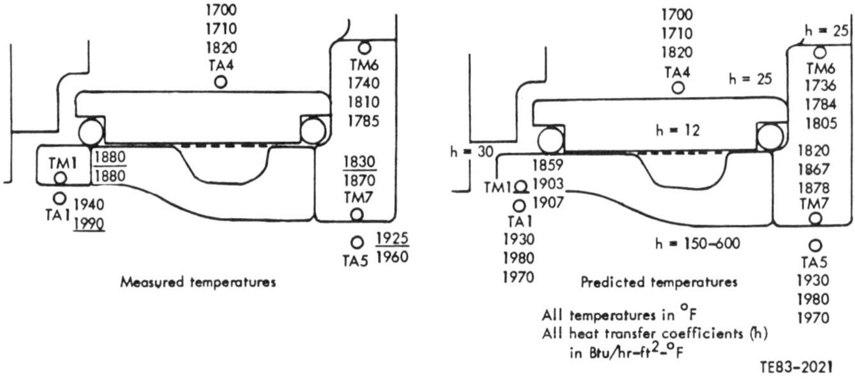

Figure 260. Metal outer vane support ring temperatures: measured versus predicted at 1038°C (1900°F) steady-state conditions without impingement on outer flange.

while the predicted axial gradient is 16.1°C to 21.7°C (29°F to 39°F). The measured radial gradient is 47.2°C to 50°C (85°F to 90°F) while the predicted radial gradient is 40.6°C to 46.7°C (73°F to 84°F). The correlation at steady-state conditions is considered good.

The heat transfer coefficients used for the prediction include the effects of radiation on the outside diameters of the ring. This is the main effect on heat loss for these surfaces of the ring since the cooling air has been sealed. The flow-path heat transfer coefficients are approximately the same as the thermal shock rig with the only change being at the trailing edge. To correlate, the coefficient was lowered from 3747.6 W/m²-°C to 3406.9 W/m²-°C (660 Btu/hr-ft²-°F to 600 Btu/hr-ft²-°F) to reduce the heat input.

Once steady-state correlation was obtained, the transients were analyzed. Figure 261 shows measured transient temperatures plotted against calculated temperatures at circumferential location "A." The highest measured gradient occurred during the snap acceleration at circumferential location "A." Figure 261 shows that correlation is good for the transients.

Having matched the measured temperatures, the revised heat transfer conditions were applied to an RBSiC model of the outer vane support ring to provide temperature data for stress analysis. The results of the heat transfer analysis for the outer vane support ring at 1132°C (2070°F) steady-state conditions are shown in Figure 262. The steady-state axial gradient in the part changed from -7.2°C (13°F) to a range of 18.3°C to 25°C (33°F to 45°F). The minus sign indicates that the leading edge is cooler than the trailing edge. The steady-state radial gradient increased from 13.3°C (24°F) to a range of 43.9°C to 51.7°C (79°F to 93°F).

The steady-state stress analysis performed indicated the lowest P_s (0.999998) at the "B" location, with a peak maximum principal stress of 47.6 MPa (6.9 ksi). This probability assumes strength characteristics of σ_{os} = 774.1 MPa (50.0 ksi), σ_{ov} = 626.3 MPa (27.0 ksi), and m = 8 for both. The location of the maximum principal stress is at the forward edge of the flange outside diameter.

Tables XLV and XLVI show the peak maximum principal stresses and P_s for the acceleration and deceleration conditions for the outer vane support ring. The probabilities shown in these tables assume σ_{os} = 774.1 MPa (50.0 ksi), σ_{ov} = 626.3 MPa (27.0 ksi), and m = 8 for both. The results show the

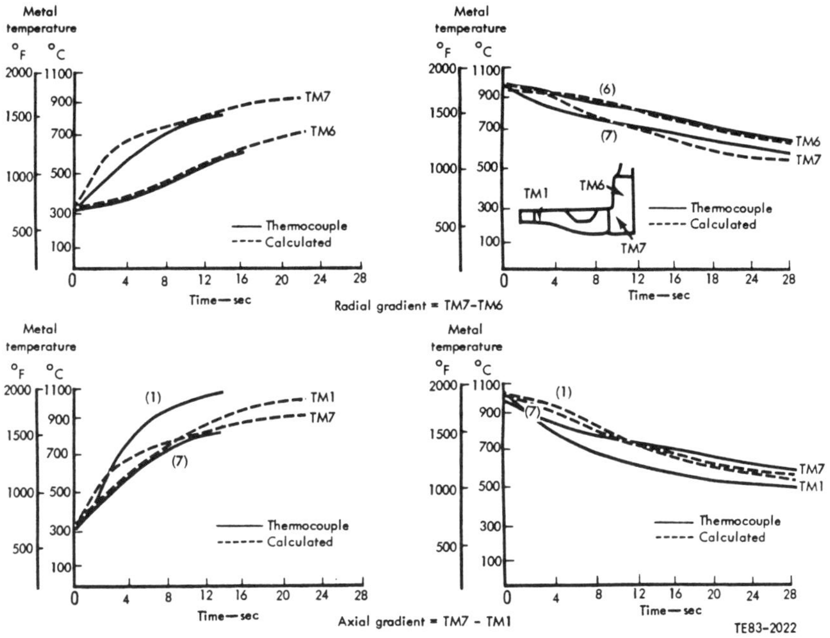

Figure 261. Outer vane support ring: comparison of measured and calculated engine transient temperatures at circumferential location "A."

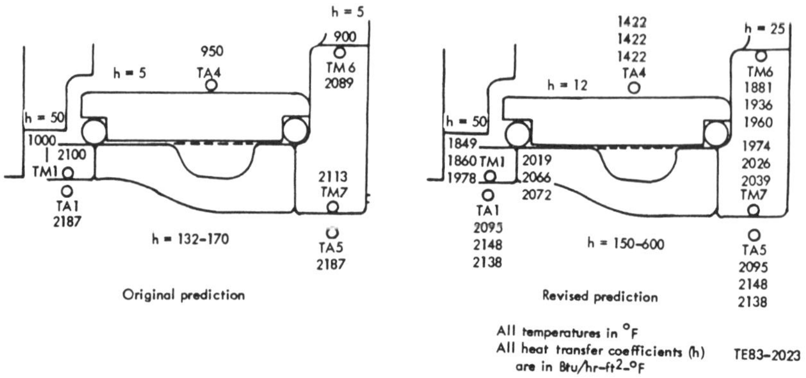

Figure 262. Ceramic outer vane support ring temperatures: original versus revised analysis at 1132°C (2070°F) steady-state conditions without impingement of cooling air.

Table XLV.
Peak maximum principal stresses and probability of survival for outer vane support ring engine acceleration.

Time—sec	Maximum principal stress MPa	(ksi)	Probability of survival
12	126.2	(18.3)	0.98982
14	132.4	(19.2)	0.98581
16	136.5	(19.8)	0.98211
18	137.9	(20.0)	0.98195
20	135.8	(19.7)	0.98438

Table XLVI.
Peak maximum principal stresses and probability of survival for outer vane support ring engine deceleration.

Time—sec	Maximum principal stress MPa	(ksi)	Probability of survival
8	94.4	(13.7)	0.999470
12	137.2	(19.9)	0.99137
16	128.2	(18.6)	0.99453
18	123.4	(17.9)	0.99602

lowest P_s (0.98195) at 18 sec into the acceleration. The peak maximum principal stress is 137.9 MPa (20.0 ksi) and is located at the forward edge of the flange outside diameter, corresponding to failure origins observed in parts. When the surface characteristics are changed to reflect as-fired and oxidized surfaces—σ_{os} = 474.7 MPa (30.66 ksi) and m = 8—the P_s is reduced to 0.93575.

These results show that diverting the cooling air behind the gasifier turbine reduces the stress level in the outer vane support ring. The steady-state stress is reduced by almost a factor of 4, but the transient stress is only reduced from 147.5 MPa (21.4 ksi) to 137.9 MPa (20.0 ksi). This indicates that the acceleration transient is the critical operating mode for the outer vane support ring.

The calculated steady-state shroud temperatures at selected locations are shown in Figure 263. The boundary conditions were modified for the revised cooling circuit. The calculated axial and radial gradients in the ring were 9.4°C (17°F) and 6.1°C (11°F), respectively.

As was the case with the outer vane support ring, the measured steady-state temperatures showed that air and cavity temperatures were lower than predicted. The measured temperatures throughout the ring were significantly lower than predicted because of the cooling air flowing across the outside diameter flange.

To match the measured temperatures, the heat transfer coefficient on the outer flange had to be increased significantly. Since the impingement of cooling air was directed more onto the shroud ring than onto the outer vane support ring, the heat transfer coefficient was increased more than the outer vane support ring value (942.6 W/m²-°C [166 Btu/hr-ft²-°F] compared with 590.5 W/m²-°C [104 Btu/hr-ft²-°F]). The flow-path coefficients were assumed to be constant along that surface.

The revised calculated steady-state temperatures are also shown in Figure 263. The correlation with measured temperatures was not as close as with the other rings. The calculated temperatures are higher than measured at most locations around the ring. The measured axial gradient was 78.3°C (141°F) while the calculated axial gradient ranged from -32.8°C to 23.3°C (-59°F to 42°F). The minus indicates the leading-edge temperature was lower than the trailing-edge temperature. The measured radial gradient ranged from 101.1°C to 177.8°C

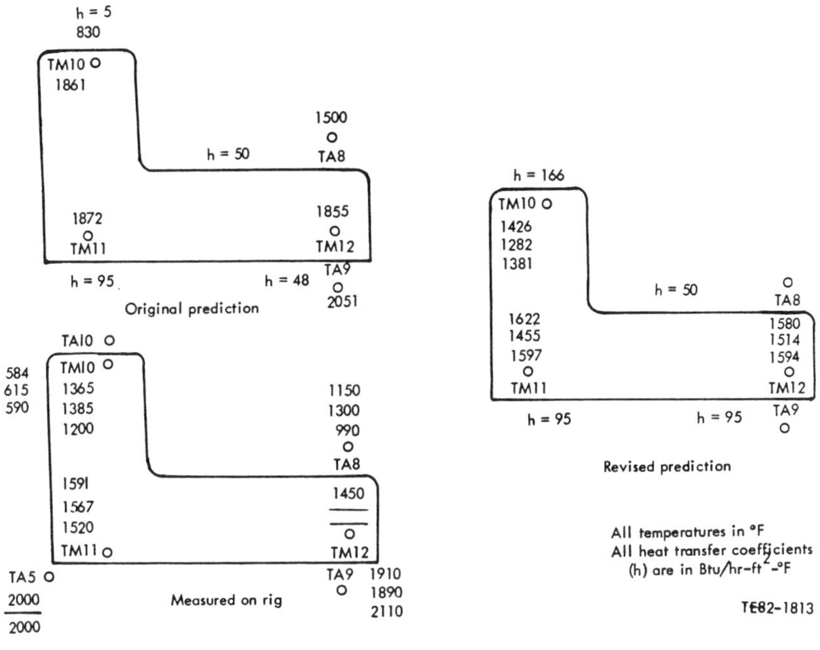

Figure 263. Metal shroud ring temperatures at 1038°C (1900°F) steady-state conditions with impingement on outer flange.

(182°F to 320°F) while the calculated radial gradient ranged from 96.1°C to 120°C (173°F to 216°F).

The shroud transient temperatures were analyzed after correlating calculated with measured steady-state temperatures. A comparison of calculated and measured transient shroud temperatures is shown in Figure 264 for rig circumferential location "A." This position was selected because the largest measured transient gradients occurred at this position. The calculated transient gradients were not as high as the calculated steady-state gradients and based on the results of the inner and outer vane support ring analyses, only the steady-state results were applied to the ceramic model.

Once the measured temperatures had been matched, the revised boundary conditions were applied to an RBSiC, nonabradable shroud model for 1132°C (2070°F) steady-state conditions. The results of the heat transfer analysis are shown in Figure 265. The axial gradient changed from 12.8°C (23°F) to a range of -27.8°C to 20°C (-50°F to 36°F). The radial gradient changed from 2.8°C (5°F) to a range of 95.6°C to 115.6°C (172°F to 208°F).

The stress analysis performed indicated the lowest P_s (0.90912) for steady state at rig position "C." The maximum principal stress—219.9 MPa (31.9 ksi)—is located at the aft edge of the flange outside diameter. This site was observed as a failure origin during testing. The P_s of 0.90912 is obtained with σ_{os} = 774.1 MPa (50.0 ksi) and m = 8.0. When the surface characteristic strength is modified to reflect as-fired and oxidized surfaces—σ_{os} = 474.7 MPa (30.66

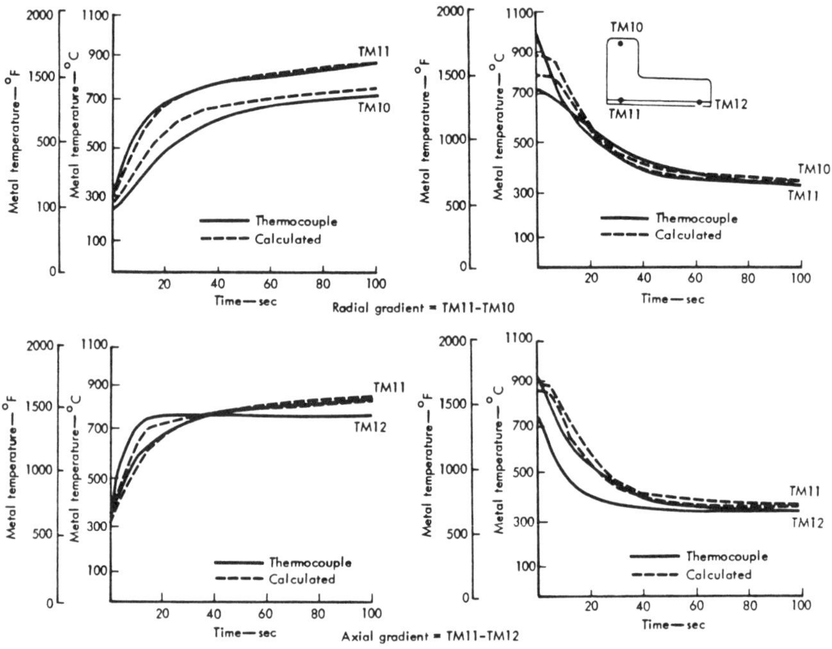

Figure 264. Shroud ring: comparison of measured and calculated transient temperatures at circumferential location "A."

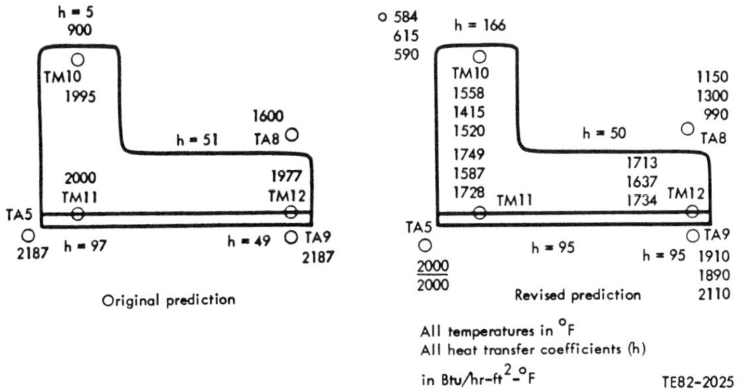

Figure 265. Ceramic shroud ring temperatures: original versus revised analyses for 1132°C (2070°F) steady-state conditions.

ksi) and m = 8.0—the P_s decreases to 0.16764. The volume characteristic strength (σ_{ov}) is 626.3 MPa (27.0 ksi), and m = 8.0 for both cases.

The cooling circuit modification designed to improve the survivability of the outer vane support ring also helped the shroud ring. The metal shroud ring was reinstrumented and installed with the reworked nozzle assembly into engine C-4. The assembly was run through the cycles described in the inner vane support ring subsection.

Figure 266 shows the measured and predicted temperatures for the shroud ring at 1038°C (1900°F) steady-state conditions. The predicted axial gradient is 28.9°C to 38.9°C (52°F to 106°F) while the measured gradient is 33.3°C (60°F). The predicted radial gradient is 10.6°C to 12.8°C (19°F to 23°F) while the measured gradient was 27.8°C to 38.9°C (50°F to 70°F). The correlation is not considered good, but the analysis was modified as much as reasonably possible.

The heat transfer coefficients used for correlation include an estimation of the effects of radiation from the outside diameter flange of the ring and conductivity through the metal shroud assembly. The flow-path heat transfer coefficient was lowered slightly—from 539.4 W/m²-°C to 454.3 W/m²-°C (from 95 Btu/hr-ft²-°F to 80 Btu/hr-ft²-°F)—to reduce the heat input along that surface.

After correlation was obtained at steady-state conditions, the transient conditions were analyzed. Figure 267 shows the measured and predicted temperatures plotted against each other at circumferential location "A." The highest measured gradients for the shroud ring occurred during deceleration with all three circumferential locations being about the same. Similar to the steady-state conditions, the correlation is better for the inner and outer vane support rings than for the shroud ring.

The revised heat transfer conditions were applied to an RBSiC, nonabradable model of the shroud ring to provide temperature data for stress analysis. The steady-state heat transfer results for the shroud ring at 1132°C (2070°F) conditions are shown in Figure 268. The steady-state axial gradient in the part changed from 12.8°C (23°F) to a range of 21.1°C to 53.9°C (38°F to 87°F). The steady-state radial gradient changed from 2.8°C (5°F) to a range of 4.4°C to 8.3°C (8°F to 15°F).

The stress analysis performed for steady-state conditions indicated the highest stress—18.6 MPa (2.7 ksi)—at the "B" location. The P_s, using strength characteristics of σ_{os} = 774.1 MPa (50.0 ksi), σ_{ov} = 626.3 MPa (27.0 ksi), and m = 8 for both, was 1.00. This P_s is a result of the low predicted gradients.

Tables XLVII and XLVIII show the peak maximum principal stresses and P_s for the acceleration and deceleration conditions. The probabilities shown assume strength characteristics of σ_{os} = 774.1 MPa (50.0 ksi), σ_{ov} = 626.3 MPa (27.0 ksi), and m = 8 for both. It can be seen that the highest stress—51.7 MPa (7.5 ksi)—occurs at 12 sec into the deceleration. The P_s at this point is 0.9999987. When

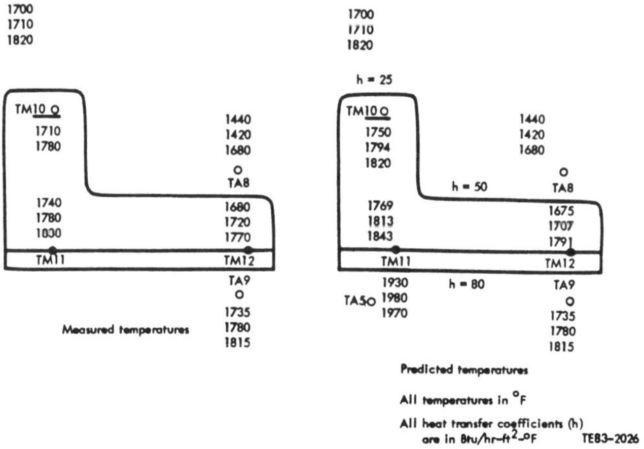

Figure 266. Metal shroud ring temperatures: measured versus predicted at 1038°C (1900°F) steady-state conditions without impingement of cooling air.

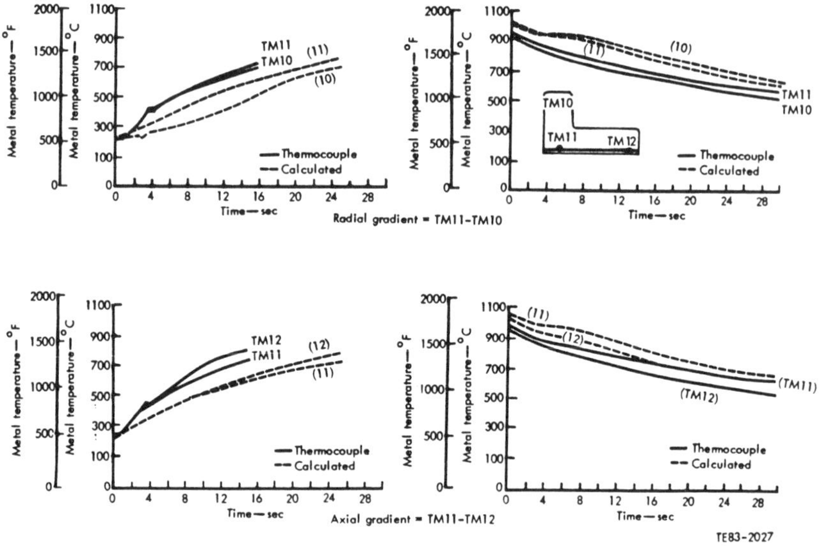

Figure 267. Shroud ring: comparison of measured and calculated engine transient temperatures at circumferential location "A."

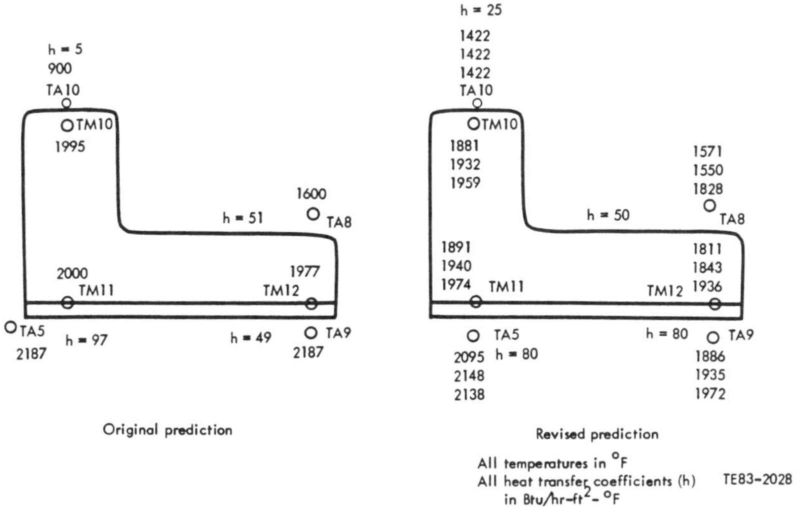

Figure 268. Ceramic shroud ring temperatures: original versus revised analysis at 1132°C (2070°F) steady-state conditions without impingement of cooling air.

Table XLVII.
Peak maximum principal stresses and probability of survival for outer vane support ring engine acceleration.

Time—sec	Maximum principal stress MPa	(ksi)	Probability of survival
8	37.2	(5.4)	0.9999996
12	37.9	(5.5)	0.9999994
16	37.2	(5.4)	0.9999995
18	36.5	(5.3)	0.9999996

the surface characteristics are changed to reflect as-fired and oxidized surfaces (σ_{os} = 474.7 MPa [30.66 ksi], m = 8), the P_s does not change.

Once again diverting the cooling air reduced the stresses in the part. The predicted stresses are reduced from 219.9 MPa (31.9 ksi) to 18.6 MPa (2.7 ksi) at steady state and from 144.8 MPa (21.0 ksi) to 51.7 MPa (7.5 ksi) during a transient. The P_s also increased by at least six orders of magnitude. As was seen with the inner vane support ring and the outer vane support ring, the transient case is the critical operating mode for the shroud ring.

The instrumented nozzle assembly testing yielded important information for finite element analysis on the ceramic rings. The thermal environment surrounding the rings has to be fully understood before proper results can be found. The effect of the ambient air-cooled block configuration on the ceramic rings was not fully understood until the instrumented nozzle assembly was tested. Table XLIX summarizes the stress results from the instrumented nozzle testing. This table shows the predicted stresses before and after modifications to the cooling air circuit and the effect of sealing the leakage underneath the flange of the inner vane support ring. The P_s calculated corresponds to the peak maximum principal stress calculated for each condition. The P_s for the modifications and as-fired and oxidized rings are considered acceptable for testing in this program.

Table XLVIII.
Peak maximum principal stresses and probability of survival for outer vane support ring engine deceleration.

Time—sec	Maximum principal stress MPa	(ksi)	Probability of survival
8	32.4	(4.7)	0.9999999
12	51.7	(7.5)	0.9999987
16	51.0	(7.4)	0.9999987
18	48.9	(7.1)	0.9999990

Table XLIX.
Calculated maximum principal stress and probability of survival for 2070°F conditions.

Component	Condition	Stress—MPa (ksi) S.S.	Trans.	Probability of survival* σ_{os} = 774.1 MPa (50.0 ksi), m_s = 8.0	σ_{os} = 474.7 MPa (30.66 ksi), m_s = 8.0
Outer vane support ring	Cooling air	167.5 (24.3)	147.5 (21.4)	0.94627	0.64199
	Without cooling air	47.6 (6.9)	137.7 (20.0)	0.98195	0.93575
Shroud ring	Cooling air	219.9 (31.9)	144.8 (21.0)	0.90912	0.16764
	Without cooling air	18.6 (2.7)	51.7 (7.5)	0.9999987	0.9999987
Inner vane support ring	Leakage	225.4 (32.7)	225.4 (32.7)	0.43285	0.01783
	Without leakage	57.9 (8.4)	118.6 (17.2)	0.99526	0.98219

* σ_{ov} = 626.3 MPa (27.0 ksi), m_v = 8 for both cases

The outer vane support ring model used for the instrumented nozzle analysis assumed there were slots at the leading edge of each vane pocket. The ring was slotted based on the original 3-D stress analysis, which indicated high stresses in the vane pockets, shown in EDR 10383 (Ref 5), pp 74-76. Subsequent testing in engine C-4 indicated the cantilevered subsection was a possible cause of the engine failure (see engine test subsection). Therefore, additional analysis, using the revised temperature predictions from the instrumented nozzle assembly testing, was performed to determine the effect of slotting the leading edge of the outer vane support ring. The deceleration case was chosen for the analysis because peak pocket stresses occur during this transient.

With slots at the leading edge, the maximum principal stress was 137.2 MPa (19.9 ksi) with a P_s of 0.99137 for 12 sec into deceleration. Figure 269 shows that the principal stress in the pocket inside diameter is approximately 62.0 MPa (9.0 ksi). This 2-D analysis does not account for a stress concentration factor in that area. With no slots at the leading edge, the maximum principal stress was reduced to 125.5 MPa (18.2 ksi) with a P_s of 0.99535 for the same conditions. Figure 270 shows that the pocket stress increased to 96.5 MPa (14.0 ksi) even though the maximum principal stress decreased. The P_s stated for these cases assumes σ_{os} = 774.1 MPa (50.0 ksi), σ_{ov} = 626.3 MPa (27.0 ksi), and m = 8. Even with a stress concentration, the predicted pocket stresses were not expected to cause failure in an unslotted ring.

Additional detailed results based on the preliminary analyses were reported in the following reports: EDR 9722 (Ref 2), pp 61-66; EDR 9951 (Ref 3), pp 125-41; EDR 10156 (Ref 4), pp 73-80; EDR 10383 (Ref 5), pp 74-76; EDR 10841 (Ref 7), pp 57-64; and EDR 11006 (Ref 8), pp 36-40.

Discussion: Process Development

The 2070°F-configuration parts were initially ordered from Carborundum under a process development contract. This was done to help Carborundum fabricate quality components.

The development involved establishing a controlled process, evaluating sample parts, and then revising the process to provide the finished parts.

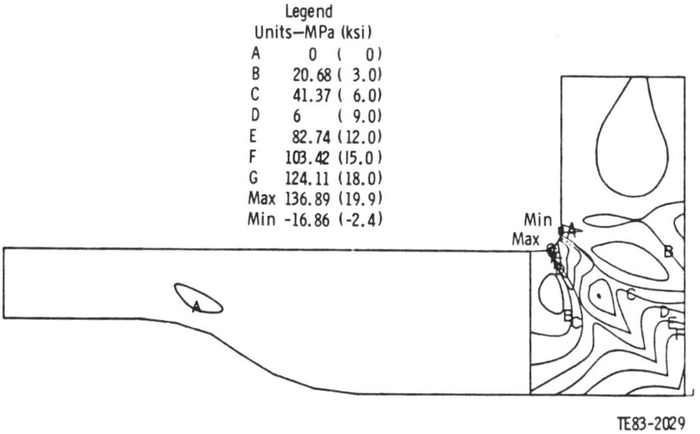

Figure 269. Plot of maximum principal stress at 12 sec into deceleration for outer vane support ring with slots at leading edge of each vane pocket.

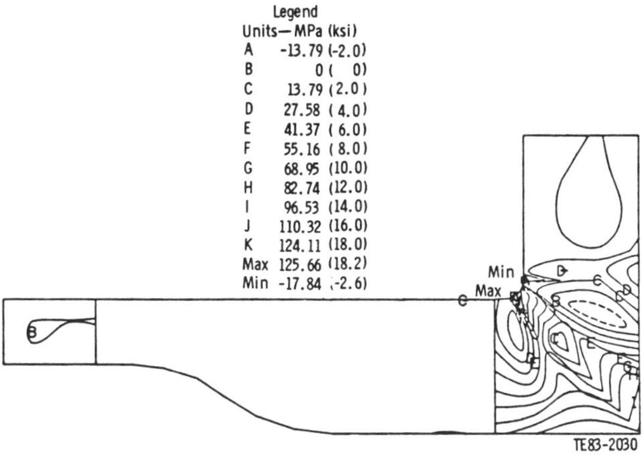

Figure 270. Plot of maximum principal stress at 12 sec into deceleration for outer vane support ring with no slots at leading edge of each vane pocket.

The development was planned for the vanes, inner vane support ring, outer vane support ring, and shroud ring. Some significant results from the process development are detailed in the following paragraphs.

The vanes were made from transfer-molded RBSiC. A significant increase in the strength of transfer-molded material was found due to the addition of a hammer-milling step in the powder processing. The average room temperature MOR increased from 314.50 MPa (45.62 ksi) to 393.16 MPa (57.03 ksi) with the addition of this step.

The rings were made from warm compression-molded, RBSiC. A significant problem identified was the existence of porosity in some test bars and pieces. This condition drastically decreased the room-temperature MOR strength from 466.65 MPa (67.69 ksi) to 247.36 MPa (35.88 ksi). The cover mix during sintering was increased to improve siliconization and therefore to reduce porosity.

The existence of silicon-filled cracks on the outer vane support ring and inner vane support ring indicated another problem: an investigation into these cracks revealed that the higher molding pressure used for these rings created residual stresses. The higher pressure was used to obtain a higher density, which was thought to give a higher strength. Further investigation revealed that the shroud rings molded to a lower density attained identical MOR strength, and, therefore, the outer vane support ring and inner vane support ring processing was changed to obtain the lower density.

Additional detailed information on NDI results, process development, and test bar results can be found in EDR 9951 (Ref 3), pp 133-51; EDR 10156 (Ref 4), pp 80-95; EDR 10383 (Ref 5), pp 76-82; and EDR 10672 (Ref 6), pp 67-69.

Discussion: Rig Testing

Similar to the 1900°F-configuration ceramic parts, all 2070°F-configuration ceramic parts were subjected to extensive rig proof testing before being installed into an engine. These proof tests were designed to identify components with a high probability of failure due to nondetectable flaws. Once the parts pass the proof tests, the probability of failure in an engine is greatly reduced.

The following rig tests simulate the engine operating conditions and qualify a ceramic component for engine test:

- cold flow rig—measures flow capacity of vane row for performance evaluation
- hoop pressurization fixture—features mechanical proof load for evaluation outer vane support ring and shroud material strength
- vibration rig—evaluates response of metal support structure and effect on ceramic components
- thermal shock rig—induces maximum thermal and mechanical proof loads experienced in engine testing

Cold Flow Rig

Similar to the 1900°F-configuration assembly, the 2070°F-configuration assembly was tested in the cold flow rig. The purpose of the cold flow capacity test was to determine the airflow at different pressure ratios for the 2070°F-configuration nozzle. The test procedure was similar to the tests conducted on the 1900°F-configuration nozzles. This test on the first 2070°F-configuration nozzle found the flow capacity to be 0.77 kg/s (1.69 lbm/sec) corrected flow at a total-to-static pressure ratio of 1.45, as shown in Figure 271. Nozzle flow capacity is 2.8% below the anticipated value but is satisfactory to operate the engine at 1038°C (1900°F).

Hoop Pressurization Fixture

The bladder proof test, initiated for the 2070°F-configuration outer vane support ring and shroud ring after several thermal shock rig failures at apparent low stress levels, led to an investigation to determine ring strength. The test was designed to subject the rings to a stress level that the part will be exposed to during engine testing. A finite element model was used to match the desired stress level with the corresponding internal pressure.

A total of 20 rings (19 outer vane support rings and one shroud ring) were tested in this fixture. Table L shows the rings that passed while Table LI presents the failure pressure and peak stress for the failed rings.

An important result from bladder testing was the discovery of a low strength as-fired surface con-

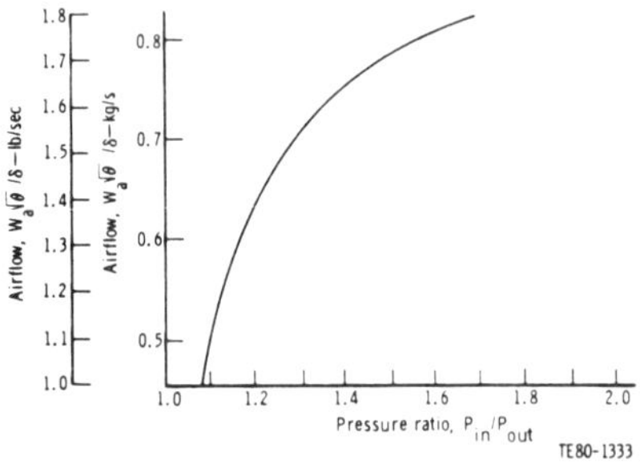

Figure 271. Cold flow characteristics of first 2070°F-configuration gasifier turbine nozzle.

Table L.
2070°F-configuration ceramic rings passing bladder proof test.

Material and ring	Supplier
6 sintered alpha SiC outer vane support rings	Carborundum
5 reaction-bonded SiC outer vane support rings	Carborundum
1 reaction-bonded SiC nonabradable shroud ring	Carborundum
1 Refel SiC outer vane support ring	Pure Carbon

hardware. Zyglo inspection of the ceramic parts revealed no damage from either the cold flow or vibration tests.

Thermal Shock Rig

Similar to the 1900°F-configuration ceramic components, the 2070°F-configuration ceramic components were tested in a thermal shock rig. The ceramic components were subjected to maximum engine condition proof load prior to actual engine testing. The test cycle included transient and steady-state operation.

The testing was conducted in two phases. The first phase consisted of a sharp acceleration out of braking to 1038°C (1900°F) rotor inlet and hold at steady-state temperature for 8 minutes to stabilize.

Table LI.
Summary of 2070°F-configuration outer vane support rings that failed bladder proof test.

		Burst pressure		Peak stress	
Supplier	Material	kPa	(psig)	MPa	(ksi)
Carborundum	Reaction-bonded SiC	3,240	(470)	45.78	(6.64)
Carborundum	Reaction-bonded SiC	2,827	(410)	39.92	(5.79)
Carborundum	Reaction-bonded SiC	3,344	(485)	47.22	(6.85)
Carborundum	Reaction-bonded SiC	3,792	(550)	53.57	(7.77)
Carborundum	Reaction-bonded SiC	4,585	(665)	64.80	(9.40)
Carborundum	Reaction-bonded SiC	3,895	(565)	55.08	(7.99)
Carborundum	Sintered-alpha SiC	15,167	(2200)*	213.0	(30.89)

* Intentional failure

dition. A relatively thick (170 μm) layer of silicon carbide crystals was observed on the outside diameter of several outer vane support rings. This layer was found to have an average room temperature MOR of only 100.03 MPa (14.51 ksi). Additional information on this surface condition and on the bladder proof test results was reported in EDR 11006 (Ref 8), pp 23-37.

Vibration Rig

The vibration test on the nozzle was intended to expose it to the excitation frequencies expected from engine operation. The nozzle was exposed to both radial and axial vibratory input, as defined in Table LII.

During the vibration test, vanes were spring loaded to simulate aerodynamic load, as shown in Figure 272. The results of this testing indicate that the 2070°F-configuration ceramic components and method of restraint are vibrationally acceptable for development engine testing. A total of 4.5 hr of vibration testing was accumulated on the ceramic

This cycle was selected to be consistent with initial engine operation on the test stand. The results of two rig builds are shown in Table LIII.

A full set of parts was qualified for engine testing after these tests. They were inserted into engine C-4 and tested for 890 hr (see engine test subsection). Additional detailed information on the initial phase of thermal shock rig testing of 2070°F-configuration parts can be found in EDR 10383 (Ref 5), pp 83-93.

The second phase of testing added the deceleration into braking from maximum gas temperature.

Table LII.
2070°F-configuration nozzle vibration test input spectrum.

Frequency range	Vibration input level
10 to 130 Hz	1 g
130 to 652 Hz	7.6 mm/s (0.3 in./sec) average velocity
652 to 2000 Hz	5 g

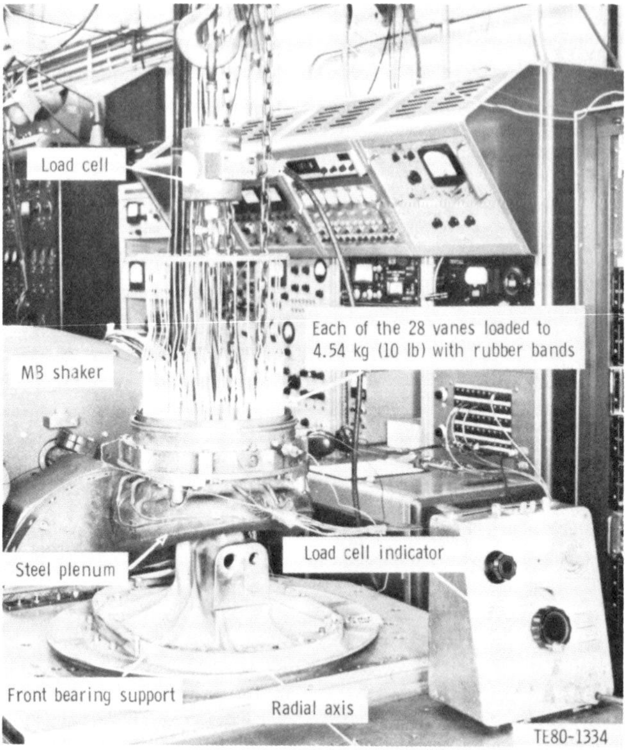

Figure 272. Vibration test rig with 2070°F-configuration ceramic gasifier nozzle assembly—radial axis excitation on an MB-C26H shaker.

The steady-state level was also raised to 1132°C (2070°F) rotor inlet. A new thermal shock rig was built to test the ceramics to the higher temperatures, 1132°C (2070°F) and eventually 1241°C (2265°F). A total of 9 builds with 2070°F-configuration ceramic nozzle components were tested in this rig. All tests were not run through the complete qualification cycle because of a failure in one or more parts. Table LIV summarizes the results of thermal shock rig testing to 1132°C (2070°F) rotor inlet.

Additional detailed information on the failure causes and modifications can be found in the following reports: EDR 10672 (Ref 6), pp 68-81; EDR 10841 (Ref 7), pp 63-72; and EDR 11006 (Ref 8), pp 39-47. The results of several builds not reported earlier will be reported in detail in the following paragraphs.

Two builds of a 2070°F-configuration ceramic nozzle assembly were attempted to qualify the ceramic parts for engine testing. Neither build was completed because of operational problems. The plan for the tests was to run a slow acceleration to 1152°C (2106°F) burner outlet, stabilize for 8 minutes, and then slowly cool down to 316°C (600°F). The nozzle assembly would then be inspected. This portion would qualify the parts for steady-state testing. To qualify the parts for transient testing, an ac-

Table LIII.
Results of thermal shock rig testing of 2070°F-configuration ceramic components to 1038°C (1900°F).

Parts	No. tested	No. passed
Inner vane support ring		
Carborundum—reaction-bonded SiC	2	1
Outer vane support ring		
Carborundum—reaction-bonded SiC	1	1
Shroud ring		
Carborundum—reaction-bonded SiC	2	1
Retaining ring		
Pure Carbon—Refel SiC	2	2
Vanes		
Carborundum—reaction-bonded SiC	50	45
Carborundum—sintered alpha SiC	4	4
Strut shells		
Carborundum—sintered alpha SiC	2	2
Pure Carbon—Refel SiC	6	6

Table LIV.
Results of thermal shock rig testing of 2070°F-configuration ceramic components to 1132°C (2070°F).

Parts	No. tested	No. passed
Inner vane support ring		
Carborundum—reaction-bonded SiC	5	1
Kyocera—sintered alpha SiC	1	0
Outer vane support ring		
Carborundum—reaction-bonded SiC	7	1
Carborundum—sintered alpha SiC	1	1
Pure Carbon—Refel SiC	1	0
Shroud ring		
Carborundum—reaction-bonded SiC	5	3
Retaining ring		
Pure Carbon—Refel SiC	3	3
NGK—sintered Si_3N_4	1	1
Vanes		
Carborundum—reaction-bonded SiC	108*	97*
Carborundum—sintered alpha SiC	8	7
Pure Carbon—Refel SiC	7	7
Strut shells		
Carborundum—sintered alpha SiC	9	8
Pure Carbon—Refel SiC	4	4

* Includes 10 vanes previously qualified to 1900°F

celeration schedule, stabilization, and a slow cool-down were to be conducted. Then a slow heat-up, stabilization, and an exponential decel were to follow. At this point the parts were to be considered qualified for 2070°F engine testing.

To conduct a slow heat-up to steady-state conditions, a manual control was needed. During this portion of the test on BU9, a switch that stops the signal from the automatic fuel controller was inadvertently left open. This caused the controller to run the preprogrammed acceleration schedule. The temperature trace resulting from this acceleration is shown in Figure 273.

Inspection of the assembly revealed the outer vane support ring cracked in two places. Examination of the fracture surfaces revealed a low strength silicon carbide layer (approximately 80 μm thick) similar to that observed during bladder-proof testing, as shown in Figure 274. The failure origins are at the leading edge of the outside diameter flange, corresponding to the peak calculated stress location. The failure at pocket No. 3 is considered the primary failure because of the larger fracture mirror. This ring had previously been bladder-proof tested to 5170 kPa (750 lb/in.^{2}), corresponding to 55.2 MPa (8 ksi) stress on the flange, which indicates that the transient stress is higher than 55.2 MPa (8 ksi). The instrumented nozzle analysis presented earlier predicted a transient stress of 137.9 MPa (20.0 ksi).

For BU10, the outer vane support ring was replaced with an RBSiC ring that had the as-fired surface removed. This was done to remove the low strength crystalline layer. All other parts from BU9 remained the same.

During BU10 of the thermal shock rig, the rig was operated through the first phase of testing (slow heat up, stabilization, and slow cool-down). On inspection of the ceramic parts, a foreign substance was seen on the vanes and rings (see Figures 275 and 276). The assembly was removed for a closer inspection.

Detailed inspection revealed that free silicon had exuded out of 11 RBSiC vanes, the outer vane

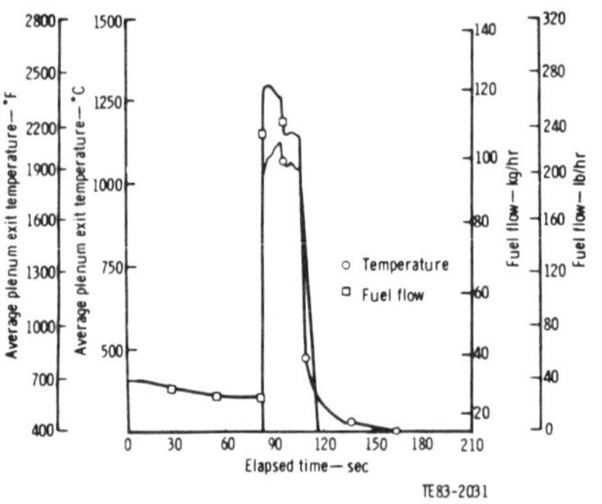

Figure 273. CATE thermal shock qualification test BU9.

support ring, and the inner vane support ring, indicating an extreme overtemperature. In the rotating probe data, the average vane outlet temperature was 1371°C (2500°F) with peak temperatures of 1649°C (3000°F). Figure 277 shows the probe data and the location of the 11 RBSiC vanes. Figures 278 and 279 show two RBSiC vanes with the silicon exuded. The vane deposits are composed of silicon and calcium. Figure 277 shows the deposits in the top right corner. A close-up of the silicon is shown in the bottom left while a close-up of the calcium is shown in the bottom right. Adjacent to the RBSiC vanes were some alpha SiC vanes that did not appear distressed, although they were subjected to the same temperatures. Figure 280 shows some metal deposits on the pressure side of one of the vanes. These deposits are composed of nickel and iron and are thought to have come from the melted portion of the combustor skirt.

Inspection of the outer vane support ring revealed some cracks between the lug slots, shown in Figure 281, and areas of superficial porosity where the silicon had exuded. Figure 282 shows a cross section of the outer vane support ring at the crack in a lug slot. Area 1 shows the SiC crystals without any free silicon present. Area 2 shows the microstructure when the free silicon is beginning to melt. Area 3 shows a typical microstructure of RBSiC. Area 4 shows the silicon bubbling out on the surface of the ring. Area 5 shows the microstructure where some free silicon had exuded and the rest was melted. Inspection of the inner vane support ring reveals a small area of superficial porosity at the trailing edge of two vane pockets. Inspection of the shroud did not reveal any damage due to overtemperature.

There was some difficulty in disassembling the ceramic nozzle assembly because the metal support structure had distorted and locked up the locating tabs on the shroud and the outer vane support ring. After disassembly, the metal outer support was found to be distorted with a crack starting at the inner diameter, as shown in Figure 283.

Investigation into the cause of the overtemperature revealed the failure of two airflow reference thermocouples. This resulted in an erroneous fuel/air temperature calculation that caused overtemperature operation. At the same time, four of eight plenum thermocouples failed, which resulted in a measured temperature average that agreed with the erroneous fuel/air temperature. Following the inadvertent overtemperature, the procedures for the next build of the thermal shock rig were reviewed and revised. The revisions included additional checks on temperature and verification of fuel/air ratio. Several

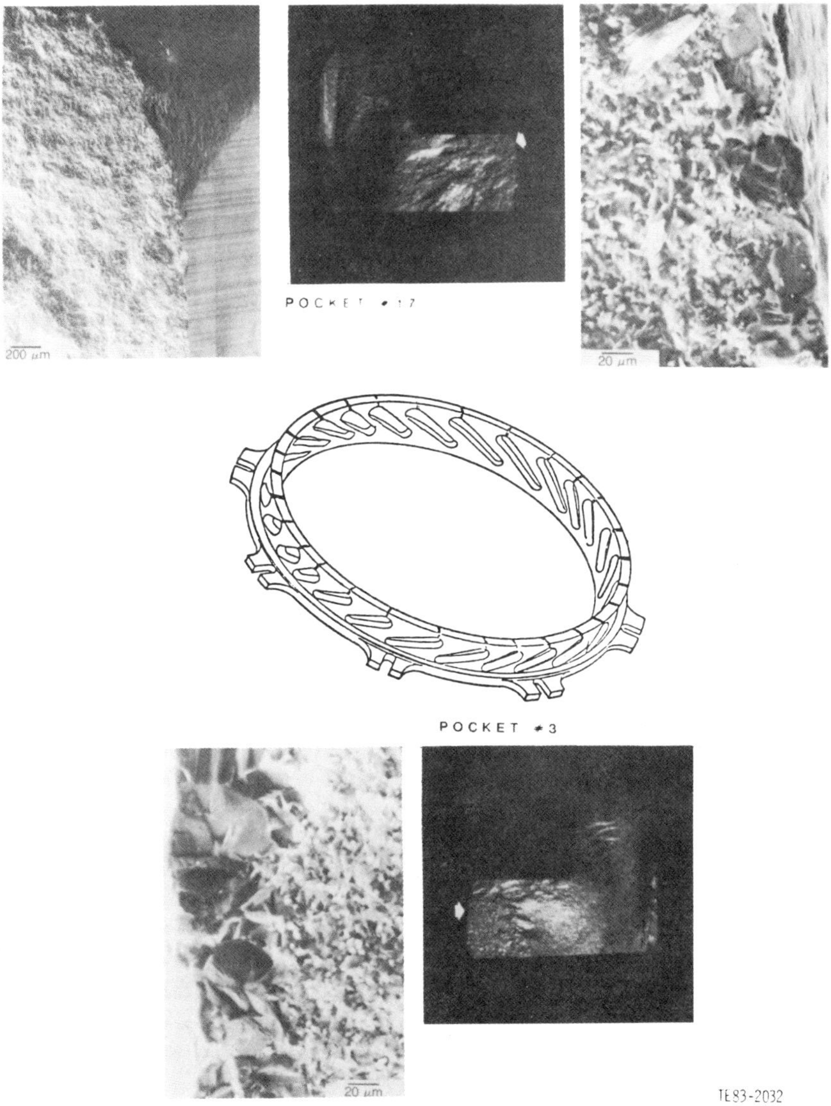

Figure 274. Failure origin of outer vane support ring (S/N FX 25830) during BU9 of the thermal shock rig.

Figure 275. View of nozzle assembly after removal from thermal shock rig BU10.

Figure 276. Close-up of silicon attached to vanes after testing in BU10 of the thermal shock rig.

temperature limiters and a fuel-flow limiter were incorporated to prevent an overtemperature condition from happening.

BU11 involved qualifying the first full set of ceramic gasifier nozzle parts for engine testing. The rig was operated through the complete 2070°F-configuration qualification cycle with all ceramic parts except for one vane passing the test. The qualified parts were installed into engine C-4 with a ceramic bladed rotor (see Section II).

The nozzle assembly that was tested in BU11 of the thermal shock rig consisted of an RBSiC inner vane support ring, an alpha SiC outer vane support ring, an RBSiC abradable shroud ring, a Refel RBSiC retaining ring, 8 alpha SiC vanes, 20 RBSiC vanes, and 4 alpha SiC strut shells. The assembly was run to steady-state conditions and then inspected. Small chips had broken away from the abradable layer in the shroud, but it did not appear to be cracked in the base material, so the testing was continued. The acceleration and deceleration conditions were run with the only distress to the parts being a few more chips in the abradable layer. The assembly was then removed from the stand for closer inspection. Figure 284 shows the assembly, minus the shroud ring, after removal from the stand. During the test, a white powder (yttrium oxide) was injected into the inlet to determine leak paths. This powder can be seen on the inner vane support ring flange in Figure 284 where the rope packing sealed the leakage.

Inspection of the shroud ring revealed that the chips did not propagate cracks into the base material. These chips were blended out and the corners radiused so the shroud was considered qualified for 2070°F testing. One alpha SiC vane had a small indication on the outside diameter trailing-edge foot. This indication has been seen on other alpha SiC vanes and is thought to be a molding flaw.

Following the failure of engine C-4, effort was applied toward building (BU12) and testing a 2070°F-configuration ceramic nozzle assembly in the thermal shock rig. The intent was to qualify the nozzle assembly for installation and testing in engine C-4 along with a ceramic-bladed rotor assembly.

The test was conducted by operating the rig through a slow acceleration (shown in Figure 273) to steady-state stabilization for 8 minutes and then through an exponential deceleration. This qualified the parts for steady-state and deceleration operation. At the conclusion of this phase, all parts appeared to be intact.

The second phase included five complete qualification slow acceleration cycles, including a sharp acceleration, the 8 minute stabilization, and an exponential deceleration. At the conclusion of this phase, the RBSiC inner vane support ring was cracked in one place, as shown in Figure 285. Tear-

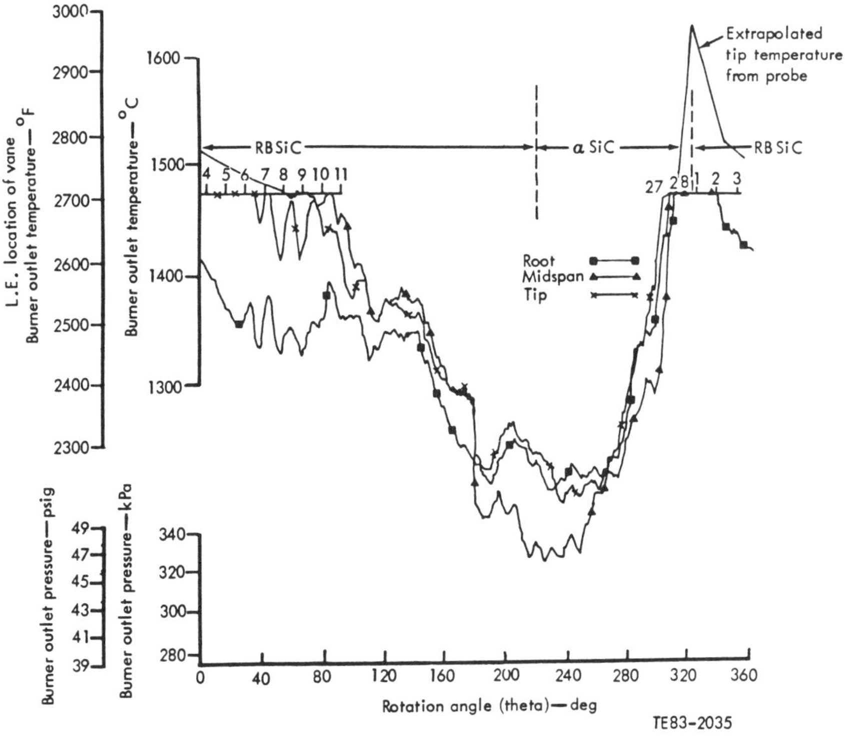

Figure 277. Rotating probe temperature trace at steady-state conditions during BU10 of the thermal shock rig.

down of the assembly revealed that two lugs were also fractured on the inner vane support ring (see Figure 286). All other parts passed the posttest inspection and were considered qualified for 1132°C (2070°F) operation. These parts included an outer vane support ring without slots at the leading edge of each vane pocket.

Closer examination of the failed inner vane support ring by the materials laboratory indicated that the crack originated at the flange inside diameter (shown in Figure 287), which is the predicted maximum stress location during an acceleration transient. Also, the inner flange is chipped in two locations (see Figure 288). One location is at the same place as the crack through the ring, while the other is 130-deg from the top (clockwise viewed from rear). These chips are thought to be caused by the large rope packing being squeezed axially at the corner of the inner flange, as illustrated in Figure 289. These chips created a stress concentration resulting in fracture during the acceleration transient.

There are two possible causes for the lug fractures. The first possibility is based on fracture dynamics. At the time of fracture, the ring separated and overloaded the lug at 270 deg viewed from the rear. The ring then rebounded and overloaded the lug at 180 deg viewed from the rear. The second possible cause is based on differential thermal expan-

Figure 278. RBSiC vane (S/N FX 23775) with silicon exuded.

sion. If the inner vane support ring was locally clamped at the inner flange by the rope packing so that no radial thermal expansion could take place at that point, then as the ring heated, it would grow away from the clamped point. This could cause the lugs to be loaded and overstressed, resulting in fracture.

Buildup 13 of the thermal shock rig was another test of a 2070°F-configuration ceramic nozzle assembly. The primary test objective was to qualify an inner vane support ring for installation and testing in engine C-4 along with previously qualified parts. To take advantage of another thermal shock rig test opportunity, several parts produced by alternate suppliers were included. A listing of the ceramic parts tested during this build is shown in Table LV.

Table LV.
Summary of parts tested during BU13.

Part	Material and supplier
Vanes	7 Refel SiC by Pure Carbon Co. 21 RBSiC by Carborundum
Outer vane support ring	Refel SiC by Pure Carbon Co.
Inner vane support ring	RBSiC by Carborundum
Turbine tip shroud	RBSiC by Carborundum without abradable layer
Vane retaining ring	Si_3N_4 by NGK
Strut shells	Alpha SiC by Carborundum

Because the large diameter rope packing on the inside diameter of the inner vane support ring

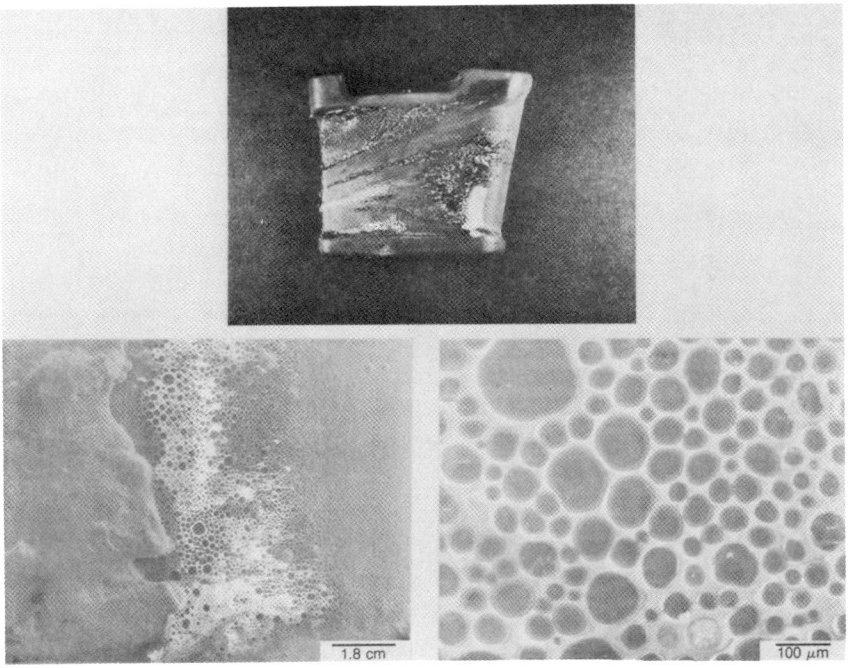

Figure 279. RBSiC vane (S/N FX 23776) with silicon exuded.

flange was thought to have contributed to the failure during BU12, it was omitted for this build. The axial squeezing of the smaller diameter rope packing on either side of the flange was considered sufficient to stop any leakage.

The ceramic parts were tested through a slow heat-up, the steady-state, and deceleration portions (see Figure 290a) of the test without any visible damage. However, following five thermal cycles, involving rapid heating, steady state, and cool-down (as in Figure 290b), the RBSiC inner vane support ring was found to be cracked in one place, as shown in Figure 291, and the Refel SiC outer vane support ring was cracked in three places. All other parts appeared to be intact.

Closer examination of the parts by the Materials Laboratory confirmed the initial observations. The inner vane support ring had failed at one location (shown in Figure 292), with the failure origin located at the flange inside diameter. A semielliptical flaw is observed at the failure origin and is indicated by the arrows in Figure 293. Two separate calculations of the fracture stress were performed based on the mirror radius and on the critical flaw size. These calculations indicated a fracture stress of 146.8 MPa (21.3 ksi) and 159.2 MPa (23.1 ksi), respectively. The predicted maximum principal stress during the acceleration for the inner vane support ring with the lip removed is 83.4 MPa (12.1 ksi). The location of this stress is also at the flange inside diameter. The discrepancy between predicted stress and calculated stress from the fracture surface cannot be explained at this time. Test bars were cut from the ring. (The results are presented in Section III.)

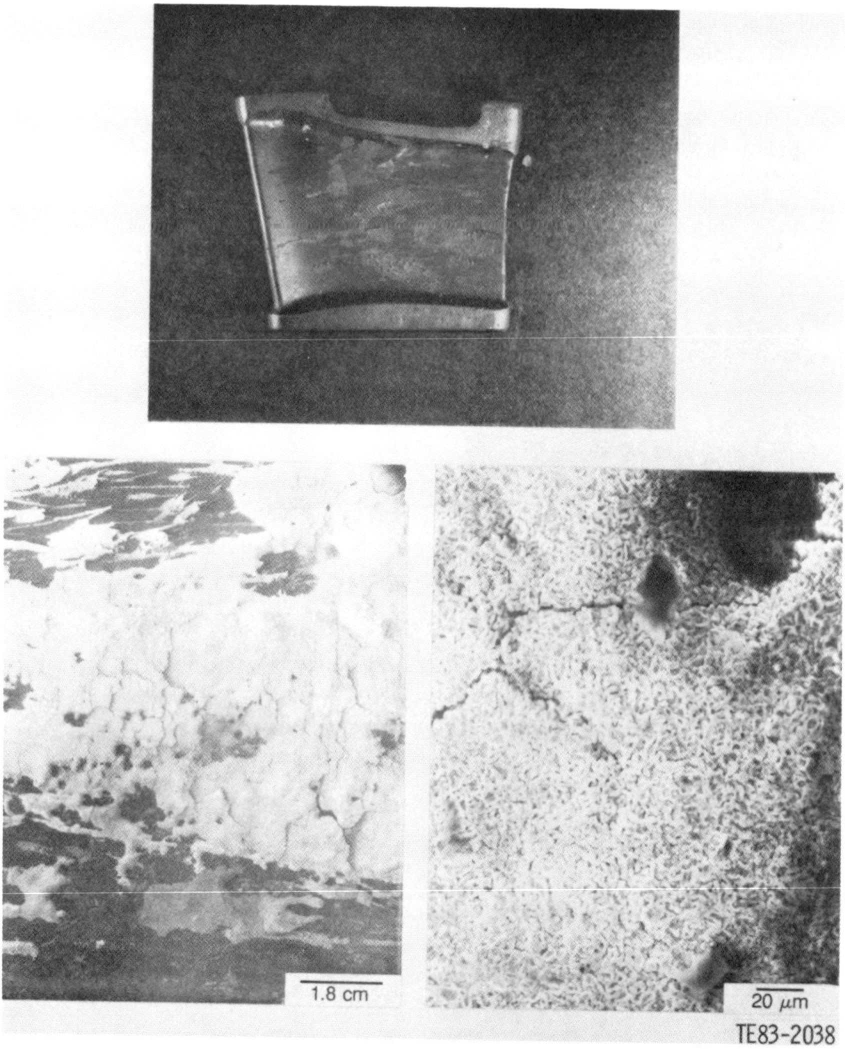

Figure 280. Metal deposits on vane S/N FX 23961 after testing in thermal shock rig, BU10.

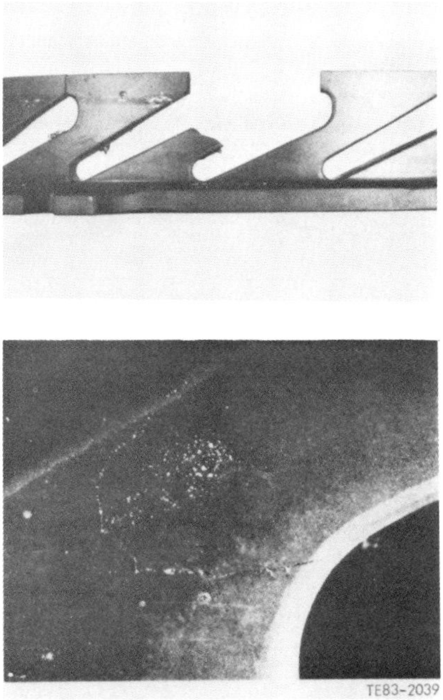

Figure 281. Outer vane support ring (S/N FX 25834) with portion of vane retention feature missing after testing in BU10 of the thermal shock rig (top); close-up of crack in another vane slot of outer vane support ring (S/N FX 25834) (bottom).

The outer vane support ring failed at three locations, as shown in Figure 294. The failure origin for each crack was at the bottom of a lug, on the ring outer diameter (o.d.). Only two of these three lugs were loaded, the bottom locating pin having been previously removed.

Figure 295 shows the fracture surface for the lug located at 270 deg viewed from the rear. A large (560 μm) pore (located by the arrow in Figure 295 was found on the fracture surface. This failure is considered primary because no other discontinuities were found at the other failure sites. The predicted peak stress during accel for the outer vane support ring is 137.9 MPa (20.0 ksi). No fracture analysis was performed because the mirror is not well defined in this material. Test results for bars, cut from the flange of this ring, are presented in Section III.

2265°F CONFIGURATION

Summary

The 2265°F configuration was the final step in the CATE program for introduction of structural ceramics into a turbine engine. The gasifier nozzle for the 2265°F configuration design was identical to the 2070°F configuration. Finite element heat transfer and stress analyses were performed for the ceramic components. Thermal shock rig and engine testing were scheduled for the parts, but funding limitations resulted in cancellation of the testing. This subsection describes the initial finite element analyses for the ceramic components.

Objective

The objective of the 2265°F configuration was to evaluate the 2070°F-configuration ceramic nozzle at elevated temperatures. The evaluation included selecting appropriate materials to be used to ensure the configuration could not need modification.

Approach

The approach used to determine the applicability of the 2070°F-configuration nozzle at 1259°C (2265°F) conditions was to perform finite element heat transfer and stress analyses for the predicted gas stream conditions. Based on these results, the appropriate material was selected to provide a desired probability of survival. Finally, thermal shock rig testing was planned to expose the parts to simulated engine operational stress levels to support the analyses and to increase the probability of survival for the engine.

Discussion: Design

The analysis procedures for the 2265°F-configuration design were the same as previously reported for the 2070°F-configuration design. Since the configuration was identical in each case, the models were the same. The material was changed from RBSiC to sintered alpha SiC for the vane to accommodate the higher temperatures. The boundary conditions were also changed to reflect the higher temperatures.

The transients selected for heat-transfer analysis are shown in Figure 296. Only the transients were analyzed because they marked when peak stresses occurred for the 2070°F-configuration com-

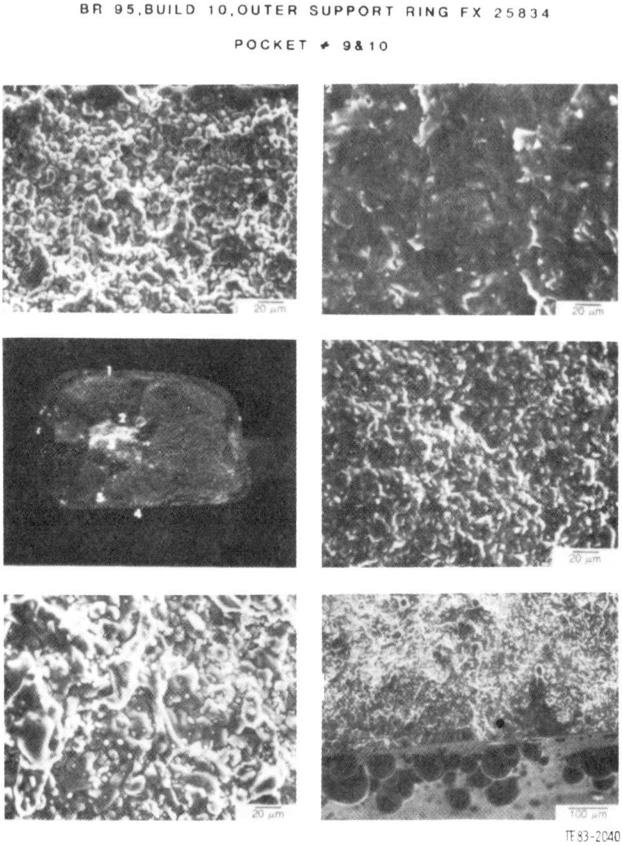

Figure 282. Cross section of outer vane support ring slot broken in BU10 of thermal shock rig.

ponents. Stress analysis was conducted at the point of maximum thermal gradient. The results of the stress analyses performed are shown in Table LVI.

Additional detailed information on this analysis can be found in EDR 10383 (Ref 5), pp 96-105, and EDR 10672 (Ref 6), pp 83-86. The shroud ring was not analyzed at 1241°C (2265°F) conditions because of the initial high probability of survival at 1132°C (2070°F) conditions.

Gasifier Turbine Blade

SUMMARY

A ceramic gasifier turbine blade was designed, fabricated, and successfully demonstrated utilizing the IGT 404 engine. The blade—an injection-molded alpha SiC material—featured a net shape airfoil and machined dovetail attachment. Approximately 9 hr

Figure 283. Close-up of crack in metal outer support due to overtemperature in thermal shock rig BU10.

Figure 284. Nozzle assembly for 2070°F-configuration after passing 2070°F-qualification test in thermal shock rig.

of engine testing of 40 of these blades, including 2 hr 39 minutes at 100% speed and temperature (36,905 rpm and 2070°F RIT), were completed. A maximum gas temperature of 1371°C (2500°F) was observed during transient operation.

The design of the ceramic blade centered on a fast fracture probabilistic technique rather than the deterministic approach typically used in design with ductile metal materials. Three-dimensional (3-D) finite element analyses combined with Weibull modeling of material strength properties—both surface and volume—made possible the calculation of the blade reliability. Extension of the probabilistic technique allowed (1) calculation of the enhancement of blade reliability via spin proof test and (2) reliability as a function of dynamic stress level after spin proof test.

The initial design effort defined the blade attachment (dovetail) configuration and the requirement for a compliant layer. To substantiate the design, dovetail coupons were fabricated and spin tested. These simulated the airfoil mass and location, and the structural soundness of the configuration was demonstrated.

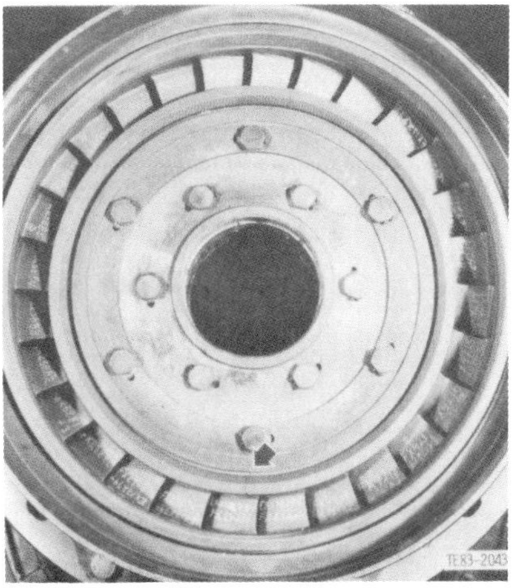

Figure 285. View of gasifier nozzle assembly from rear, showing crack in inner vane support ring after testing in thermal shock rig (BU12).

Figure 286. Ceramic nozzle assembly for 2070°F-configuration after testing in thermal shock rig (BU12), showing lug fractures in the inner vane support ring.

Concurrent with the attachment development, manufacturing process development was conducted to establish the parameters for a structurally sound, injection-molded airfoil shape. A modified IGT airfoil was used. This design and process technology was combined into the final 2070°F blade. Individual blades were proof spin tested to 117% N_G and rotor assemblies to 108% N_G. Two rotors qualified and were subsequently utilized in the engine demonstration.

Finally, an alternate configuration blade was designed that is fully compatible with the 2070°F RIT engine configuration. A factor of four improvement in the probability of failure was calculated for this blade (versus engine test blade) with the same alpha SiC material properties. Alternatively, the engine blade would require an 18% increase in material strength to achieve the reliability of the alternate design blade.

OBJECTIVE

The objective of this task was to design and develop a ceramic-bladed gasifier turbine rotor for the IGT 404 engine. The rotor was designed for operation at 1132°C (2070°F) RIT at a blade reliability goal of 0.99994 and 10,000 hr life. Also, compatibility with existing shaft-mounting features and space limitations in the IGT 404 engine was an important factor

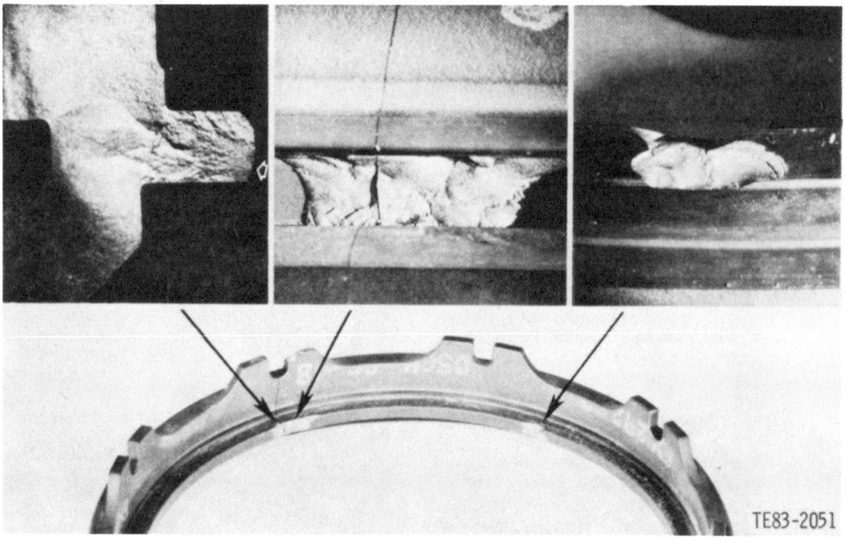

Figure 287. Failure origin of RBSiC inner vane support ring (S/N FX 29257).

Figure 288. Inner vane support ring (S/N FX 29257) after testing in thermal shock rig (BU12) showing chipped areas at inner flange.

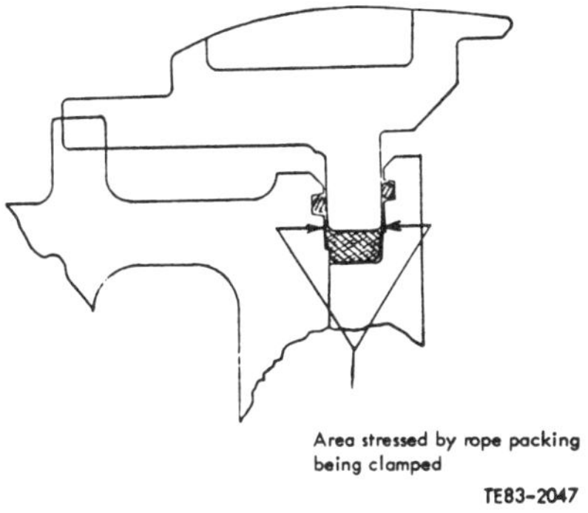

Figure 289. Axial squeezing of large rope packing.

involved in design. The design addressed high turbine efficiency airfoils designed specifically for the 2070°F engine performance cycle and minimization of gas leakage and rotor coolant usage.

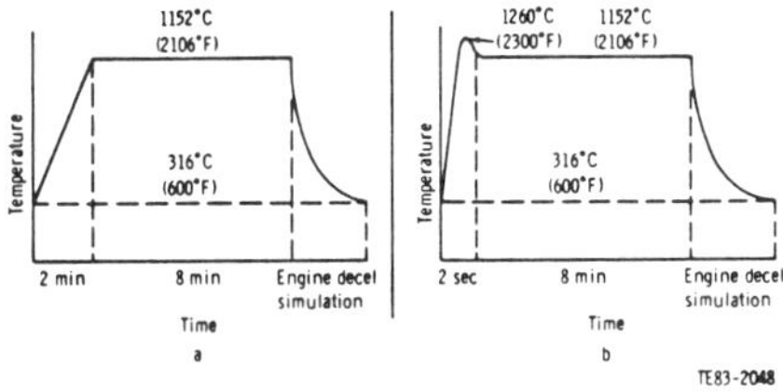

Figure 290. Test cycles conducted for BU13 of thermal shock rig.

Figure 291. View of 2070°F-configuration gasifier nozzle assembly from rear, showing cracks in outer and inner vane support rings after testing in BU13 of the thermal shock rig.

APPROACH

The ceramic-bladed turbine rotor design approach utilized a phased development program. Initial design efforts centered on establishment of an acceptable attachment configuration substantiated by spin test of dovetail coupons, as shown in Figure 297. The coupons were also used in the development of a sheet-metal compliant layer required to minimize local contact stress in the attachment.

In a parallel development phase, a prototype blade was utilized for manufacturing process devel-

Figure 292. Inner vane support ring (S/N FX 29258) after testing in thermal shock rig (BU13), showing crack through ring.

opment. The blade design for the 2070°F engine was directly impacted by experience in the attachment and prototype blade phases. This development plan is illustrated in Figure 298.

The approach in designing the successful ceramic bladed turbine rotor required an iterative process that involved consideration of the following:

- concept studies and preliminary evaluations
- aerodynamic requirements
  - flow-path geometry
  - airflow shape and thickness
  - working gas leakage
  - cooling gas flow
- material properties
  - thermal characteristics
  - elastic characteristics

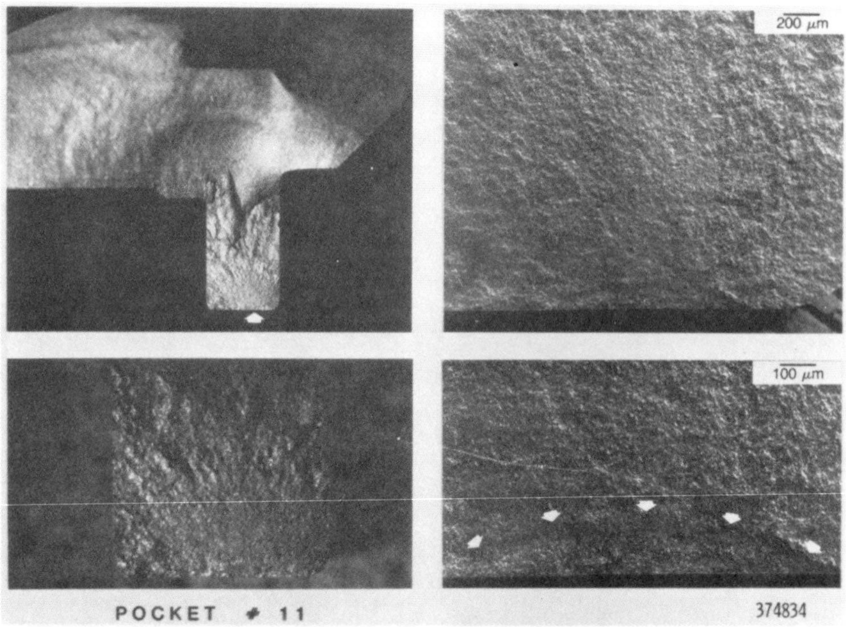

Figure 293. Failure origin of RBSiC inner vane support ring (S/N FX 29258).

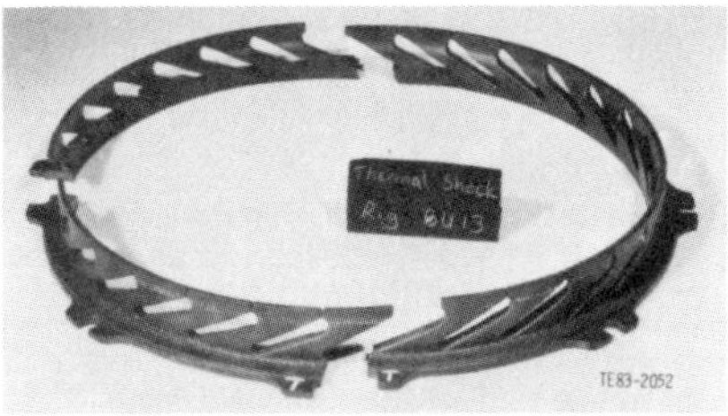

Figure 294. Outer vane support ring (S/N FX 28704) after testing in thermal shock rig (BU13), showing three cracks through ring.

  - time, temperature, and stress life characteristics
  - strength
  - oxidation (effects on strength and life)
- manufacturing requirements
  - shape and thickness
  - shape tolerance
  - finish
  - quality
  - producibility
  - cost
  - availability
- analytical appraisal of concepts
  - thermal analysis
  - stress analysis
  - vibration analysis
  - probabilistic evaluations
  - life predictions
- results of component testing
  - spin tests
  - thermal shock tests
  - vibration characterization
  - engine simulation rig
  - engine performance
  - engine durability

DISCUSSION: DESIGN

The initial task addressing the objective of a high-efficiency gasifier turbine for operation at 1132°C (2070°F) RIT was to define the design approach, criteria, configuration, and candidate materials. The definitions were made for demonstration of the hardware in the IGT 404 engine (existing shafting, etc) with potential for operation at 1241°C (2265°F) RIT. The design approach encompassed the following:

- aerodynamic requirements
- material properties
- manufacturing requirements
- probabilistic analytical appraisal of concepts
- results of component testing

The design criteria specified a 10,000 hr life in the line-haul truck, numerous transient conditions, and vibratory characteristics requirements, blade gas bending, overspeed capability (disk burst at more than 130%), and ceramic blade failure criteria.

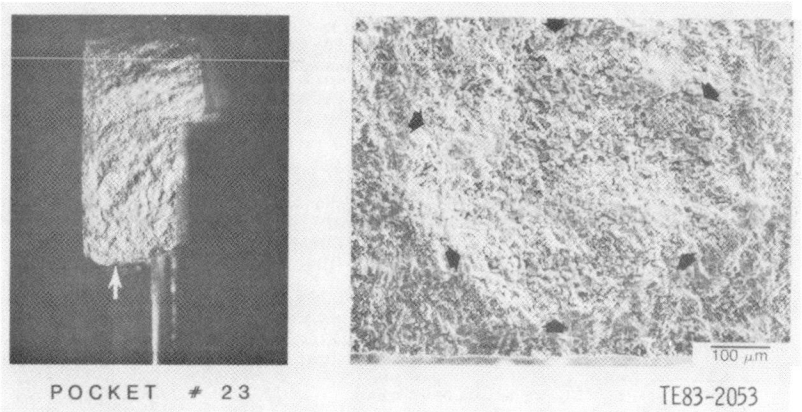

Figure 295. Failure origin of Pure Carbon Refel SiC outer vane support ring showing large pore.

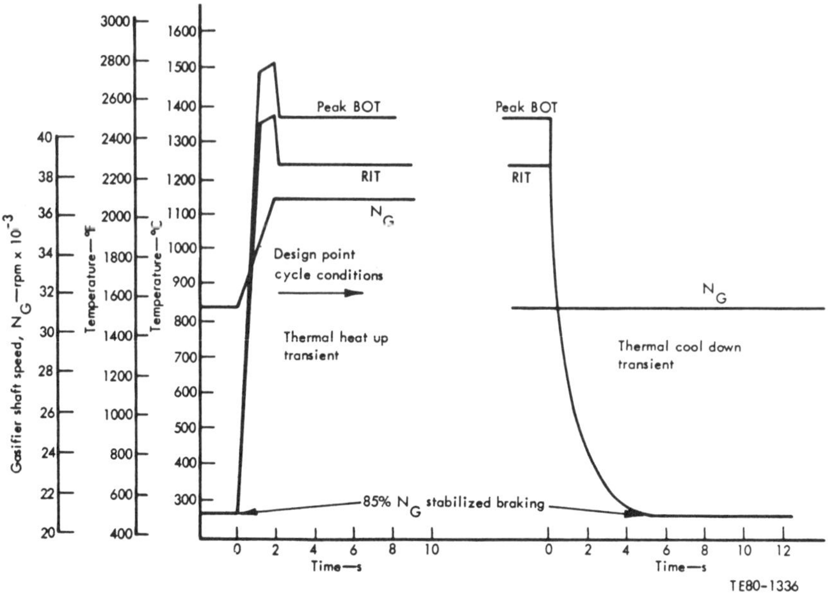

Figure 296. Gas thermal transients for 2265°F-configuration.

Table LVI.
Maximum principal stresses and probabilities of survival of 2265°F-configuration ceramic components during thermal transients.

	Maximum principal stress MPa	(ksi)	Probability of survival
Vane			
Accel at 2 sec	395.0	(57.3)	0.96101*
Decel at 4 sec	213.7	(31.0)	0.99994
Outer vane support ring			
Accel at 14 sec	200.6	(29.1)	0.99795**
Decel at 24 sec	284.7	(41.3)	0.98764
Inner vane support ring			
Accel at 24 sec	165.5	(24.0)	0.98754**
Decel at 6 sec	141.3	(20.5)	0.99642

*Probability of survival for the vane is based on alpha SiC (MOR = 342.7 MPa (49.7 ksi); m = 8.0).

**Probability of survival for rings is based on RBSiC (MOR = 366.8 MPa (53.2 ksi); m = 8.0).

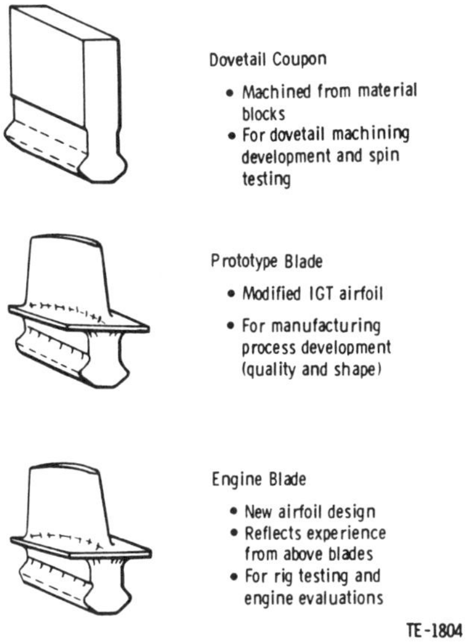

Figure 297. Ceramic blade concept and development plan.

The basic general arrangement of the gasifier turbine is illustrated in Figure 298. Note the layer of compliant material required between the blade and wheel contact surfaces to diminish load concentrations resulting from shape irregularities. The initial ceramic blade effort involved fabrication of the prototype blade, as illustrated earlier in Figure 297. (The objective of all the steps in this phased development is also noted in the figure.)

Materials selected for initial blade development were sintered silicon carbide (from Carborundum Company) and sintered silicon nitride (from GTE Laboratories). The property data used in the early heat transfer and stress analyses are presented in Table LVII. These analyses were based on two-dimensional finite element models that resulted in the blade dovetail and matching disk slot geometry also shown in Figure 297. The calculated stresses and steady-state temperature distribution are described in detail in EDR 9346 (Ref 9). The prototype blade shown in Figure 297 is a scale of the existing IGT 404-4, which has 56 blades. This scaling, constrained by ceramic strength considerations, resulted in a 40-blade design. Two-dimensional heat transfer analyses of the disk at steady-state and transient conditions with SiC and Si_3N_4 blades were performed. A transient condition with the SiC blade (conductivity of SiC is greater than Si_3N_4) was identified as the life-limiting worst case. With a blade platform forward lip seal, the disk was satisfactory at

Table LVII.
Ceramic blade material property data (1977).

	Material	
	Sintered silicon nitride	**Sintered silicon carbide**
Density — Mg/m^3 ($lb/in.^3$)	3.128 (0.113)	3.156 (0.114)
Specific heat — J/kg-°K (Btu/lbm-°F)	628 at 298°K (0.150 at 77°F) 1248 at 1273°K (0.298 at 1832°F)	678 at 298°K (0.162 at 77°F) 1260 at 1273°K (0.301 at 1832°F)
Thermal conductivity — W/m-°K (Btu/hr-ft-°F)	30.3 at 298°K (17.5 at 77°F) 16.1 at 1366°K (9.3 at 2000°F)	90.9 at 298°K (52.5 at 77°F) 34.8 at 1273°K (20.1 at 1832°F)
Modulus of elasticity — GPa ($lb/in.^2$)	306.8 at 298°K (44.5 x 10^6 at 77°F) 268.9 at 1366°K (39.0 x 10^6 at 2000°F)	409.5 at 298°K (59.4 x 10^6 at 77°F) 386.1 at 1273°K (56.0 x 10^6 at 1832°F)
Thermal expansion coefficient — m/m-°K (in./in.-°F)	1.02 μm at 294-1588°K (1.84 μm at 70-2400°F)	1.48 μm at 298-1773°K (2.67 μm at 77-2732°F)
Mean MOR — MPa (ksi)	551.6 at 1273°K (80.0 at 1832°F)	386.1 at 1273°K (56.0 at 1832°F)

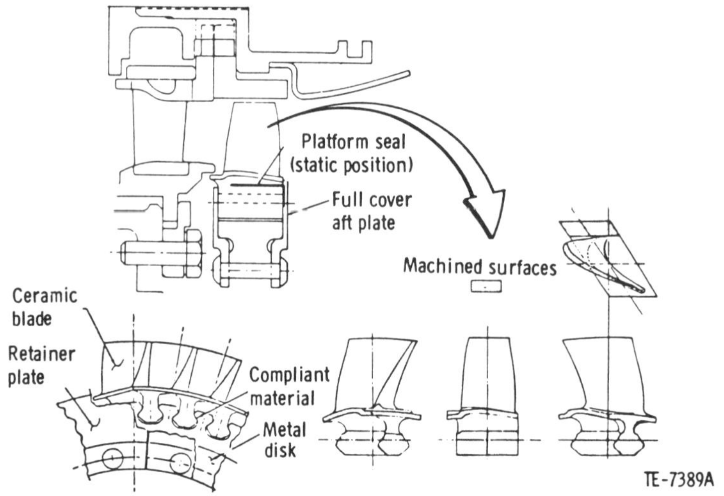

Figure 298. Development of 2070°F ceramic blade.

649°C (1200°F) and approximately 4600 low cycle fatigue life. (Additional details are found in EDR 9519—Ref 1.)

For the engine blade of Figure 297, the vibration mode frequencies analyzed for the 1132°C (2070°F) 40-blade configuration with the 28-vane nozzle were shown to be satisfactory. An interference of first mode with vane passage was predicted at transition between the idle and high time speed range, as shown by the Campbell diagram in Figure 299. IGT 404 experience indicates satisfactory operation with this type of transition interference (idle to high time range). Alternative investigations of 30 and 50 blades were not productive. (Additional information on these investigations is discussed in EDR 9519 [Ref 1]).

Using the defined airfoil shape that meets aerodynamic goals and the vibration characteristics, the basic layout was completed, as shown in Figure 300. This layout (design) features the following:

- a forward lip seal to reduce hot gas mixing
- the blade aft retaining plate, radially extended to the blade platform for elimination of leakage between blade stalks
- thin sheet metal strips attached to the aft retainer plate to provide sealing for the gaps between blade platforms

The 2070°F engine blade, whose geometry is shown in Figure 301, is injection-molded silicon carbide, and final machining is minimal. Two-dimensional finite element analysis was utilized to study the radial alignment of the airfoil section center of gravity. The stack criteria were minimization of the positive values of the maximum principal stress for steady state at design point (2070°F and N_G = 100%) and various transient conditions. The numerical values and a description (illustration) of the resulting stack is shown in detail in EDR 9722 (Ref 2).

Three-dimensional cubic finite element analysis was subsequently used as the basis for establishing the design acceptability of the 2070°F blade geometry (see Figure 302). This analysis was used to calculate temperatures, stresses, and failure probabilities for engine design point and transient conditions. The design goal is a probability of failure of 6 or fewer per 100,000. Note that the spin test coupon was also modeled. At 100% blade load, the slab type spin test coupon probability of failure was predicted at 0.0095. This was acceptable for initial spin tests addressing the attachment concept.

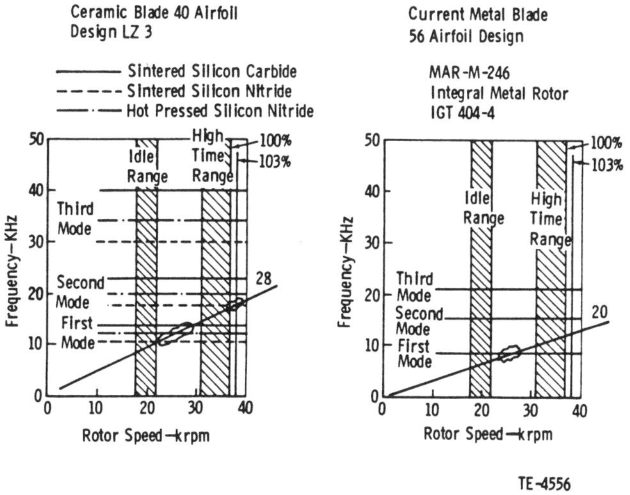

Figure 299. Turbine blade vibration characteristics.

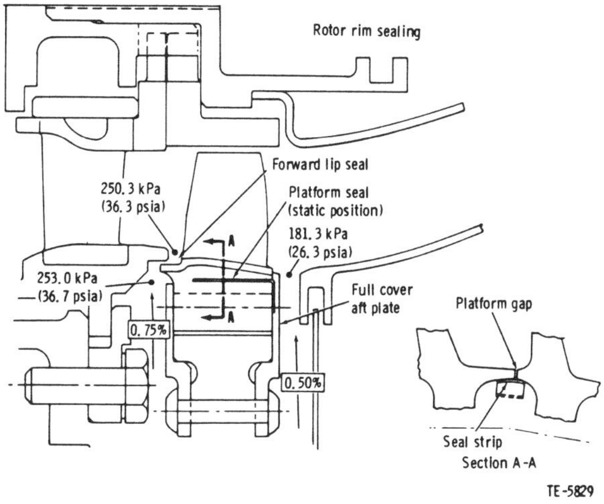

Figure 300. Gasifier turbine rotor configuration.

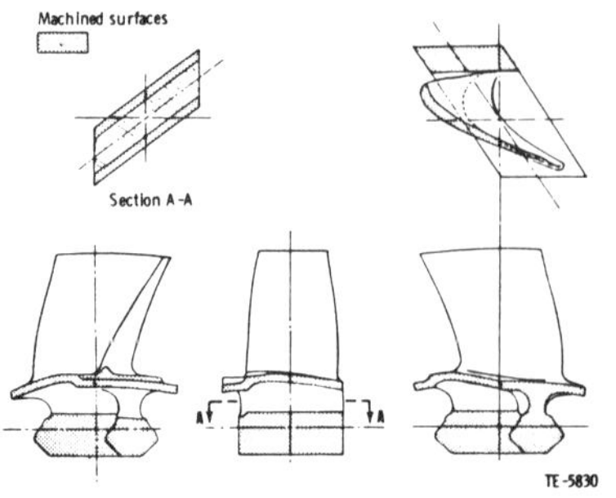

Figure 301. Geometry of 2070°F ceramic blade.

The 3-D analysis of the blade per the model shown in Figure 302 began with definition of the turbine inlet gas temperature versus time, as shown in Figure 303. The following conditions were analyzed:

1. room temperature spin at 100% rotor speed
2. steady state at 1132°C (2070°F) and maximum power, N_G = 100%
3. 3 sec into acceleration from stabilized braking to maximum power, N_G = 100%
4. 2 sec into deceleration from part power into dynamic braking, N_G = 85%
5. 5 sec into deceleration from part power into dynamic braking, N_G = 85%

Conditions 3, 4, and 5 were selected as the time points in transient operation to be stress analyzed, based on the severity of the temperature gradient. The computed maximum principal stress revealed that the blade peak stress normally occurs in the attachment. The maximum attachment stress of 326.5 MPa (47.35 ksi) occurs at 3 sec into an acceleration from stabilized braking with 100% rotor speed. The isostress plot for this case is shown in Figure 304.

Of greater interest is the predicted blade reliability for the stress distributions presented. Material strength characteristics were determined from pressed and sintered silicon-carbide test bars, which have a mean strength of 342.7 MPa (49.7 ksi), and these data have been used to predict the injection-molded blade reliability.

The blade analysis model comprises two substructures, which include the airfoil and upper portion of the platform plus the attachment, stalk, and lower portion of the platform. Because of this sub-

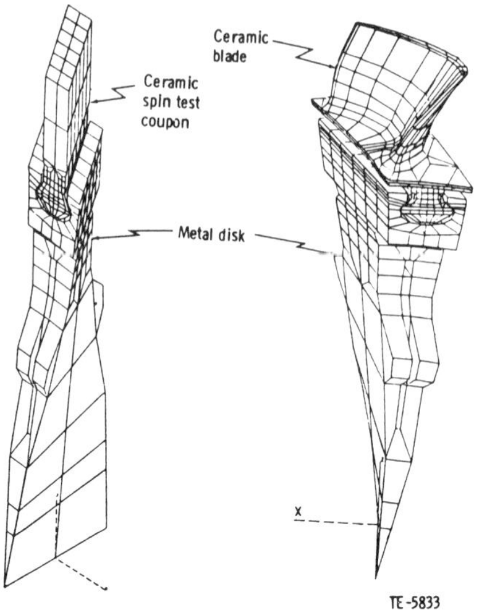

Figure 302. Three-dimensional finite element model.

structuring, it is natural to determine the probability of failure for the airfoil, attachment, and total blade. This has been accomplished for each of the selected engine operating conditions plus spin test. The results are presented in Table LVIII, which shows that failure probability in the attachment is higher than in the airfoil for all conditions except the 2- and 5-sec deceleration points, where airfoil stress and failure probability are higher. Overall blade reliability is lowest in the 3-sec acceleration condition, where a failure rate of nearly 1.8 per 100 is predicted. All failure rates were determined without a spin proof test; however, such a test would improve blade reliability in the engine. The complete report on temperature gradients and isostress plots for all the conditions analyzed is in EDR 9951 (Ref 3).

An analytic treatment of the benefit of spin proof test on component reliability was developed. The calculation of the probability of surviving a prescribed load condition after the structure was proof tested follows:

$P_S = e^{(-R_R + R_P)}$ Probability of survival

$R_R = \frac{\sigma_R - \sigma_\mu}{o}$ For load condition (σ_R principal stresses)

$R_P = \frac{\sigma_P - \sigma_\mu}{o}$ For proof test (σ_P principal stresses)

$P_s + P_f = 1$ P_f = Probability of failure

σ_o, σ_u, and m are Weibull material constants

Equation 1 is integrated over the surface and volume of the structure. For example, the P_f values shown in Table LIX were calculated for the worst-case operating condition (3 sec acceleration from stabilized braking).

A proof test speed of 43,250 (117%) was determined to provide satisfactory rotor reliability (P_f = 0.00023) for initial engine testing (rotor P_f = 0.00023/blade x 40 blades/rotor = 0.0092/rotor). That is, the risk of rotor failure in initial engine testing is approximately 1 in 100.

Another analytic procedure was developed for evaluating the effects of blade vibration on reliability. This procedure allows a dynamic stress to be combined with a static stress distribution for an engine operating condition. Using this procedure, a range of peak vibratory stress level distributions for the blade fundamental mode were combined with the static stress distribution for engine operation at N_G = 79% (29,000 rpm). This corresponds to the vane passage interference (see Figure 299). The results indicated that high vibratory stress could have a significant and adverse effect on blade reliability. The effect of a spin proof test was calculated considering a previously completed proof spin at 117% speed and engine environment static plus dynamic stress. The results are shown in Table LX. Results indicated a dramatic improvement in reliability at lower dynamic stress levels. Further review of results indicated a blade reliability of P_f = 0.00025 (rotor P_f = 0.00025 x 40 = 0.01 or the earlier noted rotor risk of 1 in 100) was met if the proof-tested blade dynamic stress did not exceed ± 153 MPa (22.1 ksi). This was sufficient vibratory stress tolerance for initial engine tests. (Details are reported in EDR 10156, Ref 4.)

A computer efficient 3-D model was described as part of the alternate blade design study. This model reduced the machine central processor time by 90% and is subsequently referred to as the

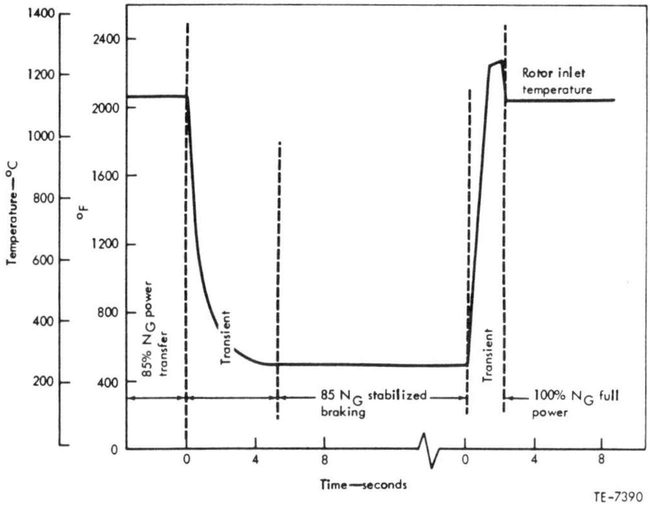

Figure 303. CATE 2070°F engine thermal transients for heat transfer analysis.

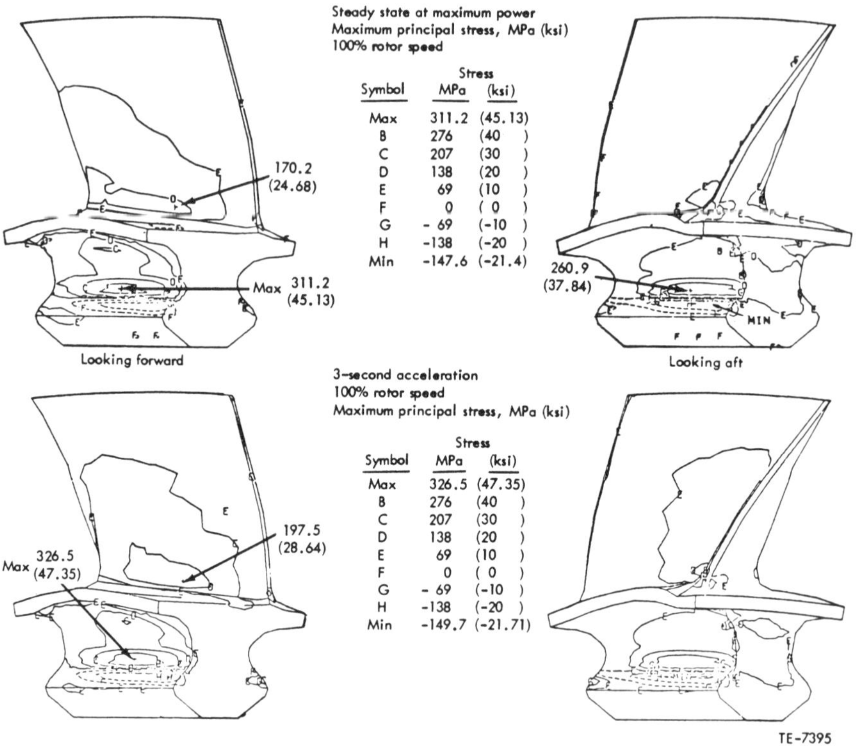

Figure 304. Stress distribution for 2070°F blade.

coarse mesh model. The accuracy of the coarse mesh model was assessed by comparing the calculated P_f of the 3 sec acceleration out of braking condition for the coarse and previous fine mesh models. The results compared favorably as follows:

Model	Total P_f
Fine mesh	0.0177
Coarse mesh	0.0159

The peak values of maximum principal stress as determined with the two models were also in agreement. Additional background is found in EDR 10382 (Ref 5).

The coarse mesh model was used to evaluate a potential alternate spin test with a skewed dovetail slot, as shown in Figure 305. The objective was to improve the effectiveness of spin proof tests by loading of the airfoil concave side. Thus, two spin tests were considered—one with a normal slot orientation and a second with a skewed slot. Analysis indicated (1) the skewed proof position would not significantly improve blade reliability and (2) further rotation in the rim was not practical. The numerical

Table LVIII.
Blade stress and reliability summary (no spin test).

	Spin test	Max power	3-sec accel	2-sec decel	5-sec decel
Rotor speed—rpm	36,905 (100%)	36,905 (100%)	36,905 (100%)	31,369 (85%)	31,369 (85%)
Avg tensile stress — MPa (ksi) (attachment neck)	82.1 (11.9)	82.1 (11.9)	82.1 (11.9)	59.3 (8.6)	59.3 (8.6)
Max principal peak — MPa (ksi)					
Attachment	276.5 (40.1)	311.2 (45.1)	326.5 (47.3)	225.5 (33.7)	228.9 (33.2)
Airfoil	201.3 (29.2)	170.2 (24.7)	197.5 (28.6)	226.2 (32.8)	263.4 (38.2)
P_f attachment	0.0057	0.0112	0.0169	0.0006	0.0008
P_f airfoil	0.0012	0.0005	0.0008	0.0012	0.0028
P_f total (blade)	0.0069	0.0117	0.0177	0.0018	0.0036

Note: Sintered silicon carbide, MOR (mean) = 342.7 MPa (49.7 ksi), Weibull m = 7.91.

Table LIX.
Probability of failure.

Proof test speed—rpm	Lost in proof—%	P_f, 3 sec accel
0 (no spin proof)	0	0.0177
40,000 (108%)	2.7	0.00053
43,250 (117%)	10	0.00023
48,280 (131%)	50	0.00017

results supporting this conclusion are tabulated in EDR 10672 (Ref 6). This completed the consideration of the proof test with a skewed dovetail slot.

The construction of the gasifier rotor with ceramic blades is illustrated in Figure 306. The ceramic blade, shown in Figure 301, features a net shape airfoil with the exception of the airfoil tip, which is machined. The following is a summary of the design features of each component:

- general configuration
  - compatible with the IGT 404 engine, gasifier rotor
  - 2070°F rotor inlet temperature
- rotor disk
  - Waspaloy material
  - maximum temperature at rim, 649°C (1200°F)
  - approximately 4600 LCF life, rivet holes
  - cooling air, front face at 0.75% and aft face at 0.50%
- blade retaining plates
  - Inco 718 material
  - centrifugal load reacted by wheel
  - finger seals for sealing the blade platform gaps
- compliant layer
  - L605 material
  - thickness, 0.203 mm ± 0.00254 mm (0.008 in. ± 0.0001 in.)
- blade (40 per rotor assembly)
  - alpha silicon carbide material
  - material properties: MOR = 383.98 MPa (55.69 ksi); Weibull slope = 7.823
  - reliability design goal: P_s = 0.99994/blade or P_s = 0.998/rotor
  - calculated reliability with spin proof test at 117% N_G: P_s = 0.99977/blade or P_s = 0.991/rotor

This latter part is for the life-limiting case, which is a 3 sec acceleration to 100% power after stabilized braking at 85% speed. A 10% failure rate is predicted for the 117% speed proof test.

The isostress plots for the 3 sec acceleration case and the room-temperature spin case are shown in Figures 307 and 308. Note that the spin condition stress values are for 100% speed (36,905

Table LX.
Effect of spin proof and dynamic stress on blade reliability.

Peak dynamic stress (dovetail)	P_f Without proof test	P_f With proof test
± 0 (steady state, N_G = 79%)	0.00025	3.0×10^{-7}
± 80 MPa (± 11.6 ksi)	0.0027	3.03×10^{-6}
± 159.3 MPa (± 23.1 ksi)	0.024	0.00034
± 318.5 MPa (± 46.2 ksi)	0.51	0.133

rpm); the values at 117% proof speed (43,250 rpm) would increase by the ratio of speeds squared.

Figure 309 presents the calculated reliability for the 3 sec acceleration condition as a function of proof test spin speed. This illustrates the selection of the 117% proof speed; above 117% there is little improvement in the blade reliability.

Blade fundamental (14,854 Hz) was calculated at interference with 28th engine order (vane passage frequency) at N_G = 79% (29,000 rpm). This was the only predicted interference in the engine operation envelope and is a transition speed (low time) from idle to the power range.

At the predicted rotor reliability of P_s = 0.991 (100% N_G) or P_f 0.009, the tolerable dynamic stress level at 79% N_G for a proof-tested blade was calculated at ± 153 MPa (22.1 ksi) at the location in the attachment of the maximum dynamic stress. The corresponding dynamic stress in the airfoil (hub, concave trailing edge location) is ± 79.3 MPa (11.5 ksi). This vibratory stress tolerance was sufficient for initial engine tests.

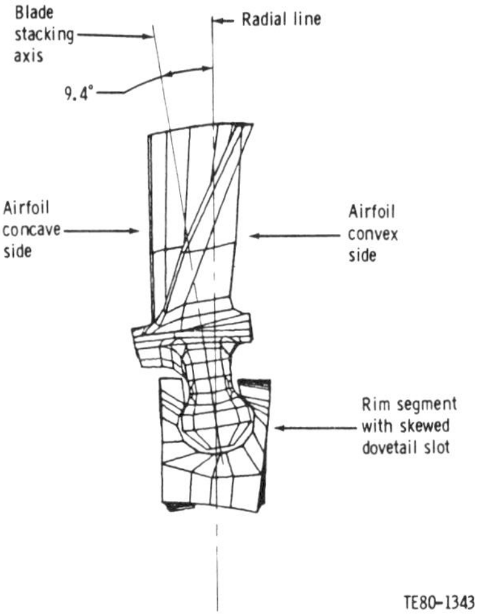

Figure 305. Spin proof test with skewed dovetail slot.

DISCUSSION: TEST

Compliant Layer Materials

Test investigation and verification of the gasifier blade system design began with bench rig tests of candidate compliant layer materials. The rig shown in Figures 310 and 311 was used to conduct tests of 347 stainless steel, L605, IN-600, and platinum compliant layers at room temperature—649°C (1200°F)—and using constant and cyclic loads. At elevated temperature and constant load, the 347 stainless steel and platinum layers bonded to the silicon carbide. Additional tests at temperature demonstrated that boron nitride lubricant applied to the load bearing surfaces of the attachment successfully eliminated the bonding.

Cyclic load tests were conducted for the four materials with the boron nitride lubricant applied to all load bearing surfaces of the attachment system. All materials except L605 showed considerable thin-out.

Room-temperature, overspeed-to-failure spin tests were conducted as the next logical step. Silicon carbide and silicon nitride slab coupons and the three candidate compliant layers were tested. The results, shown in Figure 312, included the following observations:

1. The five silicon carbide specimens with IN-600 compliant layer failed at an average speed of 38,560 rpm versus 37,889 rpm predicted (1.8% difference).
2. The use of IN-600, L605, or 347 stainless steel compliant layers does not affect failure speed in cold spin.
3. Failure initiation locations are very close to peak stress locations computed by 3-D finite element analysis.

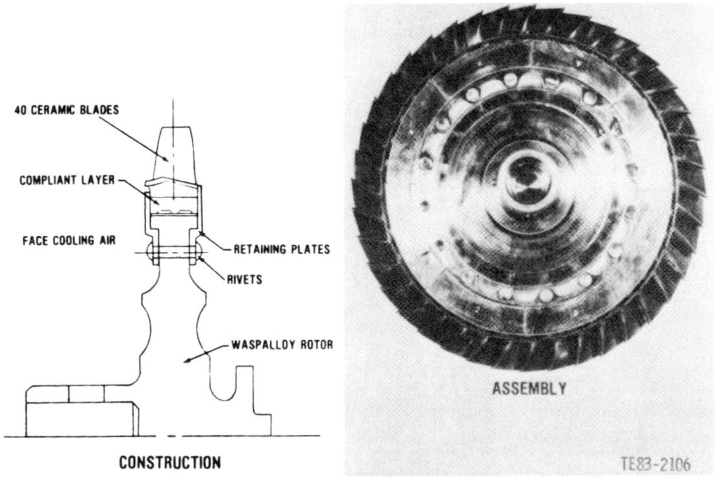

Figure 306. Ceramic bladed turbine rotor.

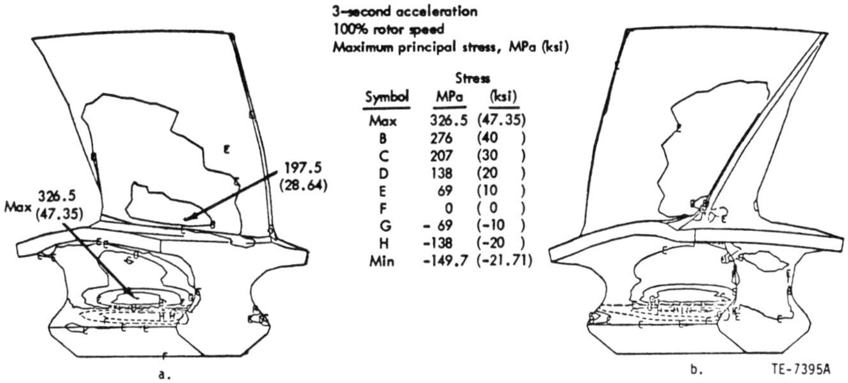

Figure 307. Contour plots of ceramic blade maximum principal stress (a) looking forward with 3 sec acceleration and (b) looking aft with 3 sec acceleration.

4. The silicon nitride specimen failed at an average speed of 49,100 rpm versus the predicted average speed of 49,600 rpm (1.0% difference).

The complete data describing the bench and cold spin rig tests are listed in EDR 9951 (Ref 3).

Elevated temperature, overspeed-to-failure blade coupon spin tests were conducted with each of three compliant layers (347 stainless steel, L605, and IN-600 material). The wheel and coupon assembly was heated to a uniform 677°C (1250°F), and the

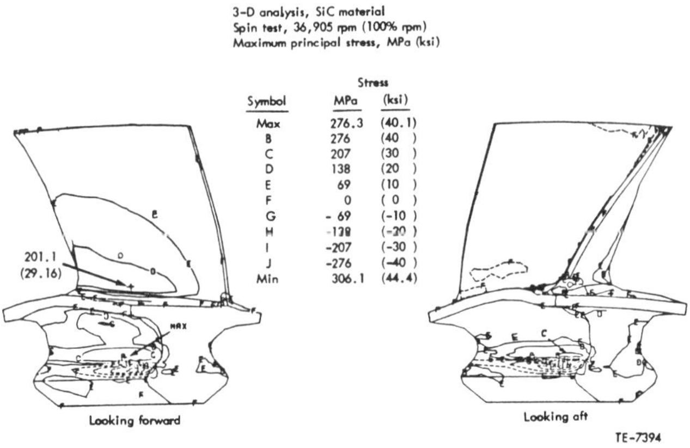

Figure 308. Contour plot of ceramic blade maximum principal stress looking forward and looking aft.

rotor speed slowly increased until failure. The average of the hot tests was shown to be within 1.7% of the average of the previously described room-temperature spin tests. None of the compliant materials seemed clearly superior.

The candidate compliant layer materials were evaluated further by hot cyclic spin tests. In this test series, the rotor assembly temperature is maintained at 677°C (1250°F), and rotor speed is cycled between 500 and 30,800 rpm (100% attachment load speed). After each test, the compliant layer and coupon were photographed and thickness change of the layer was determined. Results of the thickness change measurements are in Figure 313. The photo-

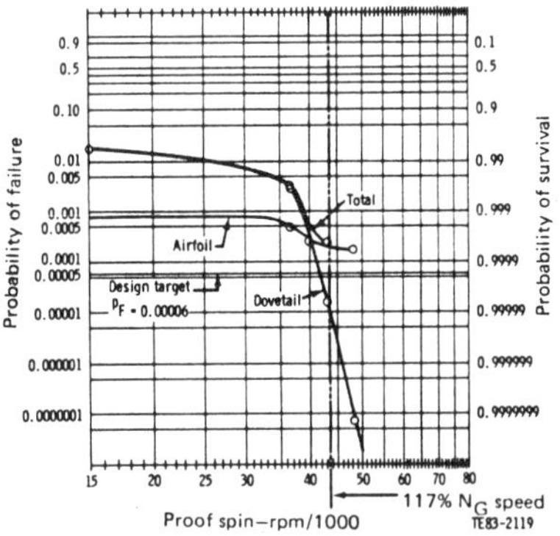

Figure 309. Calculated reliability at 3 sec deceleration condition after proof spin at room temperature.

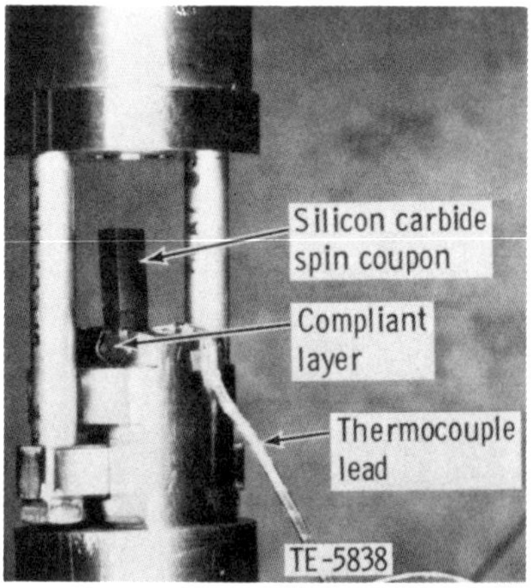

Figure 310. Compliant layer static test setup.

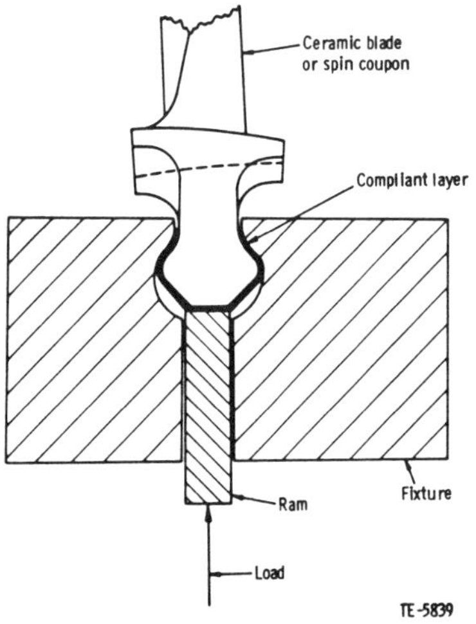

Figure 311. Compliant layer test concept.

graphs illustrating the thin-out are shown in EDR 10156 (Ref 4). Results observed with each material are as follows:

- IN-600 Compliant Layer. This material seems to be compliant, but cyclic testing resulted in complete thin-out in 250 cycles. Considerable distortion of the layer is evident at 200 cycles.
- 347 Stainless Steel Compliant Layer. This material was also compliant, but excessive thin-out again resulted. Complete thin-out was skewed at 600 cycles and continued cycling failed the ceramic coupon at 684 cycles.
- L605 Compliant Layer. This cobalt-based material used in an annealed condition demonstrated minimal thin-out. After 2500 cycles, the maximum local thickness reduction was 0.02 mm (0.0008 in.). The initial annealed condition provides compliance, and the work-hardening characteristic provides resistance to continued thin-out. Based on these results, L605 was selected for compliant layers.

Individual blade spin tests to failure continued, using the L605 compliant layer and boron nitride lubricant. The tests were on the following:

- 12 each GTE silicon nitride attachment coupon
- 4 silicon carbide prototype blades
- 14 silicon carbide 2070°F-configuration blades

GTE Spin Coupons

Primarily, this testing demonstrated a material performance degradation resulting from an axial machining lay in the attachment. Radially ground coupons failed at an average speed of 162% versus 161% predicted. The prediction was based on the corresponding case of Weibull strength parameters obtained from longitudinally ground MOR bars. Longitudinally ground coupons demonstrated 152% average failure speed, and similar coupons, hand polished with diamond paste, demonstrated 154% speed. It was concluded that machining damage (degradation) extends below the surface.

Prototype Blades

The sample of four prototype blades failed at an average speed of 135% and provided the first indication that the blade design including the attachment was successful. Furthermore, the attachment and compliant layer contact pattern was identical to patterns previously observed in coupon spin tests.

2070°F-Configuration Blades, Follow-On Order No. 1

From the start of this program, the axial machining lay selected for the blade dovetail was recognized to result in a material performance decrease. An investigation of transverse ground MOR bar strength as a function of thermal exposure treatment was conducted. A 24-hr heat treatment in air at 1250°C (2282°F) showed a 25% improvement in strength and was selected for evaluation on test blades.

The overspeed-to-failure tests of the first group of 2070°F-configuration blades are summarized in Table LXI. An average failure speed of 131% was predicted for this group, and the first sample of six blades demonstrated 116%. Past failure investigation identified a glassy layer that may have affected part strength. The blades were processed using Al_2O_3 furnace set plates rather than silicon carbide plates. All other experience has been with the silicon carbide plates.

To further assess the effect of the heat treatment, the remaining eight blades were divided in half, four with and four without post machining heat treatment. These are identified in Table LXI. Note that the heat-treated blades performed the best at 125.2% average failure speed. This was below the calculated average failure speed of 131%. Thus, the following observations were made:

- Flaws found in failed blade fracture surfaces were similar to those found in test bars.

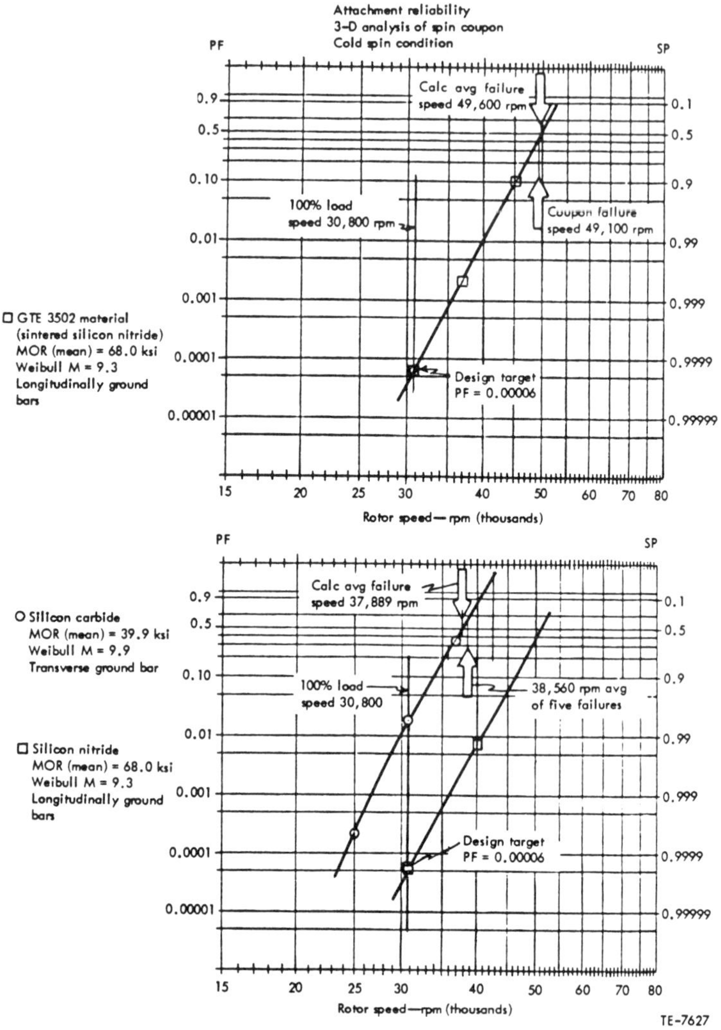

Figure 312. Spin coupon reliability versus rotor speed.

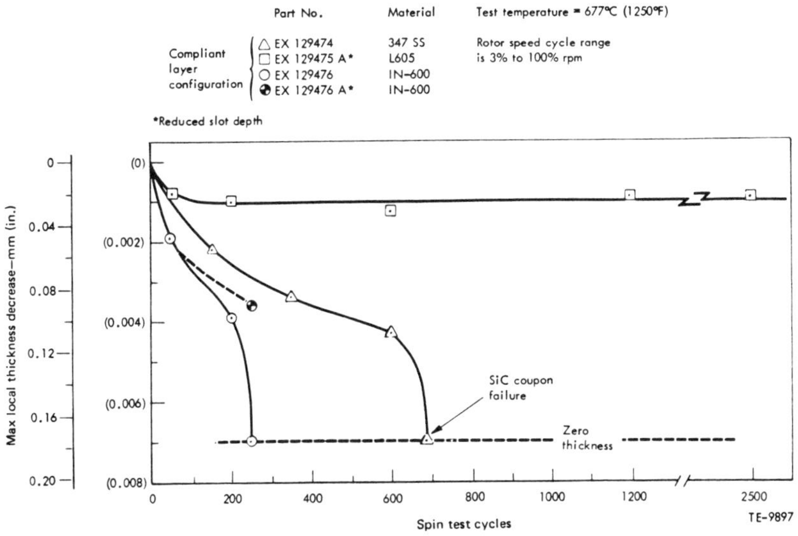

Figure 313. Hot cyclic spin evaluation of compliant layers.

- Failure initiation sites corresponded to predicted zones of peak stress in attachment, stalk, or airfoil.
- The material strength properties of injection-molded alpha SiC were not established. (A large number of bars were subsequently produced for evaluation.)
- The development blades tested did not represent the final process established for the 2070°F-configuration blade.

Photos and additional information on this group of tests are found in EDR 10383 (Ref 5).

The second half of 1980 included a major spin test effort addressing individual blades and rotor assemblies. Fifty-two finished blades were individually proof tested to 110% (40,600 rpm) in a single slot disk with L605 compliant layers and boron-nitride spray lubricant. The blades were the result of initial process optimization by Carborundum Company (CBO) with attachments longitudinally ground by Crafts Inc. Ten blades failed during proof testing for a 19.6% failure rate versus a predicted rate of 8% based on pressed and sintered MOR bars with 343 MPa (49.7 ksi) strength and Weibull m = 7.9. The failures revealed large crystals in the fracture surfaces not previously seen, thus indicating variation in the material process.

Gasifier Rotor No. 1

A rotor was assembled with the proof tested blades. The balanced rotor assembly was subjected to ambient temperature spin tests in an evacuated spin pit. During incremental speed advance from 90% to 95%, a failure occurred at 93.2% speed. (The proof speed goal was 105%.) The fracture originated in the stalk region above the attachment. The origin was on the surface and a flaw could not be found with microexamination. This location corresponds to the zone of maximum calculated stress in the blade stalk.

A failure investigation including additional spin tests of individual blades was conducted. Summary points of the investigation were as follows:

- blade quality standard—not a factor; no recorded defects in the failed blade
- dimensional perturbations—thickness variations of 0.038 mm (0.0015 in.) measured in the L605 material compliant layers
- load variations—3-D finite element model (FEM) of nonuniform loading along the attachment

Table LXI.
Overspeed-to-failure testing of 2070°F configuration blades of sintered silicon carbide development blades.

Serial No.	Heat treat*	Test temperature	Failure speed—rpm (%)	Failure location	Failure origin
12	Yes**	Room	39,000 (105.7)	Attachment	40-μm pore, volume
10	Yes**	Room	46,000 (124.6)	Attachment (cv)	100-μm pore, surface
17	Yes**	677°C (1250°F)	40,150 (108.8)	Stalk	200-μm pore, volume
FX24890	Yes**	Room	40,600 (110)	Attachment	24-μm pore, surface
FX24891	Yes**	Room	47,000 (127.4)	Airfoil at root	No flaws at initiations
				Platform at acute corner	Recorded airfoil TE flaws
FX24892	Yes**	Room	44,200 (119.8)	Attachment	Large grain, 70-μm pore, volume
		Avg 6 = 42,875 rpm (116%)			
FX24886	No	Room	45,100 (122.2)	Attachment	Surface, no flaw
FX24885	No	Room	40,500 (109.7)	Attachment	Surface, no flaw
FX24884	No	Room	45,650 (123.7)	Attachment	Surface, no flaw
FX24887	No	Room	49,450 (133.9)	Stalk	Surface, no flaw, looks like strength failure
		Avg 4 = 45,175 rpm (122.4%)			
FX24880	Yes	Room	42,700 (115.7)	Stalk	100-μm pore, volume
FX24881	Yes	Room	49,800 (134.9)	Attachment	Surface, no flaw
FX24882	Yes	Room	47,300 (128.2)	Attachment	Surface, no flaw
FX24883	Yes	Room	45,000 (121.9)	Attachment	Surface, no flaw
		Avg 4 = 46,200 rpm (125.2%)			

* 24 hr at 1250°C (2282°F) in air
** Al_2O_3 furnace set plates

contact surfaces showed high stress, low reliability (Thus, the measured L605 compliant layer thickness variation is significant.)

- effect of surface glassy layer on blade strength—spin test samples demonstrated apparent improved performance of oxidized blades with polished stalks (Based on this point, it was concluded that [1] additional spin evaluation was required and [2] a 117% blade proof was necessary for increased reliability at 100% speed.)

The additional spin evaluation was conducted, and the test failure rates (individual blades), presented in Figure 314 for comparison with the computed failure rate, were determined. The following observations were made:

- Six oxidized blades spun overspeed to failure had an average failure speed about equal to prediction.
- Fifty-two oxidized blades spun to 110% speed had a failure rate of nearly 20%, compared with 8% predicted rate.
- These two data points suggest that blade material mean strength is as anticipated, but Weibull modulus was lower than expected.
- The second 110% proof test of oxidized blades resulted in an additional 15% blade loss that cannot be explained by blade strength variability.
- The 23 oxidized and polished blades with selected-thickness compliant layers seem to have improved reliability at 117% proof speed. The achieved failure rate of 26% is still above the predicted 22% but below the 32% implied by the straight line between the oxidized-only data points.

The failure of blades in a second 110% proof test can be explained by assuming variability in test conditions; that is, inherent variability is distributed around the nominal proof-test speed.

Complete details of the rotor failure, including descriptive photographs, are found in EDR 10672 (Ref 6). An item-by-item tabulation of each spin test is also provided as a road map.

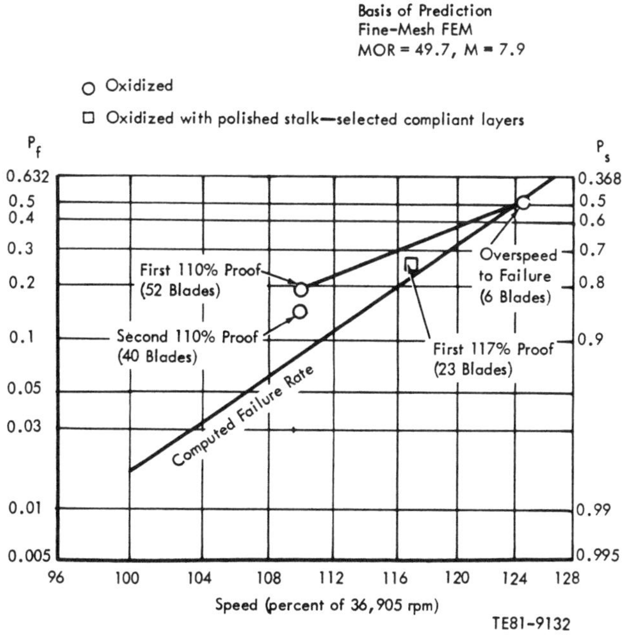

Figure 314. Reliability of 2070°F-configuration blades.

A second rotor assembly was built using the supply of blades twice proof tested at 117% speed. The test results are presented in Table LXII. Note that the combined benefits of oxidization, stalk polishing, and controlled compliant layer thickness variation resulted in 100% clearance of the repeated proof test.

Blade Vibration Testing

The initial bench dynamic characterization of ten 2070°F-configuration blades demonstrated that fundamental mode blade frequencies were 11% lower than predicted and 5.5% higher than predicted in the first torsion mode. The measured data are shown versus engine rpm in Figure 315. The interference with mean fundamental mode and vane passage frequency is at 26,000 rpm. This nominal interference is lower and preferred to the predicted 29,000 rpm. The lower test value probably indicates reduced fixity in the test case.

Gasifier Rotor No. 2

Following the 49 individual blade/proof spin tests at 117% speed (see Gasifier Rotor No. 1 subsection), a second rotor was assembled for a 5 minute proof test at 108% speed.

The rotor assembly included the following features:

- injection-molded sintered alpha SiC blades produced by CBO
- blade dovetails longitudinally ground by Crafts Inc
- blades that received a post-machining oxidation treatment of 24-hr exposure to 1250°C (2282°F) in air
- blade stalk (convex side) surfaces, which are polished after oxidation to remove glassy layer formed during oxidation treatment (Note: The polishing is done with a hand-held high-speed grinder driving a soft bristle nylon brush and lab-

Table LXII.
Blade proof testing at 117% speed.

	Test 1	Test 2
Proof speed	117%	117%
Number of blades	62	49
Condition	Oxidized and polished stalk	
Compliant layer	L605	IN-600*
Thickness variation	0.015 mm (0.0006 in.)	0.005 mm (0.0002 in.)
Results		
Number cleared	49	49
Number failed	13 (21%)	0
Failure origin	6 attachment 6 stalk 1 airfoil	

*Supply of L605 layers was exhausted at time of Test 2.

oratory diamond compound paste. The polishing begins with a 60-μm paste, followed by passes using 45-, 30-, 15-, and 60-μm pastes and finished with a 1-μm paste. The blades are visually inspected at 10X for removal of the glassy layer and, finally, cleaned by degreasing and immersion in boiling methanol.)

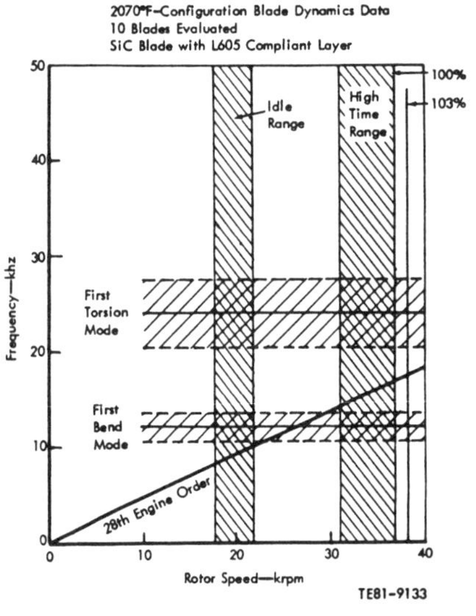

Figure 315. Measured 2070°F-configuration ceramic blade dynamic characteristics.

- uniform (tolerance of ±0.000635 mm [0.000025 in.]) IN-600 compliant layers
- each blade in the assembly twice proof tested to 117% rpm

Rotor testing was accomplished in the evacuated spin facility previously used in CATE blade testing. The rotor was visually inspected after completing 5 minutes at 85%, 93%, 98%, and 103% speed. No abnormal conditions were noted. Failure occurred at 107% in an attempt to reach 108% speed. Four blades failed and three sustained damage at the blade tip, as shown in Figure 316.

Figure 316. Rotor assembly: proof test failure at 107% speed.

The interesting failure was the attachment at the 10 o'clock position, shown in Figure 316. The origin was located at the leading edge acute corner and was associated with an abnormal compliant layer contact pattern as shown in Figure 317. Proficorder traces of the failed blade and, for comparison, traces of a nonfailure are shown in Figure 318. Records revealed the major depression in the failed blade was due to polishing to remove a surface flaw. This nonuniform surface precipitated the failure. All other blades were checked and those with nonuniform contact pattern were replaced along with failed or damaged blades for the third rotor build.

Gasifier Rotor No. 3

Proof testing of the third assembly was successful at 108% speed with a 5-minute hold. The rotor assembly included the following features:

- Blades were injection-molded sintered silicon carbide.
- Blade dovetails were longitudinally ground.
- Blades received a postmachining oxidation treatment.
- Blade stalks were diamond-polished after the oxidation treatment.
- IN-600 compliant layers of uniform thickness were used.
- Attachment contact surfaces were coated with boron nitride as a lubricant.

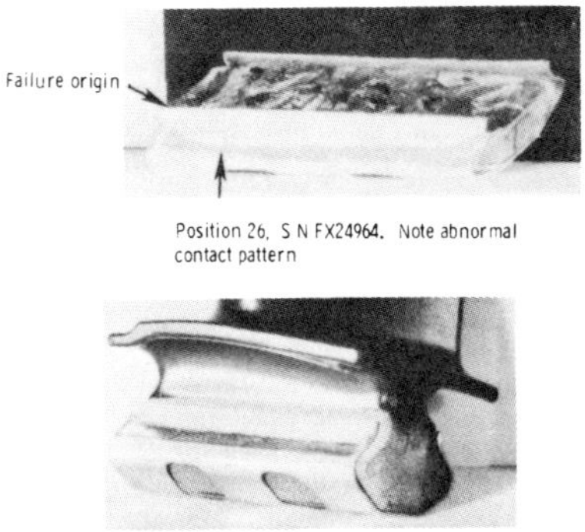

Figure 317. Attachment contact pattern.

Following completion of the 108% cold spin proof test, the rotor was carefully examined and then successfully tested in the hot engine simulation rig. Thirty minutes of test included two starts plus operation at speeds from idle to 100% rpm. At maximum speed, a maximum local temperature of 1122°C (2052°F) was reached. Additional description of rotors No. 2 and 3 can be found in EDR 10841 (Ref 7). This rotor was ultimately installed in engine C-4/BU14 for engine demonstration. A blade suspect primary failure occurred during C-4/BU14 testing. This is described in Section II of this report.

In other rotor tests, an instrumented rotor spin test was conducted with static and dynamic gages. The static strain measurements agreed very closely with values predicted by finite element analysis. Dynamic strain gages attached to blade airfoils recorded no measurable response in the spin pit. (Numerical values for the static gages are tabulated in EDR 10841—Ref 7.)

The successful rotor tests prompted the need for additional blades, proof tested for subsequent

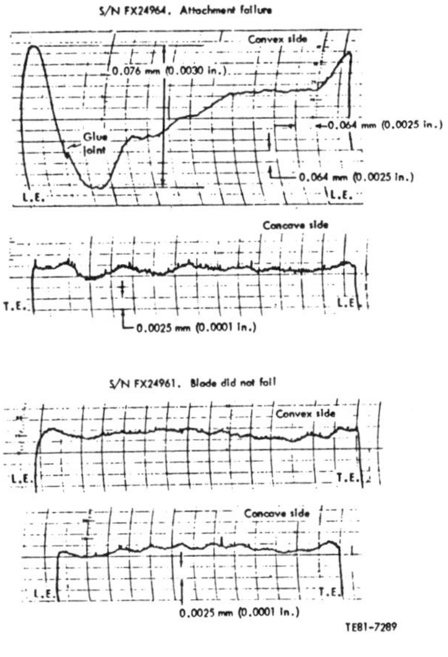

Figure 318. Proficorder data—dovetail contact surfaces.

engine tests. This ongoing investigation of blade strength was addressed by individual spin tests of the remaining sample of 41 blades from the initial production run by CBO. The results are listed in Table LXIII.

The proof test at 117% speed with 26 blades revealed the following:

- demonstrated benefits of uniform thickness layer and flat dovetail
- demonstrated general agreement with the calculated reliability, as shown in Figure 319, displaying state-of-the-art average strength characteristics based on the companion bars to the blades:
  - average MOR = 369.2 MPa (53.55 ksi), uncorrected for test bar level edges, and average MOR = 369.2 x 1.04 = 383.97 MPa (55.69 ksi), corrected for bevel edges
  - Weibull parameter of surface σ_{os} = 810.5 MPa (51.04 ksi), m_s = 7.76; Weibull parameter of volume σ_{ov} = 606.4 MPa (33.64 ksi), m_v = 10.1

The dead-soft layer investigation (platinum versus L605) revealed the following:

- The L605 compliant layer produced the higher average failure speed.
- On one sample, the platinum was extruded by bearing stress.
- The dead-soft platinum layer was determined to offer no improvement over the L605 material; this concluded the investigation of soft layer material.

Extensive detail is available in EDR 11006 (Ref 8), describing layer thickness and performance, blade dovetail finish, blade postfailure investigation, and other supporting data.

GTE Silicon Nitride Blades

A development program with GTE Sylvania Laboratories began in February 1980 to establish process parameters suitable for injection-molded, sintered silicon nitride blades in the 2070°F configuration. The program yielded 32 blades plus 100 MOR bars. Two groups of the bars, 12 each, were tested in 4-point bending with the following MOR results:

- as-fired surface—408.25 MPa (59.21 ksi)
- machined surface (longitudinal)—438.12 MPa (63.54 ksi)

Table LXIII.
Single-blade spin test—41 blades (all creep ground dovetails).

No. of blades	Compliant layer	Test	Objective
26	L605	Proof at 117% 23 cleared 3 failures for 11.5% compared with prior experience of 26% (see Figure 314)	Spare rotor blades
5 (+5*)	L605	To failure, average failure speed = 138%	Identify any influence of dead-soft compliant layer
5 (+5*)	Platinum	To failure, average failure speed = 138%	
5 HIP process	L605	To failure, average failure speed = 121%	Identify any influence of HIP process on structural integrity
41 total			

*Five blades for each test were drawn from the 23 cleared; thus, 13 spare blades cleared at 117%.

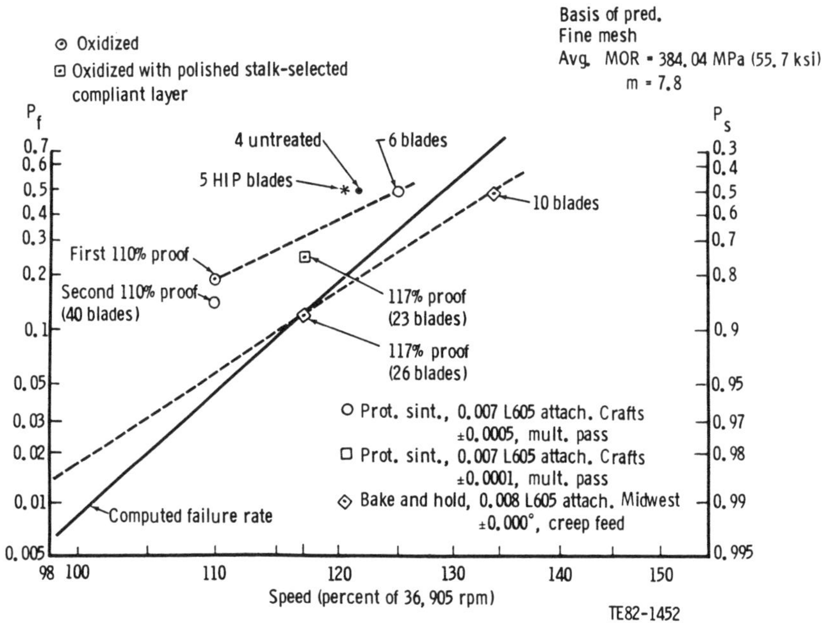

Figure 319. Cold spin testing, predicted and measured reliability.

Detailed NDE of the blades revealed the following:

- 37.5% (12) were free from any defect indications, and the geometry conformed.
- 25% (8) were suitable for rework to blend minor surface indications in noncritical areas, and the geometry conformed.
- 37.5% (12) were rejected for crack indications.

From the sample of 20 (12 + 8), 10 were finish machined by Crafts Inc, using a multiple-pass grinding technique of the attachment with handlap of the ends to eliminate spurs, and 10 were finish machined by Midwest Precision Services Inc (MPS), using a single-pass creep feed technique. A typical finished contour for the Crafts sample is shown in Figure 320. The surfaces are flat with no local spurs on high points. The contour for the MPS sample is similar, but no lapping was required to achieve the flat attachment surfaces.

The form (envelope) of the dovetail attachment conformed to the master chart in all cases. However, the attachment center plane of the Crafts sample was typically shifted 0.5 mm (0.020 in.) relative to the airfoil, as illustrated in Figure 321. This was the result of indexing from one side of the blade to locate the blade in the matrix box. This technique was modified to check both sides to ensure centering (including the MPS sample).

The overspeed-to-failure test of the GTE blades is listed in Table LXIV. Several observations are noted for these data. The failure location of the Crafts sample is erratic, and only 2 of 10 failed in the attachment. This was explained by the failure origins associated with shell-like surface cracks. These artifacts were the result of an abrasive cleaning operation (grit blasting) performed on the blades after sintering to enhance the surface finish. Subsequent to the manufacture of the sample of 32 blades, the grit blast cleaning was discontinued.

The MPS sample also exhibited some shell-like surface artifacts at the origins. However, the more typical pore feature at the failure origin was ob-

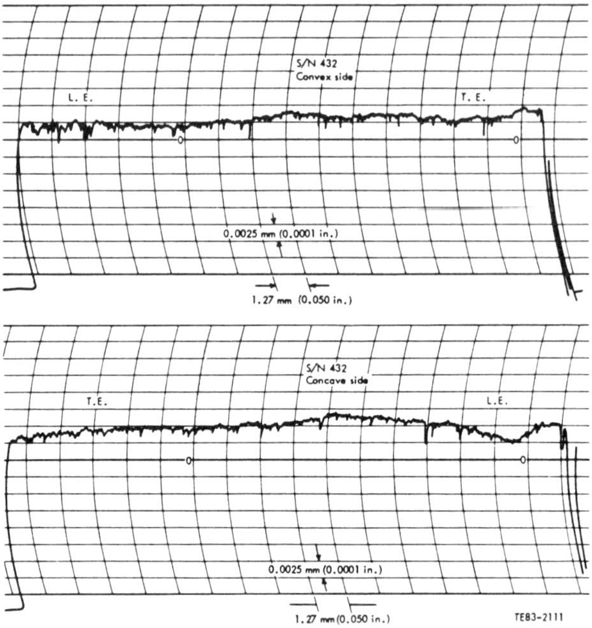

Figure 320. Proficorder trace of multiple pass ground dovetail (typical) with handlapped ends.

served. (The pore feature observed in ceramics can be corrected via processing.)

The spin test performance of the GTE silicon-nitride blades is compared with the silicon carbide blades in Figure 322. The sample machined by MPS substantially exceeded the SiC blades and was indicative of the 438.12 MPa (63.54 ksi) MOR measured for the longitudinally ground test bars. This completed the spin test evaluation of Si_3N_4 blades.

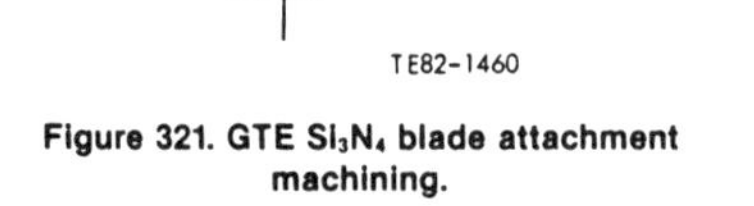

Figure 321. GTE Si_3N_4 blade attachment machining.

Based on the excellent spin test results and the goal of reduced porosity, a follow-on order for an additional 180 blades and 100 test bars was placed. An improved process specification (AY6-405) was used that achieved the highest strength to date for any GTE injection-molded Si_3N_4 material and dimensional conformance to the drawing. For example, longitudinally ground MOR bars demonstrated an average strength of 617.03 MPa (89.49 ksi). Additional strength data, including values at elevated temperature, are found in the material process development subsection of Section II of this report.

The results of the dimensional layout on a sample of three blades are summarized in Table LXV. The dimensional conformance is generally excellent. A small tilt of the airfoil stack axis toward the convex side of the airfoil is apparent, and the tip section of the airfoil exhibits 1.5 deg of additional twist above the high limit. These deviations were the

Table LXIV.
Overspeed-to-failure spin test, Si_3N_4 blades.

No.	Blade S/N	Failure	Failure speed—rpm	(%)
	(Machined by Crafts Inc, P/N EX 130190—multiple pass grinding)			
1	430	Stalk, convex side (shell)	48,500	(131.4)
2	432	Platform, convex side (shell)	49,570	(134.8)
3	433	Stalk, convex side (shell)	46,700	(126.5)
4	437	Stalk, convex side (shell)	48,600	(131.7)
5	441	Platform, convex side (shell)	49,850	(135.1)
6	445	Attachment, convex side (surface, low density)	45,200	(122.5)
7	485	Stalk, convex side (shell)	49,450	(134.0)
8	513	Attachment, concave side (surface, low density)	47,800	(129.5)
9	554	Airfoil root (shell)	49,750	(134.8)
10	831	Stalk, convex side (shell)	51,700	(140.0)
		Average failure speed	48,730	(132.0)
		One standard deviation	1,831	(5.0)
	(Machined by Midwest Precision Services—single-pass, creep-feed grinding)			
1	403	Attachment, convex side (missing)	52,250	(141.6)
2	404	Attachment, convex side (pore)	50,600	(137.1)
3	405	Attachment, convex side (pore)	52,000	(140.9)
4	412	Attachment, concave side (crack)	53,500	(145.0)
5	456	Attachment, convex side (pore)	47,400	(128.4)
6	459	Stalk, convex side (pore)	52,900	(143.3)
7	460	Airfoil, convex side (shell)	52,800	(143.1)
8	484	Stalk, convex side (shell)	57,000	(154.5)
9 & 10		(Two blades damaged in machining)		
		Average failure speed	52,300	(141.7)
			2,702	(7.3)

maximum observed. Considering the observations throughout the program, these rank on a qualitative basis as excellent dimensionally. This completed the development work on the Si_3N_4 blades.

2070°F Blades, Follow-On Order No. 2, Alpha SiC

This terminology identified an order to CBO for 180 sintered alpha SiC blades made by the established process (same as proofed blades). The objective of this effort was additional resources to support engine tests. Of 208 blades shipped to Allison, 103 cleared the combined NDE and dimensional inspection and a net of 67 cleared machining. The dovetail attachment was machined by MPS using multiple pass creep-feed grinding versus prior single-pass creep-feed grinding. Multiple pass machining was required to reduce the machining contact loads and preclude movement of the blade in the matrix box. A very slick, glasslike surface appearance was observed for this sample and may have contributed to the movement of the blade in the Cerra-bend potting metal.

Companion MOR bars were tested and Weibull properties generated. The results were similar for orders No. 1 and No. 2, as shown in Table LXVI.

Forty-eight of the 67 machined follow-on order No. 2 blades were included in a spin proof test to qualify engine blades. The proof spin test was conducted per the following established practice:

- individual blade tested at room temperature in an evacuated pit with a goal of 117% N_G speed (43,200 rpm) and a 5-minute hold at maximum speed
- L605 material compliant layer, 0.203 mm (0.008 in.) and $\pm$0.00254 mm ($\pm$0.0001 in.) tolerance
- compliant layer and blade dovetail assembled with boron nitride aerosol lubricant

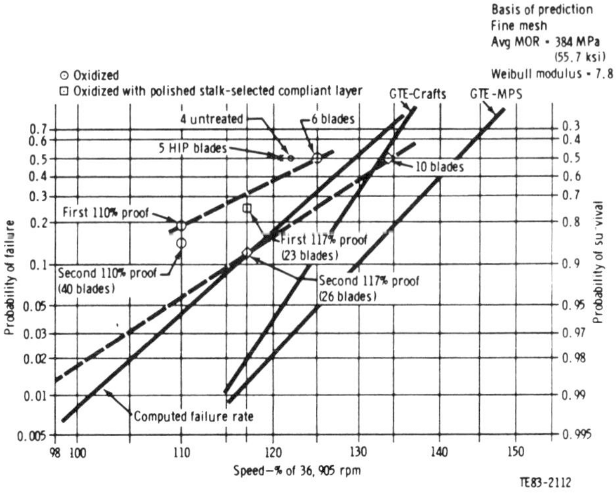

Figure 322. Cold spin testing, showing predicted and measured reliability.

The proof spin test results included the following:

- 35 of the 48 cleared at 117% N_G
- 13 failures for a 27% failure rate at an average speed of 108.7% N_G (A tabulation of the failures is shown in Table LXVII.)

A graphic interpretation of this proof-test result and prior experience is shown in Figure 323. Note that the sample of 48 blades matched the experience with the first group of blades proof tested at 117% speed. Table LXVIII is a tabulation of all spin tests. Investigation of the 13 failures revealed the following:

- satisfactory compliant layer contact pattern
- satisfactory dovetail surface finish (a typical contour shown in Figure 324)
- subsurface voids detected at the failure origin of 6 of the 13 blades
- the remaining origins not revealing

Gasifier Rotor No. 4

Following the successful proof test of an additional 35 blades, rotor No. 4 was assembled for engine C-4/BU13, the first engine built with all the CATE ceramic components.

The rotor was configured with the following parts:

- turbine wheel, EX 130186, S/N XA18047
- 40 compliant layer, EX 129475—L605 material 0.203 mm ± 0.00254 mm (0.008 in. ± 0.0001 in.)
- 8 blades retainer, front, EX 130187
- 8 blades retainer, rear, EX 130188
- 16 rivets, AN123376
- 40 blades as listed in Table LXIX

Assembly of the rotor included an investigation of the influence of the compliant layer on the dimensional scatter of the blade tip radius. Scatter in the blade platform lip seal radius was included.

The investigation addressed the seating of the compliant layers prior to spinning (centrifugal force, CF), i.e., CF was required to seat the layers and thus reduce the measured scatter in the tip radius. The blade tip and lip radius was measured with the compliant layers removed, and the resulting scatter compared with the drawing tolerance stack and data from multiple builds of rotor 1 with the compliant layers. Seating of the compliant layers was not a contributing factor to the observed dimensional scatter in the blade tip and lip seal radius. The data supporting this conclusion are summarized in Table LXX.

The measurements were repeated during final assembly of the rotor with the compliant layers, and the data are summarized in Figures 325 and 326. Within measurement accuracy, which is estimated at ±0.0254 mm (±0.001 in.), the blade tip radius

Table LXV.
Dimensional layout, GTE silicon nitride blades (order No. 2).

EX 130190—S/N 998, 997, 999

Airfoil contour
Concave and convex sides conform to drawing
(i.e., within a 0.102 mm [0.004 in.] band)

Airfoil section details—note: x conforms to drawing

Section	Chord length	Max thickness	L.E. thickness	T.E. thickness	Setting angle— +0.5 deg tolerance	Stack point location
Hub ("A")	x	x	x	x	x	x
Mid height	x	x	x	x	+0.5 deg above high tolerance	x (at maximum 0.127 mm [0.005 in.] limit toward suction side)
Tip ("F")	x	x	+0.152 mm +(0.006 in.)	x	+1.5 deg above high tolerance	0.152mm (0.006 in.) above high limit toward suction (see sketch)
Stalk	x	x				

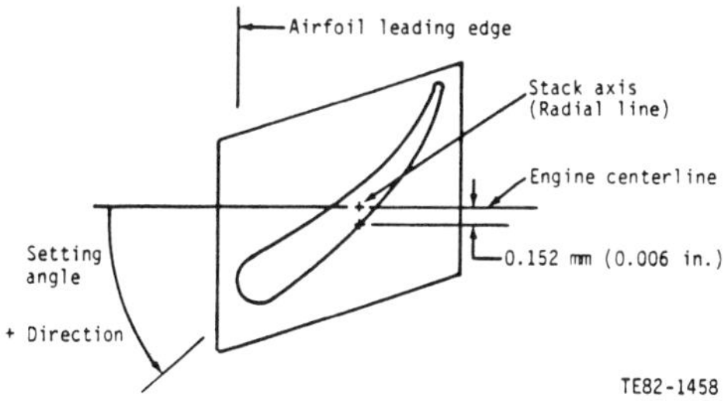

Note: Any deviations are the maximum observed.

data per Figure 325 repeat the data of Table LXX. Likewise, the blade lip seal radius data summarized in Figure 326 repeat the data of Table LXX at position A. Position B does not repeat, but the data are suspect for blade position 1; the data for position 1 do not match the trends and general level of all other blades.

The blade retention plates were riveted in place; the rotor was balanced and successfully proof tested at 108% N_G speed (39,857 rpm). The test was conducted at room temperature in a spin facility evacuated to 3.5 mm of Hg, absolute pressure. The proof test spin schedule is outlined in Figure 327. The visual inspection after the first 100% speed

Table LXVI.
Alpha SiC Weibull properties.*

	Order No. 1	Order No. 2**
Surface		
Strength	810.07 MPa (51.04 ksi)	889.27 MPa (55.24 ksi)
Modulus	7.76	7.63
Volume		
Strength	606.28 MPa (33.64 ksi)	686.33 MPa (32.29 ksi)
Modulus	10.1	8.62

*The results of the Materials Laboratory investigation of order No. 2 density, microstructure, all strength data, MOR bar failure origin, etc, are fully described in EDR 10672 (Ref 6).
**See Section II for description of the data.

Table LXVII.
Proof test failure summary.

SiC blades, follow-on No. 2
(13/48 failed—27% failure rate)

No.	Blade S/N	Failure	Failure speed—rpm	(%)
1	706-2	Attachment, convex side	42,400	(114.9)
2	705-2	Attachment, concave side	42,200	(114.3)
3	719-2	Attachment, origin unknown	39,400	(106.8)
4	766-2	Attachment, convex side	39,400	(106.8)
5	658-2	Attachment, convex side	39,200	(106.2)
6	664-2	Attachment, concave side	41,350	(112.0)
7	659-2	Attachment, origin unknown	41,150	(111.3)
8	621-2	Attachment, convex side	36,350	(98.5)
9	764-2	Attachment, convex side	33,800	(91.6)
10	682-2	Stalk, convex side	40,300	(109.2)
11	686-2	Attachment, convex side	40,500	(109.7)
12	724-2	Attachment, convex side	43,200	(117.0)
13	845-2	Attachment/stalk, concave side	42,400	(109.2)
		Average failure speed	40,116	(108.7)

step revealed no anomalies: The proof test was continued with a 5-minute dwell at a nominal 108% N_G speed. The maximum observed speed was 40,000 rpm (minor control drift) or 108.38%. At completion of the test, visual inspection again revealed no anomalies. The rotor was used in engine C-4.

DISCUSSION: ALTERNATE BLADE DESIGN

A successful study was made addressing an alternate blade design (versus current 2070°F-configuration blade) to improve reliability without compromise in aerodynamic performance. The airfoil aerodynamic design was consistent with the current 2070°F-configuration stator vane shape and does not necessitate stator alternations. The approach was (1) to reduce the hub section setting angle (wheel broach angle) corresponding airfoil revision and (2) to tilt and lean the airfoil. This reduced attachment peak stress and improved reliability.

The study began with construction of a cost-effective (coarse mesh), 3-D finite element model for evaluation of potential configurations. The details of this effort are found in EDR 10156 (Ref 4).

The initial aerodynamic study substantiated the reduced hub section setting angle with a blade airfoil design that suits the 2070°F-configuration vector diagram. Further study concluded that the most significant parameter affecting the blade setting angle and, therefore, the dovetail angle is the blade throat dimension, which is tied directly to the exit gas angle. Increasing the throat (and exit gas angle) decreases the setting angle. Figure 328

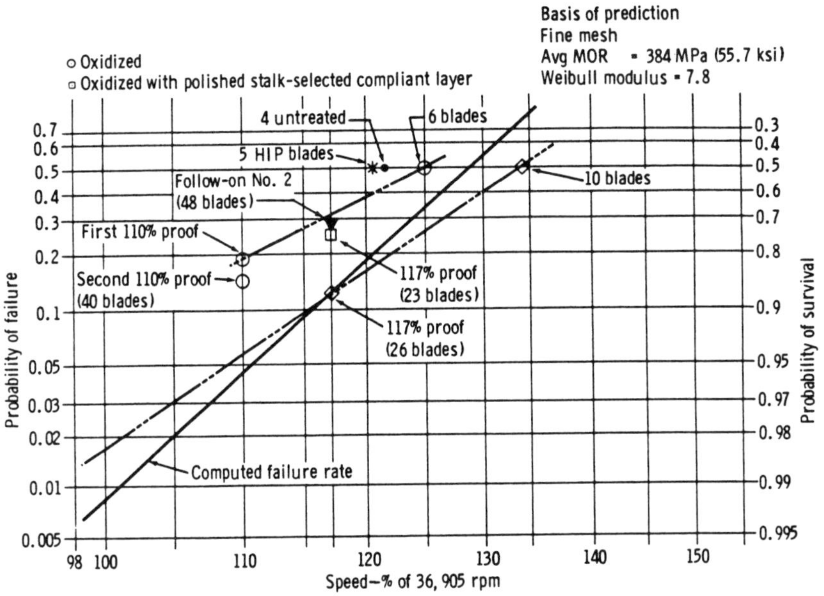

Order 1

○ L605 material thickness = 0.178 mm (0.007 in.)
tolerance = ±0.013 mm (±0.0005 in.)
Attachment machined by Crafts
Multiple pass scheme

□ L605 material thickness = 0.178 mm (0.007 in.)
tolerance = ±0.013 mm (±0.0005 in.)
Attachment machined by Crafts
Multiple pass scheme

◇ L605 material thickness = 0.203 mm (0.008 in.)
tolerance = ±0.013 mm (±0.0001 in.)
Attachment machined by MPS
Creep feed scheme

Order 2

▼ L605 material thickness = 0.203 mm (0.008 in.)
tolerance = ±0.013 mm (±0.0001 in.)
Attachment machined by MPS
Multiple pass scheme

TE83-2113

Figure 323. Cold spin testing, showing predicted and measured reliability.

Table LXVIII.
2070°F-configuration SIC blade spin test.

Condition	Test	No. tested	No. passed	Failed—%	Failure site
Oxidized	Initial 110% proof	51	41	19	6 attachment (101-110%) 4 stalk (108-110%)
	Rotor assy 105% proof	40	39	3	1 stalk (93%)
	Second 110% proof	40	34	15	1 attachment (105%) 3 stalk (106-109%) 2 airfoil (105-107%)
	Third 110% proof	5 Together	0	100	2 attachment (103%) (oxidized, 1 oxidized and polished)
	Overspeed to failure	6	0	100	5 attachment (117, 122, 128, 129, 135%) 1 stalk (116%)
Untreated	Initial 110% proof	8	5	38	3 stalk (86-96%)
	Overspeed to failure	4	0	100	3 attachment (110, 122, 124%) 1 stalk (134%)
	Overspeed to failure NASA HIP process***	5	0	100	5 dovetail (103, 119.2, 116.5, 130.7, 136.2%)
Protected center—oxidized and polished	Initial 110% proof	8	7	13	1 attachment (110%)
	Second 110% proof	7	7	0	None
	Initial 117% proof* Previous 110%, 110%	33	26	21	3 attachment (107-116%) 3 stalk (110-117%) 1 airfoil (109%)
	Initial 117% proof* (previous 110% untreated)	5	5	0	None
	Initial 117% proof*	23	17	26	3 attachment (97-112%) 3 stalk (110-112%)
	Second 117% proof**	48	48	0	None
	Rotor assy 108% proof**	40	36	10	3 attachment (107%) 1 airfoil (107%)
	2nd rotor assy 108% proof**	40	40	0	
	Overspeed to failure				3 attachment (137.5, 150, 150%)
	L605 layer***	5	0	100	2 stalk (118-119%)
	Platinum layer***	5	0	100	5 attachment (108.5, 130, (138, 143%)
	117% proof L605 layer	26	23	11.5	3 attachment (80.5, 92.5, 98.5)
Bake and hold—oxidized and polished	Overspeed to failure				
	L605 layer***	5	0	100	4 attachment (142, 133.8, 146.8, 143%) 1 low stalk (144.5%)
	Platinum layer***	5	0	100	3 attachment (133, 146.8, 138%) 2 stalk (109.7, 133%)
Blade Order No. 2—oxidized and polished	117% proof	48	35	27	2 stalk 11 attachment (91.6%-117%, avg = 108.7%)

* Selected thickness L605 compliant layers; thickness = 0.1778 mm (0.007 in.)
** Uniform thickness IN-600 compliant layers; tolerance = ±0.000635 mm (0.00025 in.)
*** Thickness = 0.203 mm (0.008 in.); tolerance = ±0.00254 mm (±0.0001 in.)

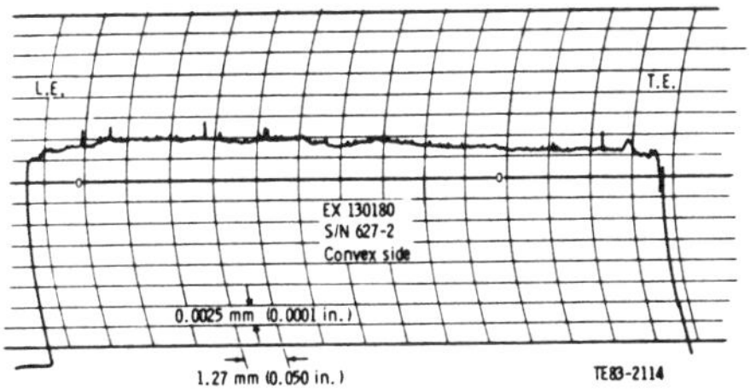

Figure 324. Proficorder trace (typical) multiple-pass creep-feed machining for order No. 2.

shows the original design throat distribution, which increases with radius until 70% span is reached and then decreases slightly in the tip region. A design was selected having a near constant throat dimension from hub to tip. This opened the hub throat 17%, resulting in a 12.5-deg decrease in dovetail angle.

This redesigned airfoil, combined with optimized airfoil tilt and blade stack, achieved improved reliability. The revised geometry presents a significant change, and the relationship of airfoil hub, platform, and attachment stalk is presented in Figure 329. Dynamic analysis of the new blade indicated characteristics similar to the current 2070°F configuration, which is considered satisfactory. (Additional details, particularly aerodynamic features, are available in EDR 10841—Ref 7.)

The aerodynamic definition of the airfoil utilized a radial stack of the airfoil sections; that is, the center of gravity of every airfoil section lies on a radial line through the center of the stalk. Utilizing this airfoil definition, an analytic study was conducted to tilt and lean the airfoil to favorably affect the probability of survival under engine and proof spin conditions. Tilt and lean consisted of shifting the center of gravity of the airfoil sections off the initial radial stack and thus changing the stress distribution (probability of survival) of the blade.

The 3-D, cubic element model noted earlier was utilized for heat transfer, stress, and probability of survival calculations for the critical conditions in the operating envelope—100% power at steady-state, spin (proof) test, and transient conditions. The transient conditions were searched for the maximum temperature gradients as a function of time in (1) snap acceleration from stabilized braking to maximum power and (2) deceleration from part power to stabilized braking modes. At each condition, a radial airfoil stack was analyzed to establish a data base, and a tilt and lean condition was analyzed to establish a trend. The magnitude and direction of the tilt and lean are defined in Figure 330.

The steady-state temperature profile was calculated for the following conditions:

- 100% N_G, 1132°C (2070°F) TIT
- 85% N_G, 1132°C (2070°F) TIT (This is the base for the transient case No. 2.)

The transient temperature profile—maximum temperature gradients versus time—was calculated for the following conditions:

- snap acceleration from stabilized braking to maximum power
- deceleration from part power to stabilized braking (i.e., dynamic braking from 1132°C [2070°F] TIT to 260°C [500°F] at 85% N_G)

A sample result of the transient temperature calculation is shown in Figure 331. Illustrated is the location and magnitude of the maximum temperature gradient in the attachment versus time. This treatment of the data makes possible the logical selection of the instants for calculation of the probability of survival. (Details of the temperature calculation results are found in EDR 11006—Ref 8.)

The results of the probability of survival calculations for the various conditions are summarized in Table LXXI. Included for comparison are the P_s data for the existing engine C-4 blade. Note that the design goal of a probability of survival of 0.99994 was analytically achieved at the 115% proof test speed with a 5% rejection rate for the alternate design blade. The design limiting case, a transient condition, was improved from 2.3 failures to 0.6 failures per 10,000—an improvement factor of approximately 4. Furthermore, the redesign P_s (0.99994) was achieved with the same alpha SiC material properties as the existing blade—i.e., demonstrated properties.

A physical description of the benefit of the proof spin test on the calculated P_s for the alternate blade is shown in Figure 332. It is seen that the proof test clears the blade for the design goal and that proof speeds in excess of 42,500 rpm (115% N_G) are inconclusive. This completed the design study of the alternate blade.

Ceramic Turbine Inlet Plenum

SUMMARY

Efforts to produce a functional ceramic turbine inlet plenum for the CATE (2070°F) program were marked by the following highlights.

Table LXIX.
Rotor blades for engine C-4 (BU13) P/N EX130180.

Group 1 (CBO follow-on order No. 1) Serial No.	Position	Group 2 (CBO follow-on order No. 2) Serial No.
978	1*	
1099	2	926-2
	3	
	4	695-2
	5	847-2
	6	647-2
	7	823-2
	8	636-2
	9	831-2
	10	628-2
1056	11	
	12	627-2
315	13	
	14	745-2
	15	835-2
	16	629-2
42	17	
	18	624-2
113	19	
	20	732-2
	21	867-2
	22	687-2
342	23	
	24	676-2
	25	834-2
	26	665-2
479	27	
	28	878-2
340	29	
	30	625-2
1107	31	
	32	632-2
330	33	
	34	643-2
	35	884-2
	36	620-2
	37	633-2
	38	743-2
	39	622-2
	40	646-2

*The No. 1 blade position corresponds to the 12 o'clock position with the hub roll pin (inlet side of wheel) at the 12 o'clock position. Increasing blade count (position) is in the clockwise direction while viewing the outlet side of the wheel.

A ceramic plenum was designed, and two major suppliers were selected. The Norton Company successfully cast seven plenums of densified RBSiC. Two of them were finish machined. Efforts by Carborundum to cast a plenum of alpha SiC resulted in parts that did not conform to print dimen-

Table LXX.
Rotor assembly, blade radial runout.

	Total radial runout—mm (in.)				
	Rotor 1*			Rotor 2**	
	BU1	BU2	BU3	BU1, machined by MPS/DDA	Drawing stack
	Machined by Crafts				
Blade tip					
L.E.	0.203 (0.008)	0.178 0.007	0.203 (0.008)	0.203 (0.008)	0.229 (0.009)
T.E.	0.152 (0.006)	0.152 (0.006)	0.127 (0.005)	0.127 (0.005)	
Platform lip					
A	0.191 (0.0075)	0.152 (0.006)	0.102 (0.004)	0.254 (0.010)	0.229 (0.009)
B	0.229 (0.009)	0.178 (0.007)	0.152 (0.006)	0.254 (0.010)	

*Compliant layer installed
**Compliant layer removed

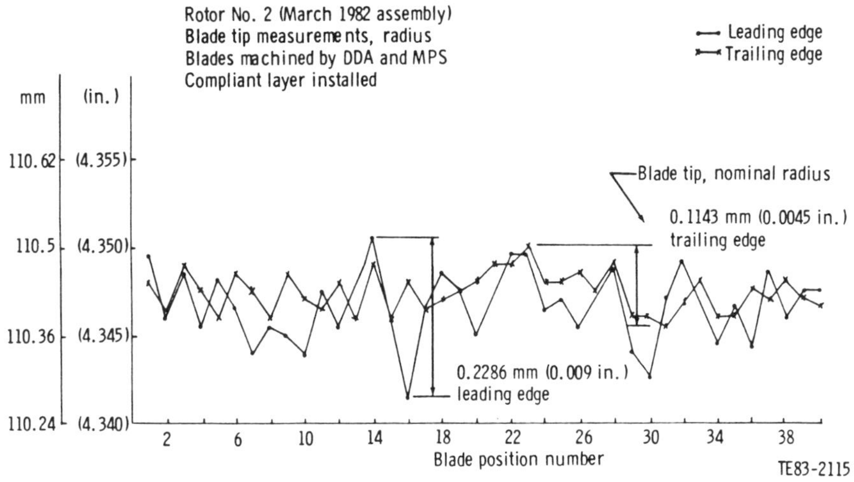

Figure 325. Blade tip radius, rotor final build.

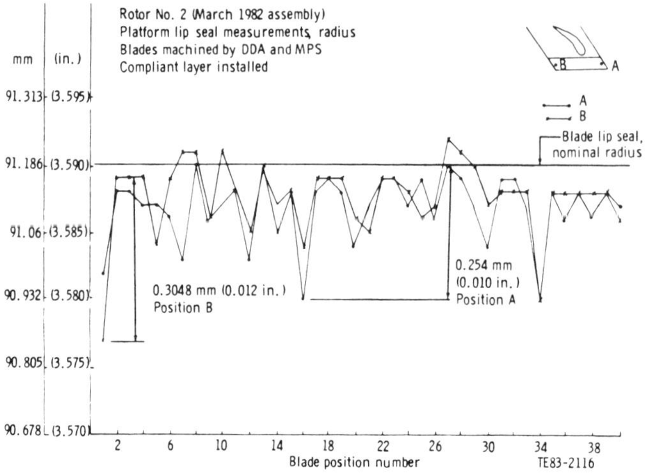

Figure 326. Blade tip seal radius, rotor final build.

sions. The chief difficulty was caused by the high shrinkage rate of the material. Nonuniform shrinkage during sintering resulted in distortion.

A detailed thermal and stress analysis indicated a probability of survival of 0.9987 for the 100% speed steady-state condition with a turbine inlet temperature (TIT) of 1132°C (2070°F). The probability of survival was calculated to be 0.9226 for the most severe transient—an acceleration from stabilized dynamic braking to 100% speed.

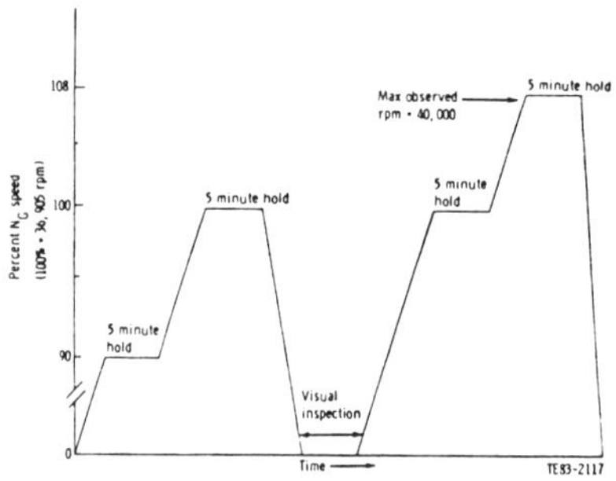

Figure 327. Rotor proof test spin schedule.

A steady-state test in the thermal shock rig resulting in a plenum failure at a burner outlet temperature (BOT) of 1043°C (1910°F) led to a review of the operating environment of the plenum. This revealed that, subsequent to the original analysis, a design change to improve engine block cooling resulted in a change in flow path of component cooling air that increased the thermal gradient near the plenum exit

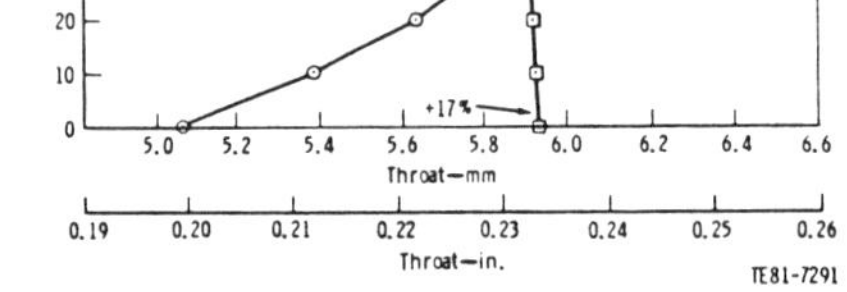

Figure 328. Original design throat distribution.

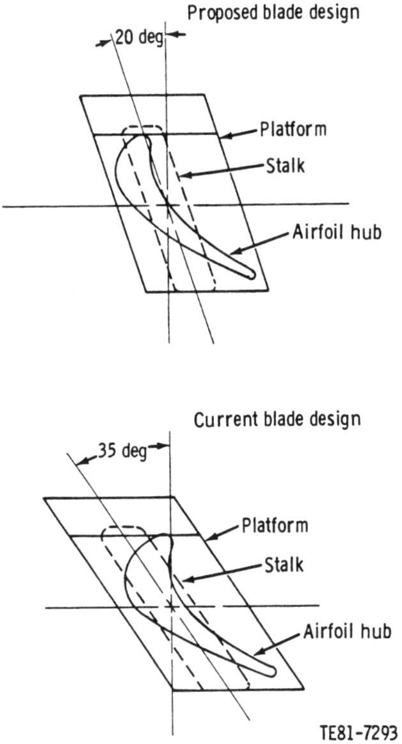

Figure 329. Current versus proposed alternate ceramic blade designs.

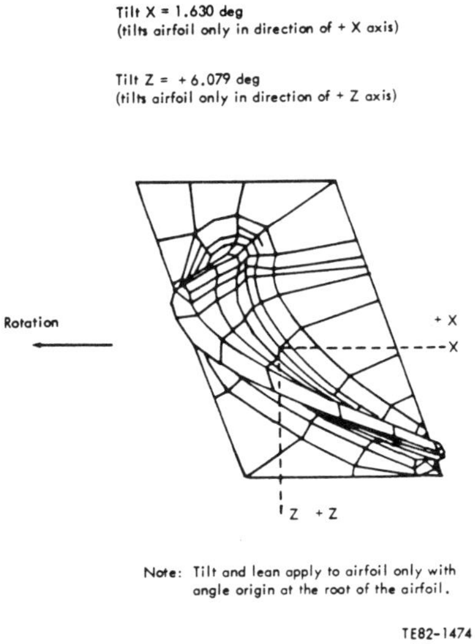

Figure 330. Definition of tilt and lean for alternate blade.

flange. Analysis of the plenum taking this into account showed that the probability of survival for 100% speed steady-state conditions with a TIT of 1132°C (2070°F) dropped to 0.384 for the rig and 0.033 for engine conditions.

A simplified axisymmetric analysis was used to evaluate proposed changes in the design of the exit flange and in the component cooling circuit. These changes were shown to significantly lower the stress and raise the probability of survival, but budget limitations prevented further detailed analysis or part modification and testing.

OBJECTIVE

The objective was to design a turbine inlet plenum that could be efficiently fabricated, machined, and assembled in the engine into a design that would minimize attachment loads at assembly and during engine operation. The probability of survival goal was 0.997.

APPROACH

Design activity on the gasifier turbine inlet plenum was initiated in early 1978. The goal was to design a plenum similar to the baseline IGT 404 gas turbine engine that would achieve comparable flow conditions to be assembled in the engine with a minimum of attachment loads. Single- and multiple-piece constructions were submitted for vendor approval. During the ensuing 12 months, a slip-cast design was developed that was one piece, except for the exit flange and the annulus that forms the inner portion of the flow path. This is shown in Figure 333. Since the exit flange was considerably thicker

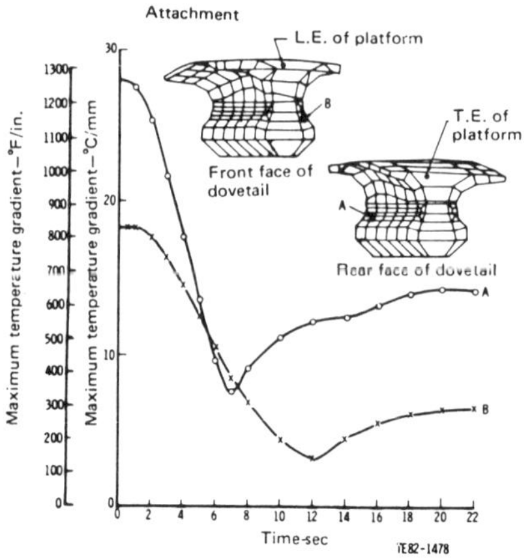

Figure 331. Maximum temperature gradient versus time for deceleration from part power to stabilized braking.

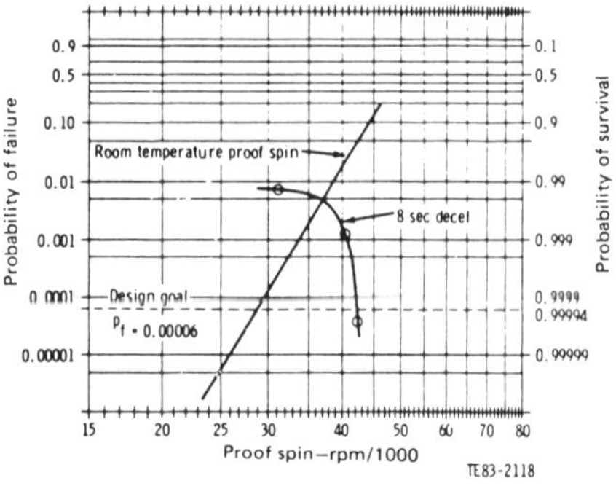

Figure 332. Calculated reliability of alternate design blade after proof spin at room temperature.

than the plenum wall, it was difficult to slip cast the flange integrally with the plenum. So it was decided to cast it separately to be assembled to the plenum in the green state and bonded at sintering. The inner annulus was cast separately to facilitate machining four slots that accommodate the struts passing through the flow path to support the outer ring of the gasifier nozzle.

DISCUSSION

Two major suppliers were chosen for slip casting. The Norton Company slip cast the plenum of Norabide NC430, a densified reaction-sintered silicon carbide ceramic. The Carborundum Company used alpha SiC, a potentially stronger material with a higher temperature capability but also a much greater shrinkage at sintering. During the remainder of the program, Norton succeeded in producing seven usable plenums and inner annuli. Those delivered early in the program were dimensionally correct except for a somewhat undersized and thin-walled inner outlet. The plenums and inner annuli were considered usable for test and two of them were finish machined. The last three plenums, delivered in late 1981 in the as-fired state, met print dimensions. Due to the high shrinkage of its alpha SiC material, Carborundum had the problem of distortion while sintering. Also attempts to bond the exit

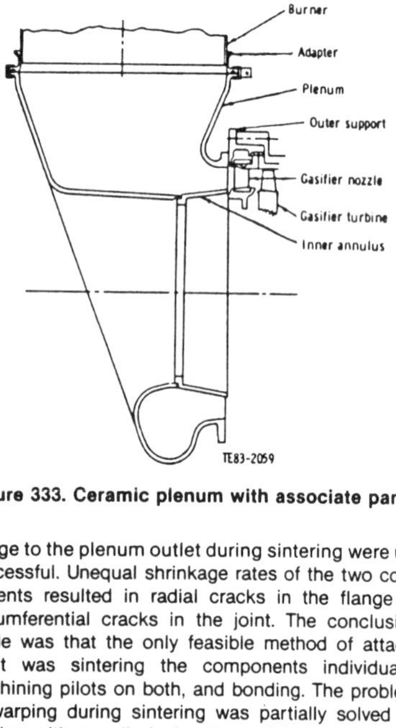

Figure 333. Ceramic plenum with associate parts.

flange to the plenum outlet during sintering were unsuccessful. Unequal shrinkage rates of the two components resulted in radial cracks in the flange or circumferential cracks in the joint. The conclusion made was that the only feasible method of attachment was sintering the components individually, machining pilots on both, and bonding. The problem of warping during sintering was partially solved by casting with a wall closing the exit to be machined away after sintering. Three parts were cast in this

Table LXXI.
Ceramic blade for 2070°F configuration.

	Existing blade (engine C-4 blade)	Alternate design
Stalk setting angle	35 deg	20 deg
Airfoil geometry	—	New aero design
Attachment profile	Identical	
Material properties	Alpha SiC	
Surface	σ_{os} = 705.9 MPa (48.7 ksi)	
	σ_{ms} = 8.71	
Volume	σ_{ov} = 557.8 MPa (25.58 ksi)	
	σ_{mv} = 8.43	
Probability of survival, P_s		
Room temperature spin	0.987929	0.995644
100% power, steady state	0.988277	0.995925
Transient (most severe), no spin proof test		
3 sec accel	0.982306	
8 sec decel		0.99326
Transient (most severe) with spin test		
3 sec accel (117% speed and 10% failure rate)	0.99977	
8 sec decel (115% speed and 5% failure rate)		0.99994

manner and delivered to Allison. These parts were found to deviate significantly from the drawing, and none were machined.

During the first half of 1979, a three-dimensional finite element thermal and stress analysis was completed and reported in EDR 9951 (Ref 3). This analysis established the combined probability of survival of plenum and inner annulus to be 0.9987 for the 100% speed, maximum power, steady-state condition and 0.9226 for the most severe transient condition—an acceleration from stabilized dynamic braking to 100% speed maximum power.

In mid-1981 the first plenum (from Norton) was tested in the thermal shock rig. After surviving steady-state tests at BOT of 316°C (600°F), 649°C (1200°F), and 816°C (1500°F) with no apparent cracks, a failure occurred at 1043°C (1910°F). The details of this test are reported in EDR 11006 (Ref 8). A review of the plenum operating environment revealed that conditions at the exit of the plenum were not the same as those assumed in the original analysis. A design change to improve the cooling of the engine block resulted in a rerouting of compressor discharge air—used to cool the nozzle supports and containment ring—that resulted in a cooler plenum exit flange. A revised thermal analysis taking into account the changed cooling flow and using some refinements in the treatment of the boundary conditions at the plenum exit flange confirmed that the thermal gradient in the exit region was much more severe than originally calculated. A subsequent stress analysis based on the actual measured thermal gradients calculated much higher stresses than had been found in the original analysis with the peak stress occurring in the region where the crack occurred in the rig failure. The probability of survival for steady-state conditions in the rig at the time of the failure was calculated to be only 0.384. For the 100% speed steady-state condition in the engine with a TIT of 1132°C (2070°F), the probability of survival was calculated to be 0.033. Detailed reports of these analyses were reported in EDR 11006 (Ref 8).

Several modifications to the plenum and its environment were proposed to increase the probability of survival of the plenum. Each of them would be expected to result in a hotter plenum outlet flange, reducing the axial thermal gradient and/or resulting in a reduction of the harmful effects of this gradient.

The plenum environment could be improved by changing the route of the compressor discharge cooling air after the containment ring is cooled. Figure 334 shows how this air could be dumped into the engine flow path downstream of the gasifier turbine. This procedure reduces the cooling of the outer support by eliminating contact of cooling air on its inner surface. This step also prevents the cooling air from scrubbing the inner portion of the plenum exit flange.

The bending moment caused by the thermal gradient between the plenum and its outlet flange can be reduced by increasing the tangential flexibility of the outlet flange; this would reduce the radial constraining force of the cool flange on the hot plenum. An increase in flexibility can be achieved by scalloping the flange—removing all of the material except the four lugs required to locate the plenum axially and the eight tangs required to locate the plenum radially and tangentially. The redesigned flange is shown in Figure 335. Part of the reason for the cool plenum flange is the conduction of heat out of the flange into the cooler outer support. This cooling could be reduced by better insulating the plenum from the outer support. In the present design a layer of zirconia 0.43 mm (0.017 in.) thick is applied to the metal outer support to prevent adhesion of the nickel-based IN-713 to the ceramic plenum at high operating temperatures. Zirconia has a conduction value about one-sixteenth of the ceramic and one-twelfth of the metal part. This zirconia layer thickness could be increased, so coating twice the present thickness is proposed.

To substantiate these proposed improvements with a thermal shock rig test, a considerable amount of rework would be required on the plenum and related parts. Due to budget reductions a second rig test to prove the proposed design changes was not made. For the same reasons the design changes were not evaluated in a detailed, three-dimensional finite element analysis of the type used to evaluate the original design. This evaluation would have required extensive model changes and large computer costs. Instead, an evaluation was made using axisymmetric finite element analysis, which requires a much simpler model and much lower computer costs. The results of this analysis are reported in the following paragraphs.

The axisymmetric analysis was not expected to produce an exact determination of the change in peak stress and probability of survival that could be obtained with the proposed modifications. Rather it was intended to determine if the modifications could bring about a significant improvement and to rate each proposal relative to the others.

To determine the effectiveness of the axisymmetric analysis for evaluating the proposed design changes, the analysis was used to analyze the base-

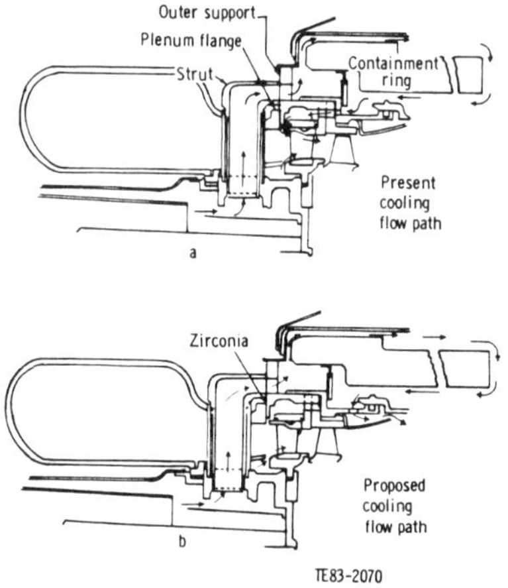

Figure 334. Proposed change in path of cooling air.

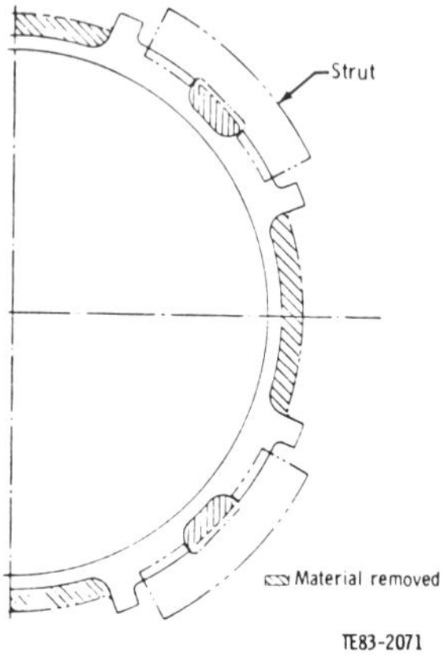

Figure 335. Proposed plenum exit flange modification.

line design under steady-state rig conditions. The results were compared with those of the 3-D analysis. The model represented a radial section through the bottom, which was the location of the maximum stress in the 3-D analysis. The calculated temperatures agreed with the temperatures in that section in the 3-D analysis within 2%. The peak stress was half of that calculated in the 3-D analysis. The pattern of the stress distribution in the two analyses, however, was identical. Because of this, the forces that are due to the 3-D effects seem to be dependent also on the magnitude of the axial thermal gradient. Therefore, the axisymmetrical analysis was determined to be a valid tool to evaluate the design changes.

The effectiveness of each of the three proposals suggested was evaluated individually and in combination with the other two. The results of the thermal analyses under rig and engine conditions are shown in Figures 336 and 337, respectively. The results of the stress analyses for rig and engine conditions are shown in Figures 338 and 339. The results are compared in Table LXXII.

All of the proposals are shown to be effective in raising the probability of survival substantially, even though the maximum principal stress is not significantly lowered in every case. This is due to the stress distribution. The stress peak is very localized and the stress is lowered over a large portion of the plenum. The scalloped flange and rerouted cooling air are shown to be the most effective, and the combined effect of all three is shown to significantly increase the probability of survival of the plenum.

Power Turbine Nozzle

SUMMARY

In the second half of 1980, a two-stage power turbine for the CATE 2265°F configuration was designed. To avoid increasing program costs, ceramics were limited to the first-stage nozzle and turbine tip shroud. The temperature of the second-stage nozzle was low enough to achieve the program life goals with a one-piece all-metal design. To draw on the experience gained in developing the ceramic gasifier nozzle, the first-stage power turbine nozzle was designed with individual vanes whose outer platforms were inserted into slots in the outer ring. Since the design configuration made it neces-

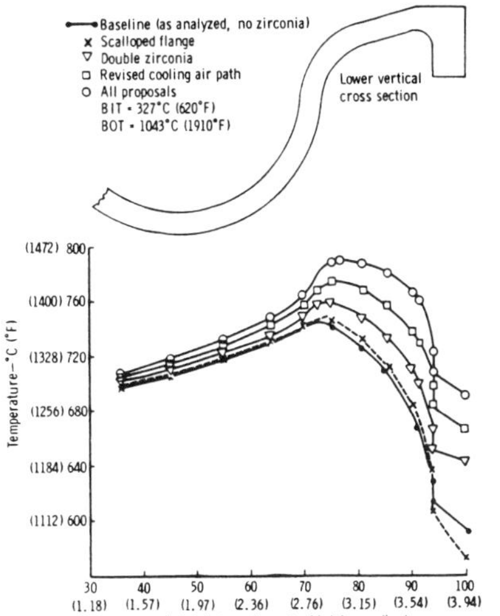

Figure 336. Plenum wall outer surface temperatures at steady-state rig conditions.

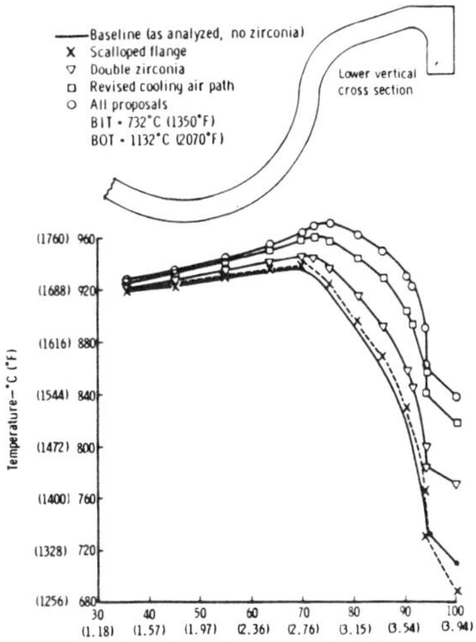

Figure 337. Plenum wall outer surface temperatures at steady-state engine conditions.

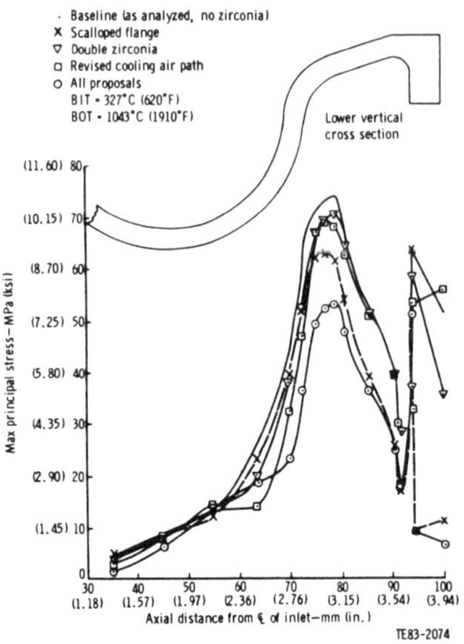

Figure 338. Maximum principal stress on plenum wall outer surface at steady-state rig conditions.

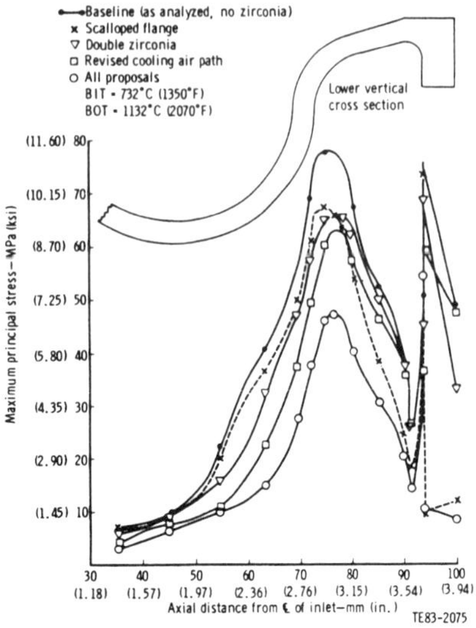

Figure 339. Maximum principal stress on plenum wall outer surface at steady-state engine conditions.

Table LXXII.
Plenum stress analysis for rig and engine conditions.

Design condition	Total axial thermal gradient—°C (°F) Rig	Total axial thermal gradient—°C (°F) Engine	Max principal stress—MPa (ksi) Rig	Max principal stress—MPa (ksi) Engine	Probability of survival Rig	Probability of survival Engine
Baseline	151 (272)	227 (409)	73.8 (10.7)	77.9 (11.3)	0.7375	0.4371
Scalloped flange	174 (314)	250 (450)	63.4 (9.2)	73.8 (10.7)	0.9252	0.8201
Double zirconia	120 (216)	176 (317)	71.0 (10.3)	71.0 (10.3)	0.8333	0.7041
Revised cooling air	104 (187)	144 (260)	67.6 (9.8)	63.4 (9.2)	0.8722	0.8368
All proposals	98 (176)	133 (240)	53.1 (7.7)	54.5 (7.9)	0.9812	0.9852

sary to cantilever the vanes from the outer ring, the vane platforms were to be bonded to the outer ring. Development of a bonding technique to achieve this was considered the primary task of the power turbine design. However, before fabrication efforts were initiated, the CATE program was revised in early 1981, eliminating the 2265°F configuration and emphasizing ceramic materials characterization and component development.

OBJECTIVE

The objective of this segment of the CATE program was to provide a power turbine section for the

2265°F configuration using ceramic components required to satisfy the performance and life criteria of the CATE program. The assembly was to fit into the IGT 404 industrial gas turbine engine without extensive changes.

APPROACH

Preliminary studies of the 2265°F configuration revealed that a two-stage power turbine was required to meet the criteria of the CATE program. To avoid increasing program costs, it was decided that a sufficient ceramic development effort would be achieved if the application of ceramics was limited to including the first-stage nozzle vanes and turbine tip shroud. The temperature of the second-stage power turbine vanes was low enough to achieve the life goals with an all-metal design. Both power turbine wheels could be all metal and still provide the life required for all proposed test stand running and vehicle demonstrations in the CATE program.

DISCUSSION

A general arrangement, shown in Figure 340, was devised for this part of the program. The design configuration makes it necessary to use vanes cantilevered from the outer ring. Design of the ceramic first-stage turbine nozzle in one piece was considered overambitious and beyond the state of the art. Instead, the experience gained on the individual vanes in the gasifier nozzle was utilized, and the cantilever vane mounting was achieved by bonding the outer platform of each individual vane into pockets in the outer ring.

Development of a bonding technique by the vendors was considered the primary development task of the two-stage power turbine design. The second-stage power turbine nozzle was designed as a one-piece casting of a cobalt-based metal. Details of the design of the two-stage power turbine vanes can be found in EDR 10672 (Ref 6).

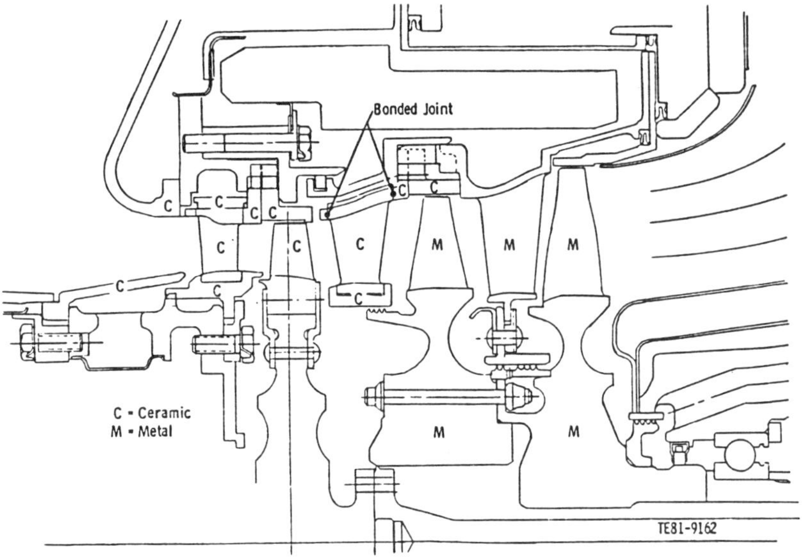

Figure 340. A 2265°F-configuration turbine.

V. CATE Ceramic Regenerator Development

Summary

Very satisfactory ceramic regenerator operation was achieved in both the 1900°F- and 2070°F-configuration engines. A number of unique features for these two engines were designed and successfully demonstrated. About 60 hr of operation at the maximum rated temperature of 982°C (1800°F) was achieved in the hot regenerator rig.

REGENERATOR DISK DEVELOPMENT

The objectives of regenerator disk development were as follows:

- improve manufacturing and inspection processes to eliminate weak areas and provide protection from handling damage
- decrease gear wear and gear attachment stresses
- improve steady-state and transient temperature tolerance
- improve effectiveness, pressure drop, and strain tolerance

A total of 1578 hr of engine testing were achieved on the 1100°C (2012°F) rated Corning aluminum silicate (AS) disks. These disks were stronger and had better performance than previous 1000°C (1832°F) disks. A total of 7814 hr of engine testing were achieved on the 1000°C (1832°F) rated disks. Testing included many miles on proving ground rough roads. Measured effectiveness ranged from 93.3% to 98.7%, and total pressure drop from 2.5% to 5.4% over the engine operating range. Tests showed disk strength highly dependent on matrix wall thickness, which varied by a factor of 3 within a core because of the wrap process. Efforts to control wall thickness showed promise but were not perfected. Failures of 1000°C (1832°F) AS disks were related to thin and porous matrix walls. The Allison regenerator design demonstrated trouble-free hub support and cantilever gear support from the cold corner of the disk. A single superior cement for both outer diameter (o.d.) coat and face fill was developed. A highly refined stress analysis explained failure location and frequency in the earlier 1000°C (1832°F) disks. Laboratory tests showed the disk more sensitive to 1177°C (2150°F) acceleration peaks than to 982°C (1800°F) steady-state exposure. Predicted life of the 1100°C (2012°F) AS disk was only 10,000 accelerations, equivalent to 300 hr of automotive operation. Three NGK-Locke magnesium aluminum silicate (MAS) disks suffered hub cracks and excessive through-wall leakage in rig tests.

REGENERATOR SEAL DEVELOPMENT

The objectives of the regenerator seal development effort were as follows:

- develop an inboard seal capable of low friction, wear, and leakage for operation at 982°C (1800°F) regenerator inlet gas temperature
- apply compressor discharge air to cool the inboard seal crossarm sufficiently to prevent creep but without introducing thermal warpage to cause leakage
- evaluate cooling crossarm wearface to provide chemical stability
- seek a crossarm wearface compound that is more chemically stable and provides a better thermal expansion match with the substrate or develop a design with a compliant isolator between wearface and substrate
- counteract thermal coning at seal rim to reduce leakage and concentrated wear

A three-piece hot side seal was developed featuring prestrained rim segments and cooled crossarm. The crossarm cooling scheme required further development to redistribute cooling flow to eliminate hot spots and seal leakage due to thermal distortion. A 90 NiO/10 CaF_2 crossarm wearface was selected for optimum friction and wear at 982°C (1800°F), whereas 70 NiO/30 CaF_2 was shown to provide less thermal distortion and consequently less leakage. Uncooled seals utilizing the 70 NiO/30 CaF_2 wearface averaged 3.4% leakage while the best cooled seals showed a 4.5% leakage rate.

Running time at high temperature was not sufficient to establish wear rate, but it was expected to be adequate. Nickel from the seal wearface was found to combine with the AS disk, producing melting at the isolated contact spots. Catastrophic reaction between seal CaF_2 and the AS disk in laboratory tests at 982°C (1800°F) without airflow did not occur in full-scale rig tests with airflow.

CONCLUSIONS

The following conclusions were reached:

1. Good ceramic regenerator disk and seal performance and endurance were demonstrated for

1038°C (1900°F) engine conditions. Additional running time at 2070°F engine conditions would help determine capabilities.
2. Laboratory tests indicate that neither AS nor MAS disk materials will tolerate an adequate number of engine accelerations because of the temperature peaks imposed by accelerations.
3. Regenerator hot seal crossarm cooling requires further development to provide even cooling for long life and low leakage.

Regenerator System Design and Analysis

EDR 9346 (Ref 9), p 30, described regenerator design objectives, configuration, loads, temperature features, and alternative designs. Two alternative gear attachment designs that were not tried were discussed in EDR 9519 (Ref 1), p 102. Plans for reducing disk bearing wear were presented.

The original heat transfer analysis was presented in EDR 9346 (Ref 9), p 57. Concerns expressed therein for high temperature damage to rubber gear mounting and seal graphite wearface proved unfounded. In EDR 9519 (Ref 1), p 94, slightly lower gear mounting rubber temperature was shown to be achievable with a thicker disk rim, a seal designed for disk and seal rim cooling (p 110), or scalloped rubber; however, none of these was necessary.

EDR 11006 (Ref 8), p 107, included the final, most refined disk stress calculations for 1038°C (1900°F) engine conditions. A maximum radial compressive stress of 324 kPa (47 lb/in.^{2}) was located at 279 mm (11 in.) radius on the cold side of the disk. Maximum tangential compressive stress was 1165 kPa (169 lb/in.^{2}) adjacent to the rubber gear mount. Maximum shear stress was 221 kPa (32 lb/in.^{2}) near the hub due to hot seal friction. Results agreed with the location and frequency of 1000°C (1832°F) disk failures. EDR 9346 (Ref 9), p 57, included initial disk stress calculations that were later judged to be in error resulting in a false stress concentration at the outer diameter (o.d.) due to the manner in which pressure drop across the disk rim, rim filler cement, and gear mount rubber were modeled.

REGENERATOR DISK DEVELOPMENT

Corning AS Disks

1000°C (1832°F) AS Disk

Effectiveness, pressure drop, and seal leakage for early thick and thin wall disks were reported in EDR 9346 (Ref 9), pp 180-91. (Drive gear deflection test versus calculation was shown on p 54.)

EDR 9519 (Ref 1), p 115, reported internal leakage of thick and thin wall disks. Then in EDR 9722 (Ref 2), p 8-12, two disk hub push-out failures due to engine block thermal distortion were discussed. Tests on other disks (p 103) showed that hub push-out failure required a load of 1180 lb and 0.406-0.432 mm (0.016-0.017 in.) deflection.

EDR 9951 (Ref 3), p 10, gave a summary of g loads experienced at the disk support in a truck driven over rail crossings and a Belgian block test road. A maximum of 3 g's and a 65-lb load on the support without damage was measured.

In EDR 10156 (Ref 4), p 133, the performances of eleven 1000°C (1832°F) disks and four 1100°C (2012°F) disks were compared. The 1100°C (2012°F) disks had higher effectiveness and pressure drop because of more passages per unit area.

EDR 10383 (Ref 5), p 126, describes four disk failures—three of which were shown to be due to thin and/or distorted walls. The failure locations agreed with maximum stress locations defined in the final stress analysis, EDR 11006 (Ref 8), p 107.

1100°C (2012°F) AS Disk

In EDR 10383 (Ref 5), p 133, 1100°C (2012°F) AS Corning disks were reported to have had 5% less open area, 11% more pressure drop, 1.3% higher effectiveness, 80% less leakage, and 85% greater strength than previous 1000°C (1832°F) disks. After further study the increase in 1100°C (2012°F) disk strength was reported in EDR 11006 (Ref 8), p 103, to be 30%, and the through-the-wall leakage decrease was recorded as 25%. Total engine test time in eight disks was 4667 hr with 1422 hr on a single disk.

Drive Gear Adapter and Gear Coating

In EDR 9951 (Ref 3), p 209, the first redesign of the disk drive gear mounting adapter to improve protection from handling damage was included. EDR 10383 (Ref 5), p 130, explained further redesign of the gear adapter to eliminate the fit problem in a rivet joint and rivet failure with a one-piece design. In EDR 11006 (Ref 8), p 103, 1680 hr of engine test time were reported to have been accumulated with the one-piece gear adapter and tungsten carbide gear tooth coating, which eliminated chipping and wear.

NGK-Locke MAS Disk Development

Three disks supplied by NGK-Locke of Japan were run in the hot regenerator rig. All three suffered hub cement joint failures attributed to thermal stresses. No failures of this type were experienced with Corning AS disks, which have much lower thermal expansion. Internal leakage in the NGK disk produced an unacceptable increase in system leakage. See EDR 10156 (Ref 4), p 123, and EDR 10672 (Ref 6), p 115, for descriptions of the first and second NGK disk failures.

Regenerator rig and internal leakage tests were completed on the third regenerator disk from NGK. This disk was composed of rectangular cemented blocks. The matrix was composed of equilateral triangles with 0.13 mm (0.005 in.) thick walls of MAS. The disk featured a revised hub design that had demonstrated greater tolerance for temperature transients in furnace tests. The disk survived a run to 807°C (1187°F) and full 290 kPa (42 lb/in.^{2}) pressure but sustained both hub cement joint cracks and radial rim cracks on a subsequent run to 896°C (1645°F). The hub joint was cracked completely around on the cold side, but only about 38 mm (1-1.5 in.) on the hot side. The hot side developed 13 radial rim cracks; three were at cement joints. The rim cracks were 1 to 1.5 in. long and demonstrated the need for rim stress relief in this higher expansion material. The cold side had two rim cracks, both at joints. Further improvement in the NGK hub joint is required.

Performance of the NGK disk was directly compared with that of a 1100°C (2012°F) Corning disk by successive runs in the regenerator rig using identical seals. Results are compared in Table LXXIII.

Engine operating conditions simulated on the rig were for the 1900°F-configuration engine. Higher temperatures for the 2070°F-configuration engine would have been run next had the the NGK disk survived. Drive torque at 100% conditions was 108 N (80 ft-lb) for the NGK disk and 136 N (100 ft-lb) for the Corning disk.

Internal leakage measurements were made on both disks with 138 kPa (20 lb/in.^{2}) imposed on a rubber sealed 3.2 mm x 152 mm (1/8 in. x 6 in.) radially oriented slot. Internal leakage for the NGK disk ranged from 0.16 to 0.21 kg/s-m^2 (0.00023 to 0.00029 lb/sec-in.^{2}), which was about 10 times greater than for the Corning disk and came largely from the cement joints. Each measurement included one joint. Calculations show that the increased internal leakage of the NGK disk over the Corning disk would more than account for the increased seal leakage. Internal leakage must be reduced to make the NGK disk a viable candidate.

Both the lower effectiveness and lower pressure drop were to be expected due to fewer and larger holes. Table LXXIV compares the physical differences of the NGK thick wall with Corning thin wall disks and with NGK extruded matrices in development.

NGK delivered samples of the 1.92:1 aspect ratio rectangular matrix to Allison for evaluation of tangential modulus of rupture (TMOR) and cyclic life. Allison had requested a rectangular matrix for its potentially better performance and strain tolerance. Higher effective surface area, comparable effective open area, and comparable hydraulic diameter resulting from the rectangular passage should produce superior heat transfer and pressure drop characteristics. Higher strain tolerance, also expected from the rectangular matrix, should reduce thermal stress. Much improvement in resistance to rapid thermal cycles of the NGK material is required because of nearly complete loss in hot face strength as shown in the following discussion on material evaluation. The NGK thin wall triangular matrix is being evaluated elsewhere prior to full-size disk fabrication.

Other Disk Materials

Proposed further development of General Electric zirconia MAS material was rejected because it

Table LXXIII.
Performance comparison of NGK and Corning regenerator disks.

Engine operating condition—%	Gas inlet temp—°C (°F)	Air inlet pressure—kPa (lb/in.^{2})	Effectiveness—% NGK	Effectiveness—% Corning	Total pressure drop—$\Delta P/P$ NGK	Total pressure drop—$\Delta P/P$ Corning	Seal leakage—% NGK	Seal leakage—% Corning
60	896 (1645)	73 (10.6)	96.8	98.2	0.018	0.025	5.4	3.2
80	808 (1487)	155 (22.6)	93.7	96.0	0.024	0.034	5.5	3.2
100	721 (1330)	293 (42.5)	89.2	93.6	0.036	0.050	7.2	5.4

Table LXXIV.
Comparison of regenerator matrix properties.

	Wall thickness—mm (in.)	Hydraulic diameter—mm (in.)	Holes per cm^2 ($in.^2$)	Open area	Effective* open area	Ratio of surface to volume—m^2/m^3 (ft^2/ft^3)	Effective ratio of surface to volume—m^2/m^3 (ft^2/ft^3)
Corning 1000°C	0.066 (0.0026)	0.549 (0.0216)	209 (1350)	0.77	0.77	5576 (1700)	5576 (1700)
Corning 1100°C	0.064 (0.0025)	0.526 (0.0207)	228 (1470)	0.77	0.77	5819 (1774)	5819 (1774)
NGK thick wall	0.130 (0.0051)	0.610 (0.024)	141 (910)	0.68	0.68	4500 (1372)	4500 (1372)
NGK Thin wall	0.089 (0.0035)	0.508 (0.0200)	236 (1520)	0.72	0.72	5800 (1768)	5800 (1768)
NGK 1.92:1 rectangle	0.114 (0.0045)	0.544 (0.0214)	208 (1340)	0.68	0.72	5035 (1535)	6658 (2030)

* Utilizing goodness factors for passage shape per A. L. London

showed no advantage over other MAS materials (see EDR 10841 [Ref 7], p 95).

REGENERATOR SEAL DEVELOPMENT

1900°F Engine Seal Configuration

EDR 9722 (Ref 2), pp 105 and 114, attributed high start and idle seal leakage to thermal distortion due to thermal expansion mismatch between wearface and substrate. Higher start and idle leakage was shown on cold fire-ups. Hold-down springs showed no benefit. Measured leakage on 14 seal sets was reported in EDR 10156 (Ref 4), p 130. Three seals with better expansion match had lower leakage (p 136). Additional examples of seal performance with the better match were reported in EDR 10383 (Ref 5), p 145, EDR 10672 (Ref 6), p 135, and EDR 10841 (Ref 7), p 103.

EDR 10672 (Ref 6), p 135, reported a 50 hr regenerator rig run at the ultimate CATE regenerator inlet temperature of 982°C (1800°F). Although the seal was not designed for that temperature and was uncooled, it suffered only light chalking of the wearface and sealing lead corrosion. Wear, however, was heavy for such a short run.

2070°F Engine Seal Configuration

EDR 9722 (Ref 2), p 105, included results of stress and buckling studies on multipiece seals intended to relieve those problems. Test data showed that lower seal temperatures resulted with ceramic regenerator disks in lieu of metal. Various rim slotting to relieve stress was also evaluated (p 108).

In EDR 10156 (Ref 4), p 126, the seal rim buckling problem and the three-piece construction devised to combat it were described. Preloading of the rim segments to counteract buckling was not discussed because patent applications were then pending. EDR 10383 (Ref 5), p 143 and 144, pictured the cooling system and described the first tests. The results of the tests showed that cooling and lack of flatness increased leakage. Leakage and cooling effects for four high-temperature seals were reported in EDR 10672 (Ref 6), p 131. Steep seal temperature gradients caused by cooling were blamed as the cause of leakage. EDR 10841 (Ref 7), p 101, showed an incremental increase in seal leakage of 3% to 4% after 254 hr of engine test. Cooling was reported as minimal at crossarm midradius, but excessive near rim and hub.

EDR 11006 (Ref 8), p 116, reported that seal cooling was still inadequate after the cooling supply passages were modified to double the flow. Parts were being fabricated to apply cooling only where needed but were not completed for the reported test.

Wearface

Early friction, wear, and thermal distortion tests on various mechanical mixtures of NiO and CaF_2 plasma-sprayed wearfaces were discussed in EDR 9346 (Ref 9), p 163. Then in EDR 9951 (Ref 3), p 91, the successful use of ultrasound to detect delamination of plasma-sprayed seal wearface was reported. EDR 10156 (Ref 4), p 127, described how the delamination problem was ultimately solved with a patented anticorrosion barrier. A seal crossarm, coated on both sides to counteract differential expansion, lost its flatness and lower leakage after 331 hr of engine test. Four seals utilizing a flexible metal layer to isolate differential expansion had slightly higher leakage than standard seals.

EDR 10383 (Ref 5), pp 140-57, reported chemical reaction between NiO/CaF_2 seal wearface and AS disk at 982°C (1800°F). The degree of reaction, friction, and wear were much more severe without airflow. Reactants were defined by microprobe and SEM. A 90 NiO/10 CaF_2 seal wearface mix produced minimum friction and wear while 70 NiO/30 CaF_2 produced minimum thermal distortion.

EDR 10672 (Ref 6), p 138, showed analysis of seal wearface glaze formed in full-scale hot regenerator testing. EDR 10841 (Ref 7), p 103, compared leakage of ten 85 NiO/15 CaF_2 seals versus six 70 NiO/30 CaF_2 seals. The latter had 0-0.8 incremental percent less leakage because of less thermal distortion.

Approximately 201 hr of engine testing were accumulated on a new type graphite seal rim wearface with 66°C (150°F) higher temperature capability. Standard graphite showed no distress after 50 hr of full-scale regenerator rig operation at 982°C (1800°F).

In EDR 11006 (Ref 8), p 118, the Ford 1112 seal wearface was reported to have spalled after 871°C (1600°F) exposure, despite Ford approval of material and application. Low friction was achieved at 982°C (1800°F) after recompounding of defective spray powder by Ford. Ford 1112 wearface applied with Allison process and barrier coat did not spall after several lengthy cycles to 871°C (1600°F).

Sealing Leaf

EDR 9519 (Ref 1), p 114, presented sealing leaf leakage for nine seals. Later, in EDR 9722 (Ref 2), p 108, seal leaf stress was shown to exceed yield strength at high temperature, and a cooling scheme, which was subsequently successful, was discussed. EDR 10156 (Ref 4), p 140, showed that leaf leakage increase due to engine block distortion was not serious.

TEST FACILITIES

EDR 9346 (Ref 9), p 163-176, described the following five regenerator test facilities with examples of data:

- seal thermal distortion (p 163)
- seal friction and wear (p 163)
- seal leaf leakage (p 170)
- disk internal leakage (p 175)
- hot regenerator rig (p 176)

In EDR 10383 (Ref 5), p 145, the first operation of a newly modified hot regenerator rig at 982°C (1800°F) was reported. The friction and wear rig screening procedure for 982°C (1800°F) wearface was also described (p 148). Then in EDR 10672 (Ref 6), p 135, a 50-hr test at 982°C (1800°F) on the hot regenerator rig was discussed.

Regenerator Core Materials Development

SUMMARY

The ceramic regenerator core material development effort can be described as an in-depth strength evaluation focused on defining the upper temperature capability of the three available regenerator disk materials: Corning AS rated at 1000°C (1832°F), Corning AS rated at 1100°C (2012°F), and NGK MAS. In addition to the strength evaluation, considerable effort has been expended to understand the mechanisms involved in loss of strength after thermal exposure. Gaining the understanding of this has been instrumental in the temperature capability increases made in the AS material.

The strength evaluations were conducted in four steps, which are summarized in the following discussion. These four steps are the evaluation of as-received strength, steady-state thermal exposure, cyclic thermal exposure, and disk-seal chemical reaction.

The as-received tests of these disk matrices showed that both the 1000°C (1832°F) and 1100°C (2012°F) AS materials were stronger in TMOR than any of the MAS disk matrices or samples. The radial compressive strength (RCS) of the precisely aligned apex-to-apex triangular channel extruded MAS was over twice that of the 1100°C (1832°F) AS matrix.

Simulated engine acceleration/deceleration cyclic thermal exposure (CTE) tests showed hot face strength levels for the 1100°C (2012°F) AS material after 10,000 cycles to 1177°C (2150°F) equal to the as-received MAS strength. Five-thousand cycles to 1120°C (2050°F) left no usable strength in the hot face of the MAS.

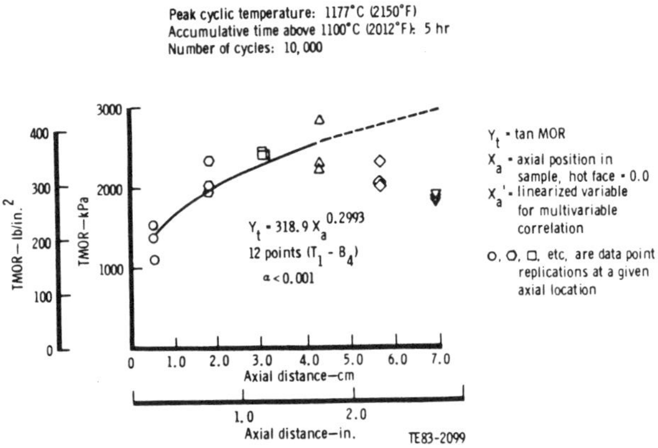

Figure 341. Disk 4, 1100°C AS, sample $R_{1\text{-}14}$ cyclic thermal exposure damage.

The following conclusions have been reached:

- The AS matrix is superior to the current MAS matrix in as-received MOR but inferior to it in RCS.
- The critical experiment for a regenerator matrix for use in a vehicular gas turbine is the CTE test simulating an engine maximum acceleration/ deceleration cycle.
- The 1100°C (2012°F) AS matrix retains a usable margin of hot face strength after 10,000 cycles and is the only current matrix that is satisfactory for further development use (see Figure 341).
- The current MAS matrix is not a satisfactory core for development (see Figure 342).
- The extrusion fabrication method produces structurally stronger and more uniform matrix than the wrapping technique.

The following are recommendations:

- The upper limits of the 1100°C (2012°F) AS matrix performance should be determined—specifically, (1) the maximum CTE temperature and (2) cycle lives of 20,000 and above to explore the keatite to mullite phase transformation under cyclic exposure.
- The possibility of improving the AS matrix resistance to CTE by controlling grain size in the ceramic should be investigated.
- A new 1500°C (2732°F) maximum matrix proposed by one of the suppliers should be evaluated for CTE resistance.
- The mass transfer and corrosion resistance of the candidate matrix materials to road salt and to hot face seal contact materials should be measured in terms of the loss in MOR and RCS.
- The role of matrix porosity in determining the as-received strength should be assessed.

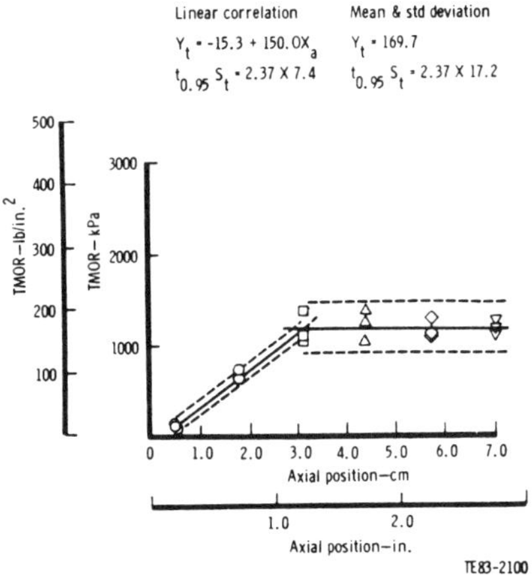

Figure 342. Disk 1, MAS, sample $R_{1\text{-}4}$, 1121°C (2050°F) for 2 hr and 30 minutes at 5000 cycles.

OBJECTIVE

The objective of the regenerator core material development effort has been to characterize the physical and chemical stability (and resulting strength) of the 1000°C (1832°F) and 1100°C (2012°F) AS matrix. These were then to be compared with an MAS candidate matrix in terms of as-received strength, tolerance of advanced vehicular turbine engine acceleration/deceleration transients, and elevated temperature levels.

APPROACH

A major program to characterize ceramic regenerator disk use in high temperature vehicular engines has been carried out. The approach to the program was based on the careful correlation of engine testing of disks with laboratory sample experiments simulating both current and advanced engine operating conditions. The interchange in information between these two parts of the program provided guidance in planning laboratory tests, verification of the laboratory experiments, and identification of the causes for the disk failures in engine tests. XRD, scanning electron microscope (SEM), EBM, and wet chemistry supported this effort. Three matrix materials were included in the program: wrapped 1000°C (1832°F) AS, wrapped 1100°C (2012°F) AS, and extruded MAS.

An experimental design for the CTE experiment was worked out using peak cyclic temperatures of 1066°C, 1121°C, and 1177°C (1950°F, 2050°F, and 2150°F) and number of cycles from 10 to 10,000. The cycle selected is a combination of a maximum CATE engine acceleration transient, with the initial ramp at approximately 1167°C/s (approximately 2100°F/sec) to the selected peak temperature followed immediately by a fully stabilized engine braking deceleration transient with an initial ramp of approximately 360°C/s (approximately 650°F/sec). This combination cycle is a worst case simulating the disk inlet gas temperature for a heavy-duty vehicle at maximum performance uphill and downhill. To achieve this cycle using the gas burner facility, controlled cold air injection is used, as shown in Figure 343. The total flow rate is controlled by the pressure ratio between supply and burner by the pressure regulating valve (PRV), the on/off by the solenoid valve, and the rate of change of flow using the pneumatically controlled valve. (The pneumatically controlled valve is operated by a control air supply system which varies both the opening and closing rates.)

The test cycles are shown in Figures 344 and 345. The lower curve in Figure 344 shows the com-

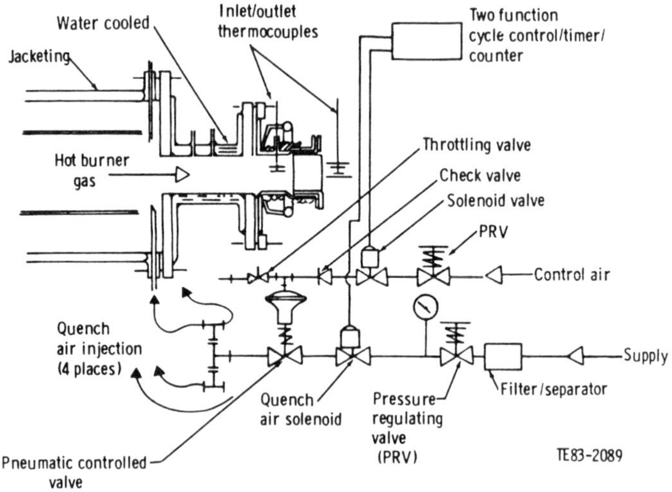

Figure 343. Cyclic thermal exposure control system.

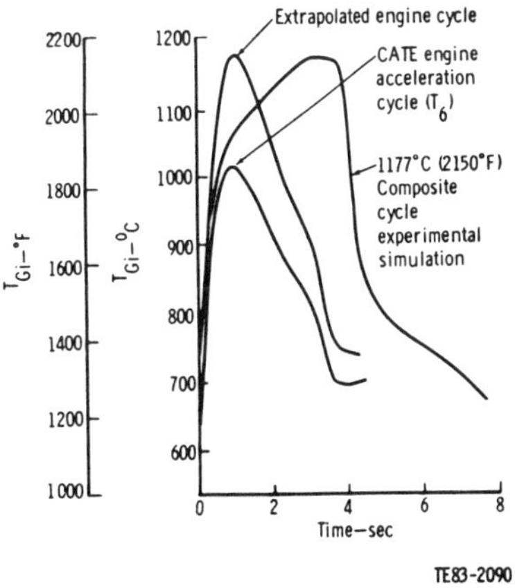

Figure 344. Comparison of engine regenerator inlet and matrix sample inlet gas temperature cycles.

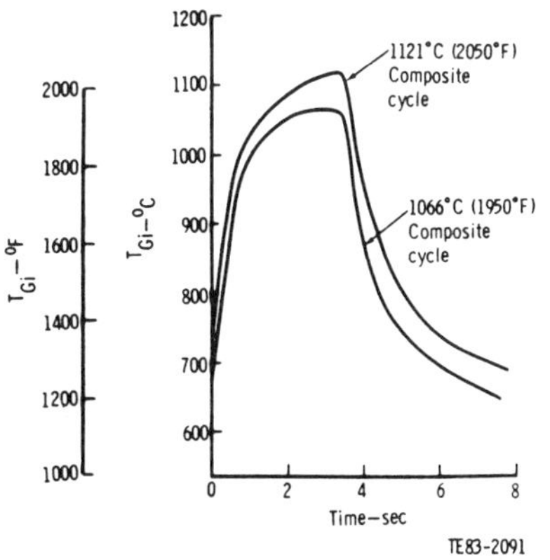

Figure 345. Simulated acceleration/deceleration cycles, sample gas inlet temperature.

posite T_6 temperatures recorded for a typical CATE engine on a test stand truck/bus, acceleration/deceleration cycle with zero steady-state time in the cycle. Immediately above is an extrapolation of this 1010°C (1850°F) peak cycle to the 1177°C (2150°F) peak anticipated for a 1121°C (2050°F) maximum temperature acceleration with a poorly trimmed engine fuel control. The test cycle finally selected to simulate the engine cycle is the third curve. This curve was selected after preliminary testing established that the CTE experiment was not sensitive to total cycle lengths from approximately 7 to approximately 16 sec. Note, however, that the axial depth of penetration of the temperature wave goes up with the cycle length, increasing the depth of CTE damage. Figure 345 shows typical curves for the 1066°C (1950°C) and 1121°C (2050°F) test cycles.

This approach to testing ceramic regenerator cores using nonsteady (or cyclic) gas temperatures was verified by examining the axial strength distribution in an 1100°C (2012°F) AS disk used in a CATE engine for 1422 hr and approximately 7800 cycles to 1010°C (1850°F) peak temperatures.

The general approach to data reduction and analysis utilized correlation of the strength level with (1) the disk and the radial location in the disk, (2) the separator wall thickness, and (3) the channel skew angle. The CTE strength distribution was first reduced by power curve correlation with the axial distance from the hot face, linearized, and then included in multiple linear correlation with the wall thickness and skew angle.

DISCUSSION

The following discussion of regenerator core materials development is arranged in four subsections: as-received strength, steady-state thermal exposure, cyclic thermal exposure, and disk-seal chemical reaction. Most of the work reported in the first two subsections was accomplished prior to 1982 and is reported in detail in previous semiannual reports. The majority of this discussion deals with cyclic thermal exposure work conducted during 1982.

As-Received Strength

Two variables affecting the strength of the as-received (and thermally exposed) wrapped AS matrix were separated and identified. The channel skew angle, defined as the local deviation from the axial direction (EDR 9951 [Ref 3], pp 111-112, 115-119) and the fracture plane separator wall thickness (EDR 10383 [Ref 5], pp 135-137) affect both the MOR and RCS. Porosity is being investigated as a third variable that will affect the strength of both the AS and MAS materials.

The Corning AS matrices are characterized by a good mean level of MOR, accompanied by wide variability, while the radial compressive strength is relatively low. The eight 1000°C (1832°F) disks tested had an average overall MOR of 1413 kPa (250 lb/in.²) and a lower 95% confidence limit of 255 kPa (37 lb/in.²). The 1100°C (2012°F) AS disks had an average MOR of 2423 kPa (351 lb/in.²) and a lower 95% limit of 1655 kPa (240 lb/in.²). The mean radial compressive strength was 633 kPa (92 lb/in.²) and the lower confidence limit was 411 kPa (60 lb/in.²). This lower 95% confidence limit has a factor of safety of 1.22. The triangular channel extruded MAS has a mean MOR of 1329 kPa (193 lb/in.²) and a lower 95% confidence limit of 1111 kPa (161 lb/in.²). However, the radial compressive strength due to the apex-to-apex configuration of the extruded matrix had a mean of 1614 kPa (232 lb/in.²) and a lower confidence limit of 1110 kPa (161 lb/in.²), as shown in Figure 346.

Steady-State Thermal Exposure

The permanent change in length, expressed as a reduction per unit length, after steady-state thermal exposure of the 1000°C (1832°F) and 1100°C (2012°F) AS matrices is shown as a function of temperature level in Figure 347. The reduction in unit length is a measurement of the reduction in volume caused by the phase transformation from the aluminous keatite structure to mullite. The larger the change in volume is, the greater the reaction rate. A

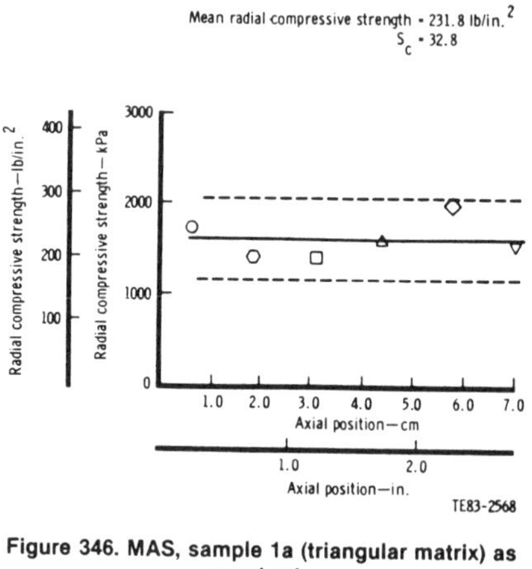

Figure 346. MAS, sample 1a (triangular matrix) as received.

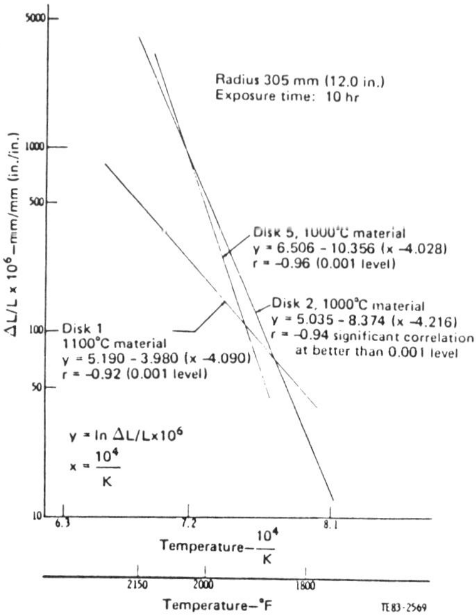

Figure 347. Comparison of phase transformation rate of 1000°C (1832°F) and 1100°C (2012°F) AS matrix materials.

major reduction in phase transformation rate is shown by the reduction in slope from the 1000°C (1832°F) data to the 1100°C (2012°F) data.

MOR tests from the steady-state phase transformation test specimens have shown a lower 95% confidence limit (loss in strength) after exposure to 1121°C (2050°F) for the 1000°C (1832°F) AS. Some test bars showed fractures after exposure to 1177°C (2150°F) for the 1000°C (1832°F) matrix. The 1100°C (2012°F) matrix retains usable strength (greater than 1034 kPa [150 lb/in.²]) after 5 hr at 1177°C (2150°F) but shows subcritical cracking. (See EDR 9722 [Ref 2], Regenerator Materials section.)

Cyclic Thermal Exposure

CTE damage to the hot face of a 1000°C (1832°F) engine disk was identified and compared with laboratory samples of 1000°C (1832°F) and 1100°C (2012°F) AS disks in EDR 10672 (Ref 6), pp 124 and 126, and correlated as MOR and RCS values against axial position in EDR 10841 (Ref 7), pp 99-

100. This work was extended to a second engine disk in EDR 11006 (Ref 8), pp 110-113. The test was then used to compare the current MAS matrix with the 1100°C (1832°F) AS (Ref 8, pp 114-115). When the hot face strength of 1100°C (1832°F) AS is plotted as a function of cyclic peak temperature and number of cycles, an overall plot of resistance to CTE damage results, as shown in Figure 348.

The plot shows the retained hot-face strength (taken from individual MOR axial position correlations) after a given number of cycles at a given peak cyclic temperature. The minimum strength point shown, after 10,000 simulated acceleration/deceleration cycles to 1177°C (2150°F), is approximately 1379 kPa (200 lb/in.^{2}), which has a usable margin of strength of approximately 345 kPa (50 lb/in.^{2}) over the suggested lower control limit level for TMOR of 345 kPa (150 lb/in.^{2}).

The CTE experimental approach and results were verified by comparison of the 1000°C (1832°F) AS engine disk sample data with laboratory samples. This model verification method was successfully extended to the 1100°C (2012°F) material using samples from an engine disk exposed for 1422 hr and some 7800 cycles. The marked similarity of the regression line shown in Figure 349 with the curve shown in Figure 350 for a laboratory sample run at 1121°C/2.5 hr/5000 cycles is readily apparent. Only the preexponential term is different. (A multiple regression equation and curve are also shown, accounting for the effects of skew angle, x_s, and separator wall thickness, x_w.)

Analysis of the CTE data suggests that the probable mechanism causing the damage is anisotropic expansion of one or more of the phases present in the matrix. If future work can demonstrate that this is indeed the case, then reduction of the grain size may reduce or eliminate the problem of CTE damage.

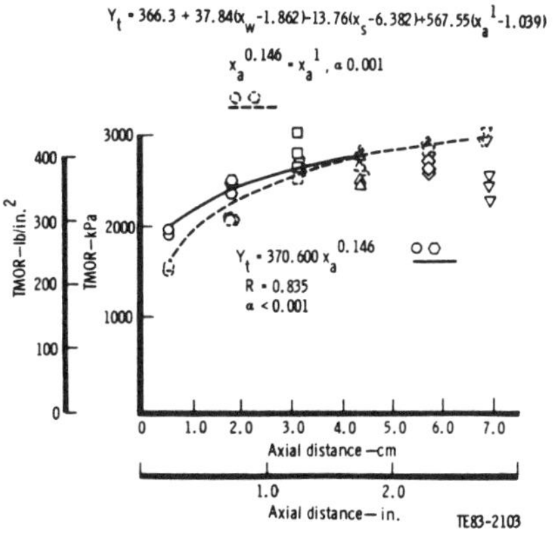

Figure 349. Disk 5, 1100°C (2012°F) AS, GM 16-04 for 1422 hr engine/7800 cycles (sample $R_{3\text{-}1}$, radius = 8.0 in.).

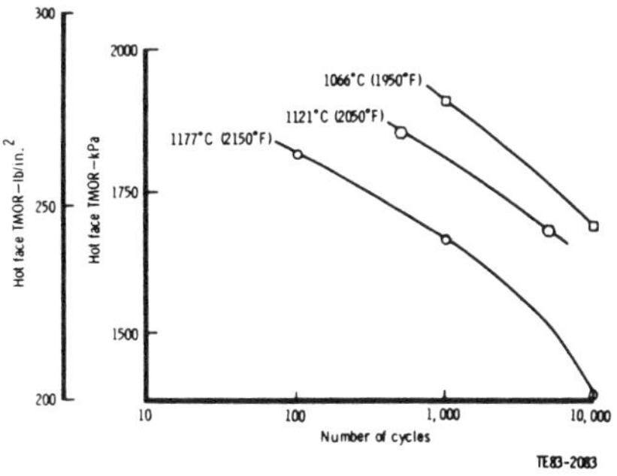

Figure 348. Power curve correlation estimate of matrix hot face strength after CTE Disk 4, 1100°C (2012°F) AS.

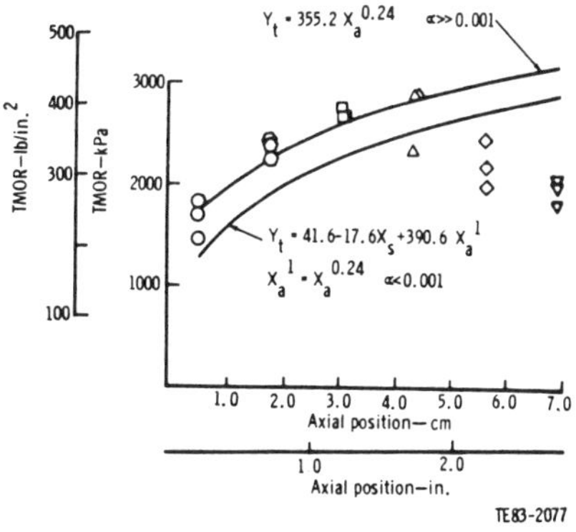

Figure 350. Disk 4, 1100°C (2012°F) AS, sample $R_{1\text{-}23}$, cyclic thermal expansion, 1066°C (1950°F) for 5 hr, 10,000 cycles.

The NGK-Locke MAS extruded matrix lost virtually all hot face strength after 2000 cycles to 1093°C (2000°F). A 50% loss in strength at the 0.02 mm level was experienced with 5000 cycles to 1121°C (2050°F).

The following is a description of the cyclic thermal exposure effort conducted during 1982. The previous discussion summarized the work conducted prior to 1982.

To simply separate the effects of the number of cycles and peak cycle temperature on the CTE damage to the 1100°C (2012°F) AS matrix, the data taken were limited to samples taken from a single radius of Disk 4. The amount of data required was reduced by approximately an order of magnitude by setting aside the effects of disk-to-disk and radial variation in strength. However, note that the information gained is limited in the same way. Multivariable correlation was used to account for the effect of channel skew angle. A single variable power curve correlation is run first to establish the correlation of the selected strength parameter with the loss in hot face strength (shown as X_a, the axial position). The axial position effect is then linearized by setting $X' = X_a^n$ and entered in the multivariable correlation of channel skew angle, X_s, and X' with the MOR or RCS. The significance of the skew angle effect, once established, can be ignored in the data plots, which are intended to show the gradient in strength, starting at the hot face ($X_a = 0.0$ mm) and extending to $X_a = 40.6$ mm (1.6 in.). The drop-off in strength accounted for by the channel skew angle effect is shown by the data points at $X_a = 57$ mm to 70 mm (2.25 in. and 2.77 in.). Figures 350, 351, and 352 give both the single and multiple correlations and curves for each as examples of the general data reduction method. Figures 353, 354, 355, and 356 show only the single correlation results for clarity.

The CTE data at 1066°C (1950°F) cyclic peak temperature for 1000 and 10,000 cycles are shown in Figures 350 and 353. A clear-cut decrease in hot face strength, 1910 kPa to 1689 kPa (277 lb/in.² to 245 lb/in.²), is shown. Several points are clear from this part of the experiment. For example, some loss in strength occurs at only 1000 cycle exposure, and the loss is progressive with number of cycles. Also the mechanism of attack is not dependent on exceeding the rated temperature of the material since the 1066°C (1950°F) peak did not approach or exceed the 1100°C (2012°F) rating.

The 1121°C (2050°F) CTE axial strength distributions for 500 and 5000 cycles are correlated and plotted in Figures 351 and 354. The loss in hot face strength is significant in both cases once again,

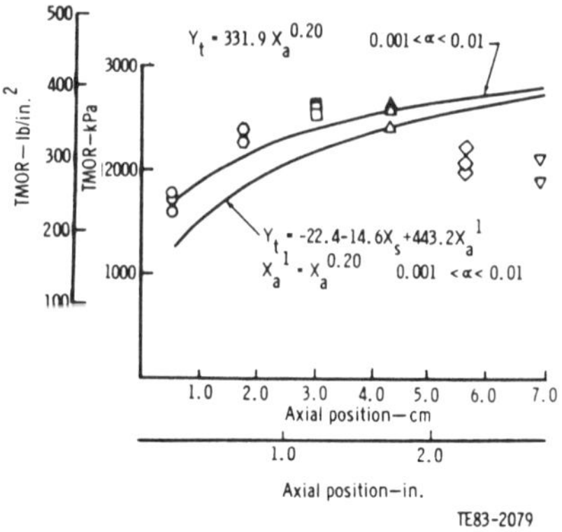

Figure 351. Disk 4, 1100°C (2012°F) AS, sample $R_{1\text{-}10}$, cyclic thermal expansion, 1121°C (2050°F) for 2 hr and 30 minutes, 5000 cycles.

dropping to 1682 kPa (244 lb/in.²) observed after 10,000 cycles at 1066°C (1950°F). The loss in strength after 500 cycles is smaller—1855 kPa (269 lb/in.²)—than the 1000 cycle loss at the lower temperature.

The final set of three runs at a peak cyclic temperature of 1177°C (2150°F) were planned to cover

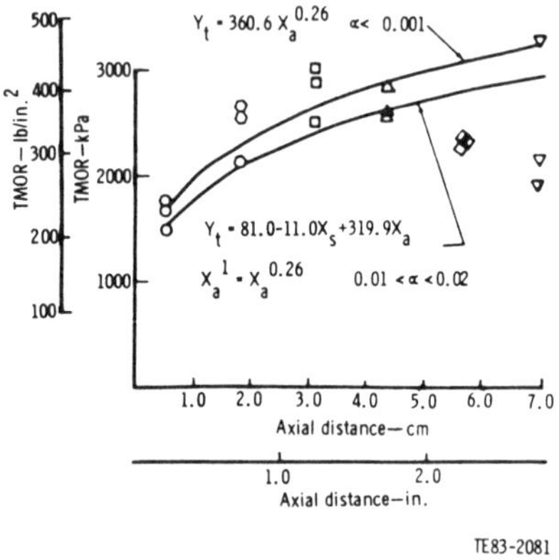

Figure 352. Disk 4, 1100°C (2012°F) AS, sample $R_{1\text{-}1}$, cyclic thermal exposure damage, 1177°C (2150°F) for 30 minutes, 1000 cycles.

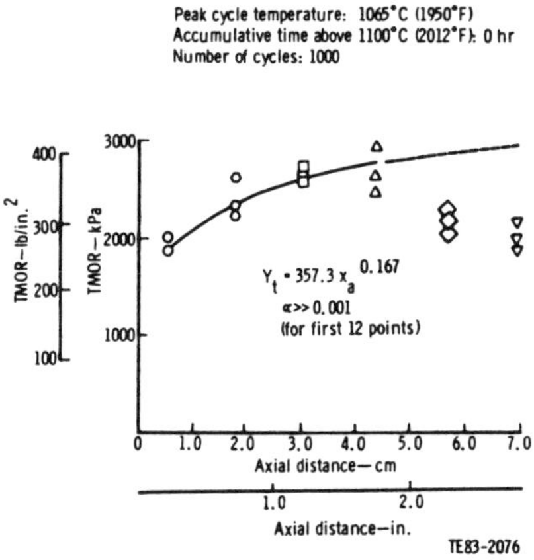

Figure 353. Disk 4, 1100°C (2012°F) AS, sample $R_{1\text{-}21}$, cyclic thermal exposure damage.

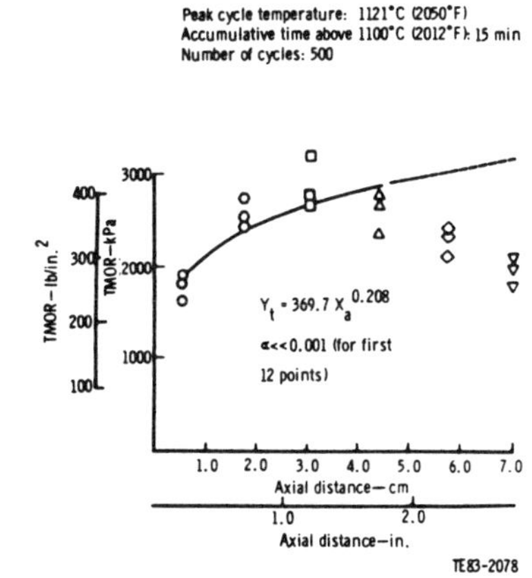

Figure 355. Disk 4, 1100°C (2012°F) AS, sample $R_{1\text{-}3}$, cyclic thermal exposure damage.

the three orders of magnitude from 100 to 10,000 cycles. The 100 cycle data were viewed as essential for developing a damage mechanism theory for the CTE experiments. The 10,000 cycles represent something less than an accumulation of 10 hr above 1100°C (2012°F), where phase transformation from aluminous keatite to mullite began to change the thermal expansion characteristics of the matrix discussed in detail in Conclusions—and represented 10% of the engine acceleration/deceleration goal of

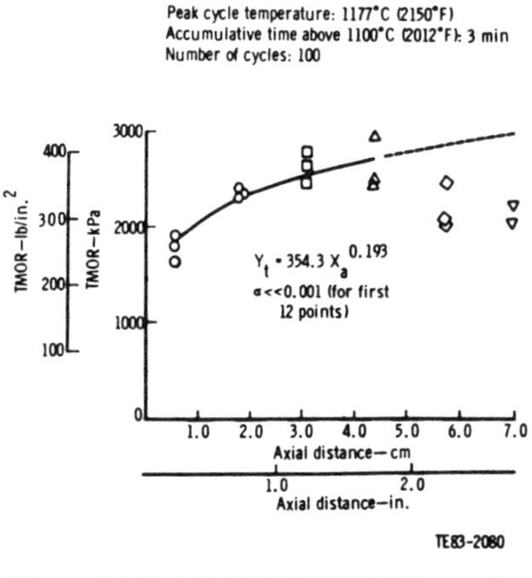

Figure 354. Disk 4, 1100°C (2012°F) AS, sample $R_{1\text{-}22}$, cyclic thermal exposure damage.

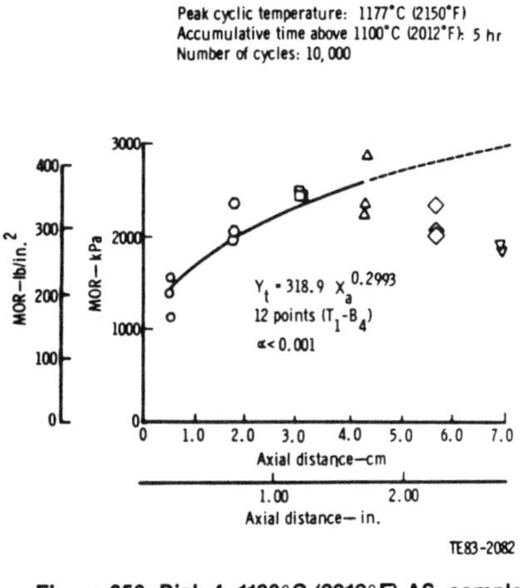

Figure 356. Disk 4, 1100°C (2012°F) AS, sample $R_{1\text{-}14}$, cyclic thermal exposure damage.

100,000 cycles. Figures 352, 355, and 356 plot the data for these runs. Once again the hot face losses were progressive with increasing number of cycles. The 100-cycle run clearly established that a small but statistically significant reduction in strength occurred in just 100 cycles to 1177°C (2150°F). However, the primary result of these three runs was that the 10,000 cycle exposure left the hot face with a usable margin of strength. (Disk 4 was below the average in as-received strength for the 1100°C [3012°F] disks tested.) The strengths measured for the three runs were 1820 kPa, 1622 kPa, and 1393 kPa (264 lb/in.^{2}, 241 lb/in.^{2}, and 202 lb/in.^{2}).

Several methods were tried to present the CTE data, including both dimensional and nondimensional plots. The approach settled on was a semilog plot of the hot face MOR plotted against the number of cycles with the peak cycle temperature as a parameter. The resulting curves, shown in Figure 348, show a superficial resemblance to fatigue curves for metals that plot stress to failure versus number of cycles. In this case the matrix hot face strength remaining after a given number of thermally induced stress cycles is the dependent variable. The data appear to be internally consistent since the curves are nearly parallel. Further, the loss is greater in going from 1121°C to 1177°C (2050°F to 2150°F) than from 1066°C to 1121°C (1950°F to 2050°F). The data, seen in this form, are consistent with a mechanism of slow crack growth (probably of preexisting cracks) that is driven by both the number of cycles and the peak temperature.

The effect of disk-to-disk variation on the CTE damage was investigated, in a preliminary way, by running a single sample from Disk 1 of the 1100°C (2012°F) series, which has the highest as-received strength of the four disk samples. As can be seen in Figure 357, the Disk 1 axial strength distribution is greater than that for Disk 4 at all points. The exponential term is smaller, indicating that the damage is less for the Disk 1 sample; however, more data are needed to prove this.

During this period, two more sets of samples of MAS matrices were obtained and tested. The first of these sets was a batch of triangular channel 0.125 mm (0.005 in.) thick wall material reported in the previous six-month report. The second set was a rectangular channel configuration with a 1.75/1.00 aspect ratio, the same wall thickness, and identical MAS material.

Figures 358 and 359 plot the mean and approximate distributions of the MOR and RCS of the triangular configuration matrices. The mean MOR of 1360 kPa (197.3 lb/in.^{2}) and sample standard deviation of 176 kPa (25.5 lb/in.^{2}) for Disk 1, reported previously, are essentially identical with the sample 6078 values. However, the level of 1358 kPa (197 lb/in.^{2}) for a 0.127 mm (0.005 in.) thick wall matrix in as-received condition is very close to the 10,000 cycle, 1177°C (2150°F) CTE value for the 1100°C (2012°F) AS matrix at a wall thickness of 0.053 mm (0.0021 in.). Figure

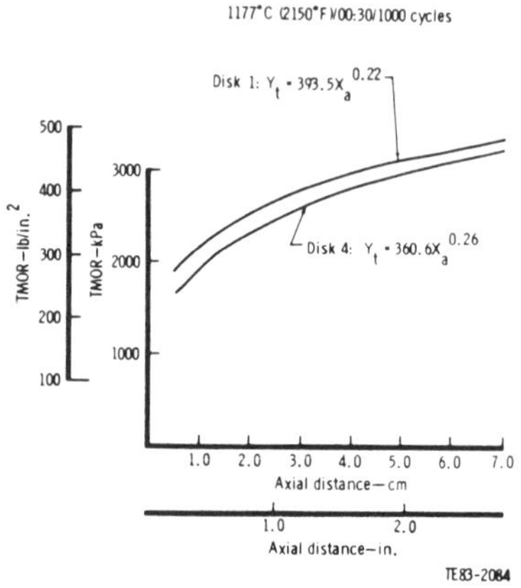

Figure 357. Comparison of cyclic thermal exposure effect on hot-face strength of 1100°C (2012°F) AS disks.

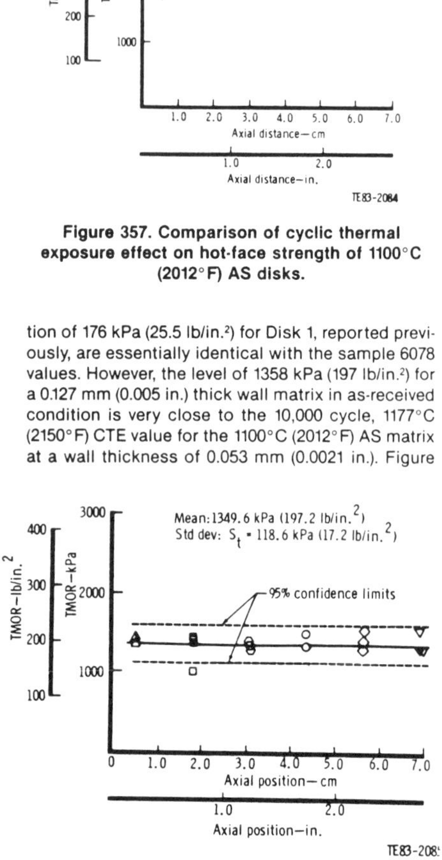

Figure 358. Triangular matrix—sample 6078, MAS, sample 1, as received.

359 plots the RCS values for both sample 6078 and Disk 1. No difference is apparent between the two batches of material; but the nearly identical values seen for the MOR tests strengthen the assumption made in the plot that the RCS values belong to the same population. Most important, however, is the very high level of the RCS when compared with the AS values 407 kPa to 703 kPa (59 lb/in.^{2} to 102 lb/in.^{2}). This is accounted for by the precisely aligned apex-to-apex triangles of the extruded MAS versus the random apex to separator strip alignments of the wrapped AS material. The separator strips fail in bending in the wrapped matrix.

Figures 360 and 361 include the corresponding MOR and RCS values for the rectangular matrix MAS. The MOR values were replicated at a mean of 980 kPa (142.2 lb/in.^{2}) and a standard deviation of 211 kPa (30.6 lb/in.^{2}) illustrating the excellent quality control in these MAS matrices. Approximately 365 kPa (53 lb/in.^{2}) is lost, however, in switching to the thermally more efficient rectangular configuration. The already high values for the triangular channel RCS values are slightly increased in the rectangular matrix data, demonstrating again the strong effect of configuration on as-received strength.

Since the RCS values have the smallest factor of safety relative to the mechanical stress in the engine, part of the RCS data available were used to

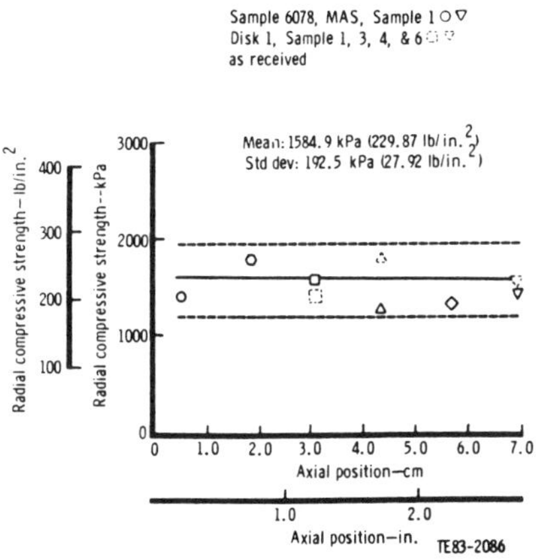

Figure 359. Triangular matrix—sample 6078, MAS, sample 1 and Disk 1, samples 1, 3, 4, and 6 as received.

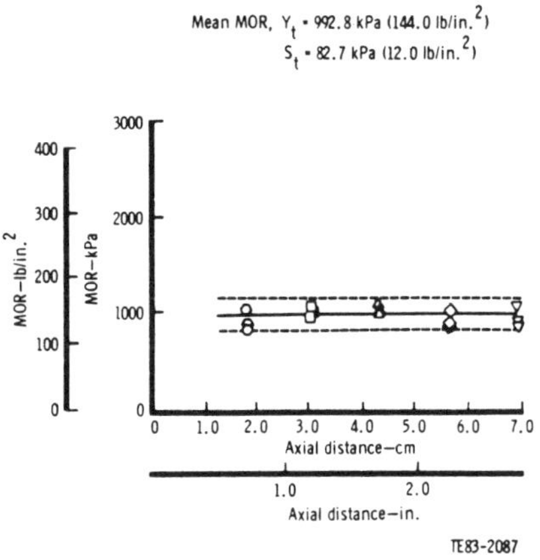

Figure 360. MAS, sample 1, rectangular matrix as received.

calculate a first approximation of the modulus of elasticity in radial compression for use in stress calculations. Two disks were selected—Disks 4 and 1 of the 1100°C (2012°F) AS disks tested. The linear part of the time-deformation-load curve was analyzed. The results are given in Table LXXV. The tabu-

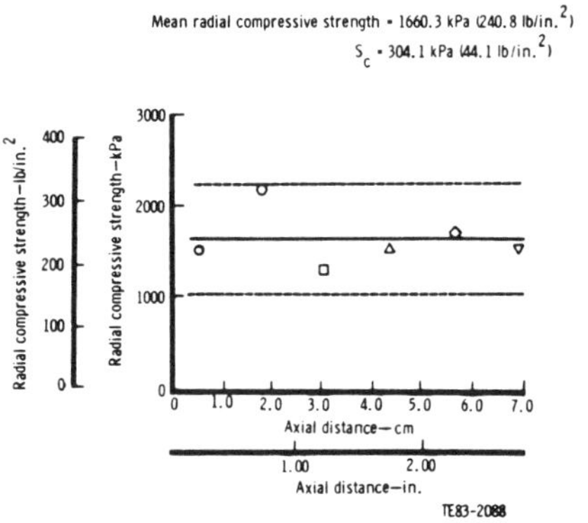

Figure 361. MAS, sample 2, rectangular matrix, as received.

Table LXXV.
Approximate modulus of elasticity in radial compression for 1100°C AS matrices.

	Sample	Modulus of elasticity kPa	(lb/in.²)	Mean kPa	(lb/in.²)	Range kPa	(lb/in.²)
Disk 4	R1-1	48,265	(7,000)				
	R1-2	61,366	(8,900)				
	R1-3	65,503	(9,500)	60,398	(8,760)	10,342	(1,500)
	R1-4	64,813	(9,400)				
	R1-5	62,055	(9,000)				
Disk 1	R4-1	70,329	(10,200)				
	R4-2	79,982	(11,600)	79,290	(11,500)	9,653	(1,400)
	R4-4	86,877	(12,600)				

lated values show that the modulus is low—fully commensurate with the low RCS values—and variable from disk to disk and within each disk.

The effects of cyclic peak temperature and the total number of cycles on the strength of the 1100°C (2012°F) AS matrix have been separated and identified for a typical disk. A small but usable margin of hot face strength remains after exposure to 1177°C (2150°F) and 10,000 cycles. This finding is particularly important in its implications for demonstrator engine reliability since the remainder of the sample axial thickness is less affected by the CTE leaving a relatively strong section to support the mechanical stresses. Further, the axial gradient in strength is consonant with the hypothesis of temperature level controlled differential thermal expansion stress as the driving force causing the CTE damage.

Time precludes an investigation of the mechanism causing the CTE damage observed. But the data and information available suggest that grain boundary sliding (Ref 10) caused by the presence of small high expansion mullite grains in the grain boundaries of the low expansion aluminous keatite structure produce the shear stress to open preexisting cracks. (Mullite is present in the as-received AS matrix.) This stress is not a function of temperature difference in the matrix walls or fillets (the 63% response time of the typical wall at maximum flow is approximately 1 ms) but is a function of the volume fraction of the mullite. The progressive nature of the damage can be related to the glassy phase, also present in the grain boundaries, which determines the relaxation time of the stress.

The short-term reliability of the AS matrix, as determined by the chemical and physical stability of the aluminous keatite material, is a function of the CTE damage. At high temperatures and longer times, the growth of the high expansion mullite phase begins to affect the internal stress and ultimate life of the regenerator disk. In isothermal testing this growth of the mullite phase can be followed by $\Delta L/L$ measurements (see EDR 10156, Ref 4). However, in CTE only the hot face is seriously affected by the high temperature inlet gas so that $\Delta L/L$ measurements are very low level and difficult to interpret while direct measurement of mullite at these levels by X-ray diffraction has not been successful to date.

The possibility of calculating the rate of transformation from keatite to mullite has been assessed. Since mullite is already present in the grain boundaries, the nucleation step is not the rate determining step. In addition, the alumina concentration in the bulk ceramic is approximately 11% to 12% and must rise to 60% at the precipitation site so that the mullite phase transformation reaction is very probably diffusion-rate controlled by movement of the Al^{+3} ion in the SiO_2. The time-dependent rate of diffusion can be calculated using equations given in Kingery (Ref 11), if the necessary diffusion coefficient data are available. A first approximation calculation was made using the data of Frischat (Ref 12), an activation energy of 25 KCAL/MOL, and a diffusion length of 3 μm as half the thickness of the mean grain size. Fewer than 1000 cycles between 1177°C and 593°C (2150°F and 1100°F) were required to transport Al^{+3} from the center of the grain to the boundary. This implies that a secondary reaction, perhaps the transformation from the keatite structure to cristobalite, determines the time to fracture since 10,000 cycle exposure produces a drop in strength but not fracture.

The results of the testing of the MAS matrices allow some important deductions about the role of channel configuration and processing methods on the strength of regenerator disks to be made. Since the material and the wall thicknesses of the triangular and rectangular channels are the same, direct comparison of the MOR and RCS values can be made. A significant loss in MOR will result by switching to the thermodynamically more efficient rectangular channel from the conventional triangular grid. No loss occurs in the RCS by adopting the rectangular grid.

Comparison of the triangular channel 1100°C (2012°F) AS and MAS matrices shows the superiority of the 1100°C (2012°F) AS material to the MAS reported here. The MOR of the AS, at less than half the wall thickness of the MAS, is superior to the MAS both as received and after CTE. However, the random orientation of the triangular apex of the wrapped AS matrix drastically reduces the RCS of the AS. The extrusion process is clearly superior to the wrapping process, provided that the wall thickness can be reduced.

Comparison of the data for the as-received AS and MAS regenerator matrix materials leads to three conclusions. The inherent material strength of the AS is from 1.5 to 2 times as great as the MAS, although the variability of the AS is about 2.5 to 3 times that of the MAS. However, the extrusion process is superior to the wrapping technique, setting aside the cementing required for disk assembly.

The intrinsic strength difference is demonstrated by comparing the MOR for the AS, shown in Figure 362, with that for the MAS, shown in Figure 363. The difference in the mean MOR of 2410 kPa and 1360 kPa (350 lb/in.² and 197 lb/in.²) is typical of these matrices, but it should be noted that all of the 1100°C (2012°F) AS disks tested were stronger than the MAS. This is so despite the fact that the MAS wall thickness ranged from 0.102 mm to 0.127 mm (0.004 in. to 0.005 in.) while the AS thickness was between 0.043 mm and 0.066 mm (0.002 in. and 0.003 in.). The mean MOR for the MAS dropped from 1360 kPa to 993 kPa (197 lb/in.² to 144 lb/in.²) in switching from the triangular to the rectangular channel configuration, shown in Figure 364. The thermally more efficient rectangular channel has a severe penalty in strength attached to it.

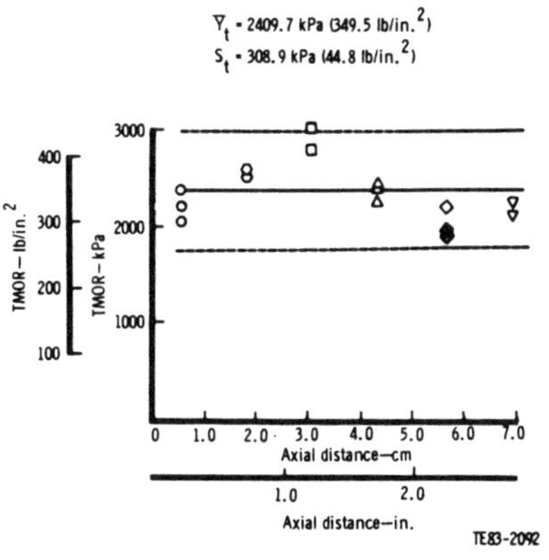

Figure 362. Disk 4, 1100°C (2012°F) AS, sample $R_{1\text{-}2}$, as received.

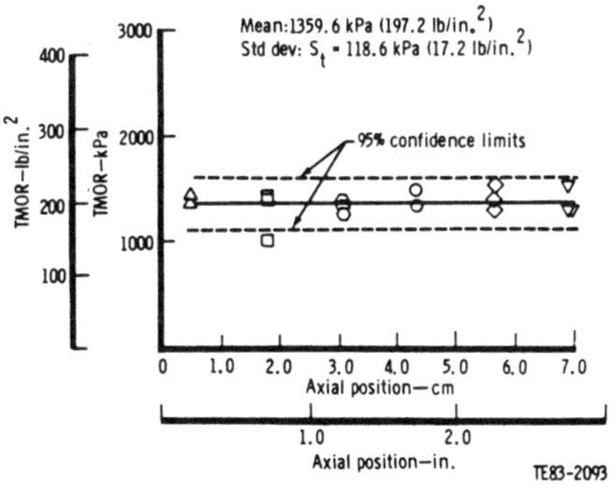

Figure 363. Triangular matrix—sample 6078, MAS, sample 1, as received.

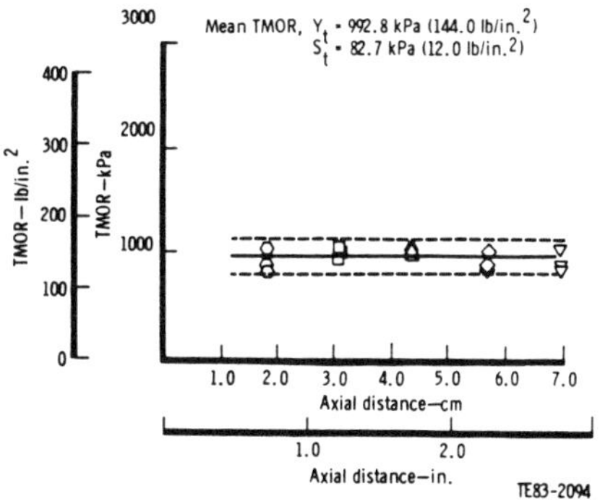

Figure 364. MAS, sample 1, rectangular matrix, as received.

When the RCS values for the two candidate materials are compared, the strong interaction of the fabrication process with the measured strength is shown. The mean value of the RCS for the sinusoidal AS material is 497 kPa (72 lb/in.²); the comparable value for the triangular MAS is 1585 kPa (229 lb/in.²) or more than three times as strong. This is accounted for by the precisely aligned apex-to-apex triangles of the extruded MAS versus the random alignments of the wrapped AS which fail the separator strips in bending. The already high values for the triangular channel RCS distribution are slightly increased in the rectangular matrix, demonstrating again the strong effect of fabrication on as-received strength. Figures 365, 366, and 367 show the mean and the axial scatter patterns for the three materials. (No significant variations are shown in the axial patterns.)

During the collection and analysis of data for this report, the variability introduced by disk-to-disk variation, radial or axial variation caused by channel skewness within the disk, and local variation caused by separator wall thickness were accounted for either by selection of blocks of data (all one disk or a single radius, for example) or by regression analyses. However, when very large groups of data are handled, another variable is very probably unaccounted for in the analysis. For example, when skew angle regression coefficients, which are intended to account for an eccentric loading effect on strength, are grouped as 1000°C (1832°F) and 1100°C (2012°F) materials, two different patterns are apparent. The 1100°C (2012°F) group coefficients have a mean of +3.8 while the 1000°C (1832°F) group mean is -9.9.

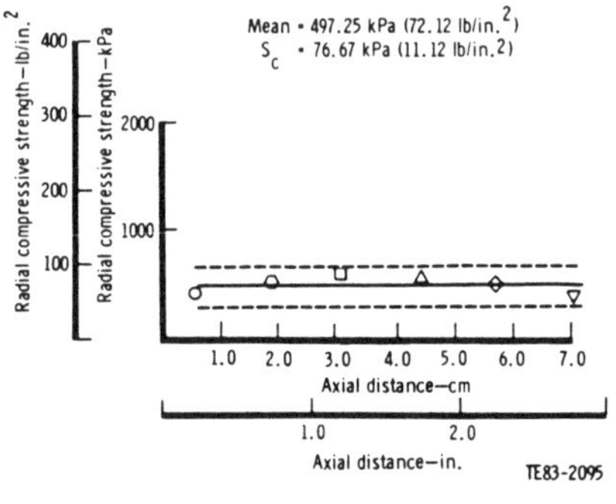

Figure 365. Disk 4, 1100°C AS, sample $R_{1\text{-}20}$, as received.

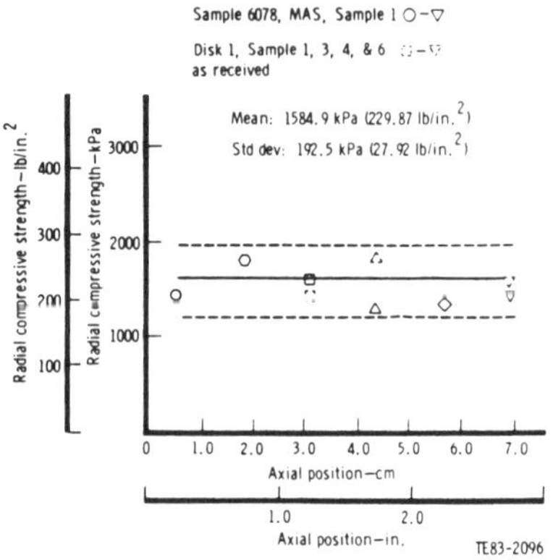

Figure 366. Sample 6078, MAS, sample 1 and Disk 1, sample 1, 3, 4, and 6 as received.

Two other things should be noted: the presence of positive regression coefficients for skew angle should have either zero or negative values only and the range of skew angles for the two groups are essentially the same. It is strongly indicated that the correlation of matrix strength with skew angle does not always link cause and effect, but is affected in the 1100°C (2012°F) group by another variable capa-

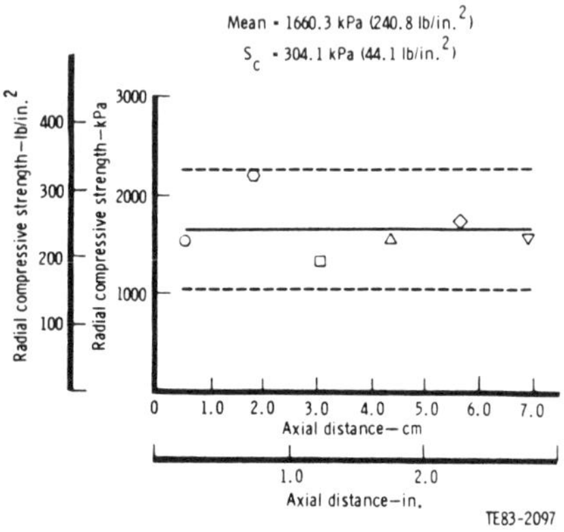

Figure 367. MAS rectangular matrix as received.

ble of overwhelming the eccentric loading effect of channel skewness.

As suggested in earlier reports, the variable that fits the requirement is porosity. Further, if porosity is the cause of the marked difference in the two groups, then the 1100°C (2012°F) porosity should be lower than the 1000°C (1832°F) material. This is confirmed by data on internal (through-wall) leak rate data accumulated by the engine project group as a disk leakage quality control measure. The data are plotted as frequency histograms in Figure 368 and show the difference in level and distribution of the through-wall leak rates of the two groups. The two groups have markedly different levels and distributions of the regression coefficients. Note that the mean wall thicknesses of the groups are 0.058 mm and 0.054 mm for the 1100°C (2012°F) and 1000°C (1832°F) groups respectively. Finally, if the effect of through-wall leak rate (or porosity) is to reduce the strength of disks to a critical level in use, there should be some correlation of leak rate with disk failures in engine service. The location in distribution of disks that failed in engine service is shown in Figure 368; these disks represent failure rates of 0.11, 0.14, 0.33, and 1.00 in the subgroups. Direct measurement of individual, sample, and test bar porosity is needed to verify and extend these findings.

A total of eight 1100°C (2012°F) AS samples were run during the CTE experiments, culminating in 1177°C (2150°F) peak cycle temperature test to 10,000 cycles and a total of 5 hr above 1100°C (2012°F). The hot face strength decreased from 2410 kPa (349.5 lb/in.²) to 1390 kPa (202 lb/in.²), which still provides a usable margin of strength in the hot face. A total of three MAS samples were tested with a final run at a peak cycle temperature of 1121°C (2050°F) at 5000 cycles and a total time above 1100°C (2012°F) of 2.5 hr. The resulting hot face strength of 103 kPa (15 lb/in.²), which started at 1170 kPa (169.5 lb/in.²), left no usable hot face strength. The other two MAS runs gave almost identical results.

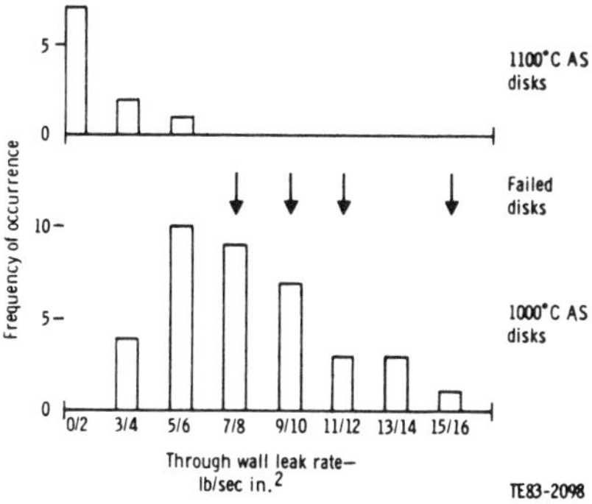

Figure 368. Leak rate frequency distributions for 1100°C (2012°F) and 1000°C (1832°F) AS disks.

The results for the two runs cited are shown in Figures 341 and 342. In both plots the preliminary data reduction step is shown. The power curve correlation shown in Figure 341 is considered the best estimate of axial strength distribution, although multiple linear regression using the separator wall thickness and the skew angle can be used to separate and identify these variables if needed. This curve should be compared with the two branch distribution that resulted from the MAS data reduction shown in Figure 342. The best representation of these data is a linear regression for the data from the hot face to the 30 mm (1.2 in.) axial slice with a mean and 95% confidence limit for the remainder. This particular set of data was selected from the three available because it shows that the CTE damage in MAS is probably progressive since the data at 18 mm (0.70 in.) show an intermediate level of damage dropping off to only approximately 620 kPa (90 lb/in.²).

Several methods were tried to present the overall CTE data, including both dimensional and nondimensional plots. The approach settled on was a semilog plot of the hot face MOR against the number of cycles with the peak cycle temperature as a parameter. The resulting curves, shown in Figure 348, resemble fatigue curves for metals that plot stress to failure versus number of cycles. In this case the matrix hot face strength remaining after a given number of thermally induced stress cycles is the dependent variable. The data are internally consistent, with the curves approximately parallel and the loss in strength greater in going from 1121°C to 1177°C (2050°F to 2150°F) than from 1066°C to 1121°C (1950°F to 2050°F). The data, in this form, are consistent with a mechanism of slow crack growth, probably from preexising microcracks, driven by both the number of cycles and the peak temperature. Note that these data apply to a single disk—Disk 4, which was selected since it lies in the center of the four 1100°C (2012°F) AS disk distribution in strength. Some evidence shows that there are significant disk-to-disk differences in response to CTE.

An approach to controlling or eliminating the problem of CTE damage in regenerator matrix has been identified as a result of a calculation of the nonsteady heat transfer in the thickness of the matrix wall during CTE. The calculation showed that the characteristic time of the 0.076 mm (0.003 in.) wall is approximately 2 milliseconds with a maximum ΔT of approximately 10-15°C (18-27°F), which is far too small to damage either AS or MAS materials. The damage must occur, then, at the microstructural level rather than the macrostructural level. Accordingly, the current literature is being monitored and has produced three papers that provide a mechanism which fits the data and background available at this writing (Ref 13-15). Briefly, the mechanism of CTE damage is believed to be due to anisotropic thermal expansion, either between two crystal phases or between the a and c axes of a single-crystal phase. The damage level is also a function of crystal size and grain boundary second phase constituents. This proposed mechanism, if it can be verified, offers the possibility of reducing or eliminating CTE damage by controlling fabrication variables.

To this point in time, the axial gradient in strength, believed to be caused by acceleration/deceleration cycles, has been seen in 1000°C (1832°F) disks with engine exposure time from 460 hr to 3050 hr. Similar hot to cold face axial gradients have been demonstrated for the 1100°C (2012°F) AS matrix over a broad range of conditions. To verify the laboratory sample CTE data, an engine exposed 1100°C (2012°F) AS disk with 1422 hr and 7800 acceleration/deceleration cycles to 1010°C (1850°F) was cut into samples and the radial and axial strength distributions were run. This peak cycle temperature is 56°C (100°F) lower than the lowest temperature investigated during the laboratory experiments; no problems occurred during the engine run.

Four radii were sampled: (1) 254 mm (10.0 in.) designated R_2, (2) 203 mm (8.0 in.) R_3, (3) 152 mm (6.0 in.) R_4 and R_5, (4) 103 mm (4.0 in.). Radii R_3, R_4, and R_5 produced very similar axial strength gradients, correlated by a power curve regression, and are in complete agreement with the laboratory CTE experiments verifying the experimental approach and the 1100°C (2012°F) AS data. The outer radius R_2, however, showed no gradient at all. These data are of particular importance since they indicate there is a cyclic peak temperature lower limit that produces no detectable reduction in strength and that this occurs at approximately equal to or less than 1010°C (1850°F).

Figures 369 and 349 show the preliminary data reduction statistical best estimates as solid lines

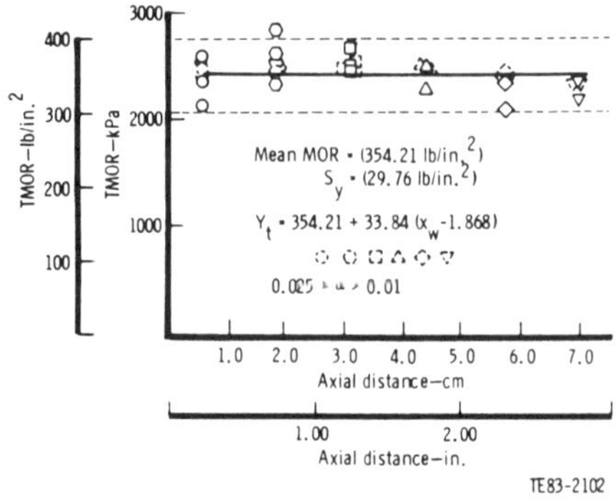

Figure 369. Disk 5, 1100°C (2012°F) AS, GM 16-04 at 1422 hr engine, 7800 cycles (sample $R_{2\text{-}3}$, radius = 10.0 in.).

and points. The dotted lines and points take into account the effect of separator wall thickness, using multiple linear regression, which reduces the scatter in both plots and shifts the regression line in the power curve plot.

A comparison of the data and regression equations shown in Figures 349 and 370 provide the clos-

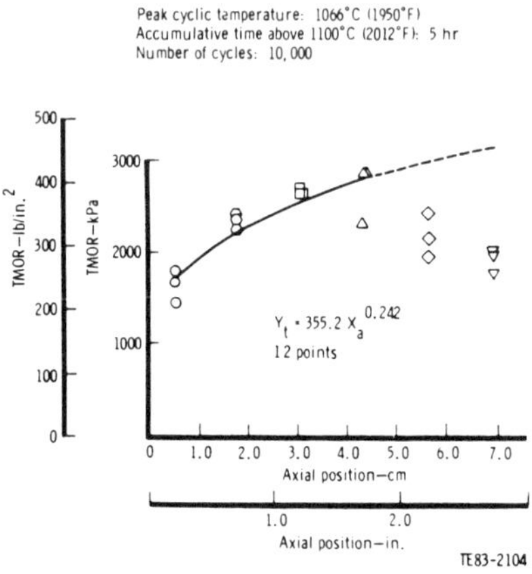

Figure 370. Disk 4, 1100°C (2012°F) AS, sample $R_{1\text{-}23}$, cyclic thermal exposure damage.

est comparison of the laboratory CTE data and the 1100°C (2012°F) AS engine disk data. The simple power curve regression line for the 1010°C (1850°F)/1422 hr/7800 cycle engine sample shows higher strength than the 1066°C (1950°F)/5 hr/10,000 cycle laboratory sample. This is consistent with the trends shown in Figure 340 and gives a rough first estimate of the locus of the 1010°C (1850°F) line on the plot.

Disk Seal Chemical Reaction

Seal contact points represent less 10% of the disk surface, are smaller than 0.13 mm (0.005 in.), and are confined primarily to matrix junctions. These contact points appear to have been molten in SEM photos (Ref 5, pp 140 and 141) and show a very high nickel content from contact with the seal NiO/CaF_2 wearface. No Ca is present. Ref 6, p 125, shows that Ni penetration is only 6-7 μm deep.

VI. General Engine Design and Component Aerodynamic Research

Summary

In support of ceramic component development, several general engine modifications were required. These are discussed in this section and include the engine block cooling and insulation, microprocessor engine control system, two-stage metal power turbine rotor, and metal combustor development.

In addition, component aerodynamic research work was planned but not implemented due to program redirection. Initial turbine aerodynamic rig design work was accomplished and is also reported in this section.

General Engine Design

ENGINE BLOCK COOLING AND INSULATION

Summary

With the higher operating temperatures of the CATE engine, it became necessary to carefully consider how the components that were not to be made of ceramic would be affected. Since the engine block is made of cast iron—a material with poor high-temperature capabilities—this portion was given detailed consideration from the beginning of the program. The following is a summary of items investigated and proposed.

Early in the CATE program a three-dimensional finite element model was used to determine whether the block, as insulated and cooled in the baseline IGT 404 industrial gas turbine, was adequate to withstand the higher CATE operating temperatures. The conclusions were as follows:

1. The baseline block was adequate up to a turbine inlet temperature (TIT) of 1038°C (1900°F).
2. At a TIT of 1132°C (2070°F) the baseline block would be subject to excessive creep distortion at the middle of the block crossarm and have a potential low-cycle-fatigue (LCF) problem at the end of the crossarm. Excessive creep deflection occurred when the crossarm temperature exceeded 538°C (1000°F). A design goal of 482°C (900°F) was established to provide an adequate margin.
3. The baseline mesh enclosed insulation blanket used to isolate high-temperature areas of the block was not performing properly due to circulation of hot gas between insulation and wall and/or due to poor coverage.
4. An improved block cooling system was required.

The following were considered to make the block capable of withstanding the CATE temperatures:

1. substitution of a moldable type of insulation that could be formed in place
2. distribution tubes to better direct the compressor discharge air used to cool the block
3. double the flow of compressor discharge cooling air
4. a sheet metal heat shield to reduce radiation of heat from the regenerator disk to the crossarm surface
5. a sheet metal heat shield with compressor discharge cooling air flowing between it and the block crossarm surface
6. substitution of ambient cooling air supplied by an external blower replacing the compressor discharge cooling air in the baseline block cooling system

The finite element thermal and stress analysis indicated that items No. 1 and 6 had the best chance for success. The decision was made to use a blanket-type insulation that was applied and molded in place while wet and subsequently cured in an oven. A modification was made to the block cooling system. The cavity adjacent to the block crossarm regenerator seal surface was isolated from the rest of the block by means of sheet metal shields and a cooling-air flow path established in which ambient air could be blown into the top of the block by an externally mounted fan and exhausted from the bottom. Engine tests using shop air to supply the ambient cooling air established that, with the aid of the air-cooled regenerator seals proposed for the CATE engine, a production automotive centrifugal heater blower could supply enough air to keep the block temperature in the safe range.

Objective

The objective of this part of the CATE program was to make the necessary modifications to the baseline IGT 404 industrial gas turbine engine block to ensure that the higher operating temperatures in the CATE engine would not result in a block failure or in block distortions that would prevent efficient operation of the CATE engine.

Approach

In the initial phase of the CATE program, there was concern that the block of the IGT 404 engine, which was to be modified into the ceramic CATE engine, would be unable to withstand the higher CATE operating temperatures without modification. Its material, cast iron, has poor elevated temperature capability, and although the hottest portion, the rear block, is made of a molybdenum-alloyed ductile cast iron, its strength and elastic modulus fall sharply above 482°C (900°F). An existing three-dimensional finite element model of the block was improved in critical areas to give better stress definition and used to analyze the block. Efforts early in the program were directed toward establishing convection coefficients by trying to match calculated temperatures to measured baseline temperatures. Results indicated that the mesh-encased blanket type of insulation used in the baseline engine was not utilizing its full insulating capabilities. An alternate type that could be molded in place was evaluated. Also investigated was a method of directing the compressor discharge air used to cool the block crossarms to locations that would optimize the cooling through the use of distribution tubes.

Discussion

This work was begun in the period from mid-1976 to the end of 1977, EDR 9346 (Ref 9), and continued through the first half of 1978, reported in EDR 9519 (Ref 1). The stresses in the block were acceptable at a TIT of 1038°C (1900°F) but at a TIT of 1132°C (2070°F), where the CATE engine would be running, there was an LCF problem. It was also established that the cooling air distribution tubes did not provide a significant improvement in cooling and that a moldable type of insulation to prevent circulation of hot air around the insulation blanket would be desirable.

In the middle of 1978 the CATE development engine block was discovered to have a permanent set at the middle of the crossarm of 1.00 mm (0.039 in.). Since the regenerator seal locates against the crossarm, excessive distortion of the crossarm results in seal leakage and inefficient engine operation. Creep data for the block material indicated potential excessive creep at temperatures above 538°C (1000°F). Since extrapolation of the curve relating block temperature to engine operating temperatures indicated that the worst running condition of the CATE engine would result in a block temperature of 704°C (1300°F), it was evident that without a significant improvement in block cooling the CATE block would have excessive distortion due to creep. Several modifications were evaluated analytically during this period in addition to the cooling air distributor tubes and the improved insulation. These include changes in the compressor discharge cooling air path, doubling the amount of cooling air, redesigns of the crossarm seal land surface, and a sheet metal zirconium-oxide-coated crossarm heat shield with and without cooling air flow between it and the block crossarm. The results of these analyses are reported in EDR 9722 and 9951 (Ref 2 and 3). Only the last of these analyses appeared to have the potential of keeping the block temperature at the required level, and its complexity was a deterrent to its adoption.

In mid-1979 a scheme was devised to cool the crossarms with ambient air instead of compressor discharge air. The block's design made it possible to separate the cavity adjacent to the crossarms from the rest of the internal block with sheet metal shields and to blow air from an external source into existing core plug openings at the top of the block and out core plug openings at the bottom. This proposal was analyzed analytically and reported in EDR 10156 (Ref 4). With a reasonable airflow, the block temperature could be kept below 482°C (900°F) and the stress kept in a range that would allow an adequate LCF life. An engine was modified and this concept tested with shop air used as an ambient cooling air supply. The method was proved feasible but the pressure drop developed in flowing an adequate amount of air through the crossarm flow path was higher than could be handled by blowers available at a reasonable price. A flow analysis was made that showed the block could be reworked to remove the restrictions so that an adequate amount of air could be delivered by a production, centrifugal, 12 Vdc heater blower. This test and analysis are reported in EDR 10383 (Ref 5). During the next year the design modification was accomplished in drafting and a development engine built incorporating the changes. It was tested extensively with shop air supplying the ambient cooling air with the capability of varying the airflow. The block was instrumented to measure pressure drop through the cooling-air flow path and block temperature. The engine could not be run at CATE operating temperatures because the ceramic gasifier turbine was not yet available but it was shown by extrapolation of data that, when running with the air-cooled regenerator seals that were designed for the CATE engine, the centrifugal heat blower would produce enough air to adequately cool the block. This engine testing is reported in EDRs 10672 and 10841 (Ref 6 and 7).

ENGINE CONTROLS

Summary

A control system for the CATE engine was developed and satisfactorily demonstrated through the 2070°F BOT engine configuration. The system was designed to permit operation up to and including the 2265°F BOT engine configuration.

Objective

The objective of the engine control system development program was to design a control system for the CATE engine that was to incorporate the following general features:

1. operate the engine up to 1241°C (2265°F) BOT for improved fuel efficiency
2. provide control modes for operation of the engine in a highway vehicle (truck or coach)
3. provide a state-of-the-art digital electronic control to control the engine
4. provide for improved fault diagnostic capability
5. provide responsive power transients without excessive temperature transients

Approach

To accomplish the objectives, the following engine control features were adopted:

1. Because present state-of-the-art thermocouples cannot tolerate steady-state temperatures at 1241°C (2265°F) with any reasonable life, it was decided to measure turbine outlet temperature (TOT) and thereby synthesize the proper BOT. This requires a carefully defined algorithm involving turbine speeds (N_1, N_2) and compressor inlet temperature (T_1) along with TOT to determine BOT, which is the critical temperature parameter.
2. A digital electronic control assembly was designed for this application using control modes for vehicle operation derived from previous experience on industrial gas turbine engines. The digital control was chosen because of its state of the art and because of its versatility in a development program to permit software changes as required to accomplish program modifications.
3. Improved fault diagnostics were made possible by the memory capability of the digital control such that interrogation of the engine parameters after a fault occurs was made available on a test connector of the digital control.

Discussion

The engine control system using a digital electronic control operating on TOT thermocouples was successfully developed and demonstrated. For detailed information regarding this effort, refer to EDRs 9722, 9951, 10156, 10383, and 10672 (Ref 2-6).

TWO-STAGE METAL POWER TURBINE ROTOR

Summary

In the last half of 1980 a power turbine was designed for the high temperature environment of the CATE 2265°F configuration with the following criteria considered:

- performance
- stress rupture
- low-cycle fatigue
- high-cycle fatigue
- rotor dynamics

The final design was a two-stage, all metal, integral blade power turbine running at a reduced speed to increase the stress rupture life of the Mar-M246 blades. The calculated life was adequate for all proposed test stand running and vehicle demonstrations in the CATE program. In early 1981 the CATE program was revised eliminating the 2265°F configuration, and activities were redirected to emphasize ceramic materials characterization and component development.

Objective

The objective of this part of the CATE program was to design a power turbine for the CATE 2265°F configuration to meet the CATE program goals regarding performance, stress rupture life, fatigue life, and burst margin that can be assembled in the basic IGT 404 turbine with a minimum of change.

Approach

Initial design studies for the power turbine for CATE 2265°F were begun in mid-1980. Ceramic and metal turbines of single-stage and two-stage designs were considered with regard to life, performance, and program cost. A two-stage, all metal, integral blade turbine rotor running at reduced speed was selected. The speed had to be reduced from 34,117 rpm to 25,000 rpm to obtain adequate blade life at the higher temperature. Two stages were then necessary to obtain the required performance.

Discussion

Even with the reduced speed the stress rupture life of the Mar-M246 material was below the design goal but adequate for all proposed test stand running and vehicle demonstrations in the CATE program. Engine cycles were established at design, and several off design points, a secondary flow scheme set up for turbine disk cooling, and concept sketches were made to use in stress and life studies.

Preliminary disk stress studies were made to establish a disk geometry compatible with the desired cyclic life and burst margin. Blade frequencies were computed to establish the number of vanes for each stage. Rotor assembly dynamics were found to be satisfactory using the existing bearing location and support system.

Near the end of 1980, a final layout and limit stack study was made setting up the cold flow path. In early 1981 the CATE program was revised eliminating the 2265°F configuration, and activities were redirected to emphasize ceramic materials characterization and development of the 2070°F configuration components. A more detailed presentation of the development of the 2265°F configuration power turbine is given in EDRs 10383 and 10672 (Ref 7 and 8).

METAL COMBUSTOR DEVELOPMENT

Summary

The metal combustion systems for the 2070°F CATE engine and the hot simulator rig were designed, developed, and satisfactorily performed for both duty cycles. The metal burner for the 2070°F engine demonstrated satisfactory performance and endurance characteristics.

Objective

The objective of the combustion system development program was to design and develop combustors for the following engine cycle:

- 2070°F CATE
- hot simulator rig

These systems, while operating under entirely different engine cycles, were designed and developed to meet the following criteria:

1. good overall performance—uniform temperature pattern, carbon-free operation, and low pressure loss
2. combustion efficiency high at low speed conditions to ensure good acceleration characteristics
3. lean burning limits well below fuel scheduling requirements

Approach

To accomplish the noted objectives, the following design and development procedures were adhered to:

- design
  - concept analysis
  - aerodynamic design
  - heat transfer analysis
  - stress analysis
  - material selection
  - detail drawings
- development
  - component testing
  - engine testing

Discussion

The combustion systems for the 2070°F CATE engine and the hot simulator rig were successfully designed, developed, and demonstrated. For detailed information regarding this effort, refer to EDRs 9722, 9951, 10156, 10383, and 10672 (Ref 2-6).

Component Aerodynamic Research Gasifier Turbine Rig

A gasifier turbine rig was designed for installation in the small turbine research facility. The rig was not built because of funding reduction.

Drawing EX 131182 defines the rig installation. A special turbine wheel was designed to mate with the turbine rig shaft and to use the ceramic turbine blades. The actual ceramic vanes were also employed as were the other ceramic components defining the gas flow path from the combustor to the turbine. Stainless steel cylindrical exit ducts were employed. Controlled and measured secondary cooling air circuits were provided for the following:

- vane hub
- vane tip
- hollow struts
- rotor face

The instrumentation provided for the following measurements:

- turbine inlet temperature, pressure, and flow rate
- shaft speed, torque, and power
- interstage pressures
- exit pressures, temperature, and yaw angle
- vane exit angle measurement at any radial position and 360 deg circumferentially was included

References

1. "Ceramic Applications in Turbine Engines: Progress Report for 16 January 1978 to 30 June 1978," Detroit Diesel Allison EDR 9519.
2. "Ceramic Applications in Turbine Engines: Progress Report for 1 July 1978 to 31 December 1978," Detroit Diesel Allison EDR 9722.
3. "Ceramic Applications in Turbine Engines: Progress Report for 1 January 1979 to 30 June 1979," Detroit Diesel Allison EDR 9951.
4. "Ceramic Applications in Turbine Engines: Progress Report for 1 July 1979 to 31 December 1979," NASA CR-159865, Detroit Diesel Allison EDR 10156.
5. "Ceramic Applications in Turbine Engines: Progress Report for 1 January 1980 to 30 June 1980," NASA CR-165197, Detroit Diesel Allison EDR 10383.
6. "Ceramic Applications in Turbine Engines: Progress Report for 1 July 1980 to 31 December 1980," NASA CR-165494, Detroit Diesel Allison EDR 10672.
7. "Ceramic Applications in Turbine Engines: Progress Report for 1 January 1981 to 30 June 1981," NASA CR-165559, Detroit Diesel Allison EDR 10841.
8. "Ceramic Applications in Turbine Engines: Progress Report for 1 July 1981 to 31 December 1981," NASA CR-167946, Detroit Diesel Allison EDR 11006.
9. "Heavy Duty Gas Turbine Engine Ceramic Component Development: 1 July 1976 to 15 January 1978," Detroit Diesel Allison EDR 9346.
10. Amar Pal Sigh Rana, "Sliding Mechanisms for Grain Boundaries Containing Preexisting Cavities in Al_2O_3-SiO_2," Ph.D. thesis, University of California (Berkeley), DOE contract No. W-7405-ENG-48, 1980.
11. W. D. Kingery, H. K. Bowen, and D. R. Uhlman, *Introduction to Ceramics,* 2nd edition, John Wiley and Sons, New York, p 443, 1976.
12. G. H. Frischat, "Zeitschrift Fur Glaskunde," 45 Jahrgang, Heft7, S.315, Bild 5, July 1972.
13. A. P. S. Rana, "Sliding Mechanisms for Grain Boundaries Containing Preexisting Cavities in Aluminum-Oxide Silicon-Dioxide Ceramics," Ph.D. thesis, University of California (Berkeley), DOE contract No. W-7405-ENG-48, 1980.
14. J. Gangas, "Application of Mossbauer Spectroscopy in Studies of Thermal Fatigue of Ceramic Materials," Final Report, 15 April through 31 October 1980, International Atomic Energy, Vienna, Austria.
15. J. J. Cleveland and R. C. Bradt, "Grain Size/Microcracking Relations for Pseudobrookite Oxides," *Journal of the American Ceramic Society,* Vol 61, No. 11-12, pp 478-81, November-December 1978.

Other Noyes Publications

CERAMIC CONTAINING SYSTEMS
Mechanical Aspects of Interfaces and Surfaces

Edited by

A.G. Evans
Department of Materials Science and Mineral Engineering
University of California, Berkeley

This book describes mechanical aspects of interfaces and surfaces in ceramic containing systems. Various multicomponent materials were investigated. Systems studied include ceramic matrix composites, metal/ceramic bonded systems and polymer particulate composites, as well as monolithic ceramics.

Fracture behavior has been observed, to develop criteria that identify fracture parameters pertinent to structural design. Micromechanics models of mechanical behaviors have also been developed whenever possible. Matrix cracking in composites and edge effects in metal/ceramic bonded systems are examples of modelled phenomena.

CONTENTS

ISBN 0-8155-1056-X (1986) **367 pages**

Other Noyes Publications

CERAMIC HEAT EXCHANGER CONCEPTS AND MATERIALS TECHNOLOGY

by

C. Bliem, D.J. Landini, J.F. Whitbeck, R. Kochan, J.C. Mittl, R. Piscitella, J. Schafer, A. Snyder, D.J. Wiggins, J.M. Zabriskie, B.A. Barna, S.P. Henslee, P.V. Kelsey

EG&G Idaho, Inc.

J.I. Federer, E.S. Bomar

Oak Ridge National Laboratory

Ceramic heat exchanger concepts and ceramic materials for use in high temperature applications are assessed in this book. Potential new concepts are compared with emerging and state-of-the-art concepts, in terms of corrosion/durability, fouling, performance, operation and maintenance, and economics. Design, functional, and cost requirements are established.

Large amounts of industrial fuel and energy consumption are lost annually due to discharge as waste heat in flue gases. Industry has attempted to decrease the quantity of lost energy with waste heat recovery systems; however, the high temperatures and corrosive constituents involved in waste heat streams now limit the recovery potential, and affect the durability of the materials of construction for heat recovery systems.

This book, then, compares and proposes heat exchanger concepts employing advanced ceramic materials, which may provide the solution to the heat loss problem. Part I of the book provides design, functional, and cost requirements of heat exchangers for high-temperature corrosive environments, a description of the state-of-the-art and emerging high-temperature heat exchanger technologies, and how additional heat exchanger concepts were obtained, selected, and developed. Part II assesses the data base on structural ceramic materials, including the concerns related to the design and use of ceramics for high temperature waste heat applications, the availability of a variety of ceramic materials and materials property data, and the applicability of current product reliability techniques.

A condensed table of contents listing **part and chapter titles and selected subtitles** is given below.

ISBN 0-8155-1030-6 (1985) **385 pages**

Other Noyes Publications

CERAMIC MATERIALS FOR ADVANCED HEAT ENGINES

Technical and Economic Evaluation

by

D.C. Larsen, J.W. Adams
IIT Research Institute

J.R. Johnson, A.P.S. Teotia, L.G. Hill
Argonne National Laboratory

Ceramic materials for advanced heat engines are described in this technological and economic evaluation. Advanced power systems for vehicles can potentially produce fuel efficiencies greatly exceeding those of today's gasoline and diesel engines. Current heat engine technologies, however, are limited by problems involving mechanical strength at high temperatures. Structural ceramics, if they can be reliably mass-produced, can make possible improved vehicle fuel efficiencies through higher-temperature operation and reduced vehicle weight.

In Part I of the book state-of-the-art technical ceramics, mainly silicon carbide and silicon nitride, that have potential as structural components in advanced heat engines, are evaluated. Thermal and mechanical property data were generated on candidate materials, and the results were interpreted with respect to microstructure, purity, secondary phases, environmental effects, and processing methods. Properties measured include: flexure strength, elastic modulus, stress-strain, fracture toughness, creep, oxidation, thermal expansion, thermal diffusivity, thermal shock, and stress rupture.

In Part II the macroeconomic impacts of structural ceramics are modeled for two scenarios, one in which the U.S. dominates the commercialization of ceramics in heat engines throughout the 1990s, and the other in which Japan dominates.

A condensed table of contents listing **part titles, chapter titles, and selected subtitles** is given below.

ISBN 0-8155-1029-2 (1985) **380 pages**